Minimax Theorems and Qualitative Properties of the Solutions of Hemivariational Inequalities

Nonconvex Optimization and Its Applications

Volume 29

Minimax Theorems and Qualitative Properties of the Solutions of Hemivariational Inequalities

by

D. Motreanu

Department of Mathematics,
University of Iasi, Romania

and

P. D. Panagiotopoulos

Department of Civil Engineering,
Aristotle University, Thessaloniki, Greece
and
Faculty of Mathematics and Physics,
RWTH Aachen, Germany

SPRINGER-SCIENCE+BUSINESS MEDIA, B.V.

Library of Congress Cataloging-in-Publication Data

Motreanu, D.
Minimax theorems and qualitative properties of the solutions of hemivariational inequalities / D. Motreanu, P.D. Panagiotopoulos.
p. cm. -- (Nonconvex optimization and its applications ; v. 29)
Includes bibliographical references.

1. Hemivariational inequalities. 2. Maxima and minima. 3. Nonsmooth optimization. I. Panagiotopoulos, P. D., 1950-
II. Title. III. Series.
QA316.M68 1999
515'.64--dc21 98-45178

ISBN 978-1-4613-6820-5 ISBN 978-1-4615-4064-9 (eBook)
DOI 10.1007/978-1-4615-4064-9

Printed on acid-free paper

Dedicated to
J.L. LIONS

Hoc tibi, quod potui, confectum carmine munus
pro multis aliis redditur officiis

Catulli Carmen LXVIII

Preface

Boundary value problems which have variational expressions in form of inequalities can be divided into two main classes. The class of boundary value problems (BVPs) leading to variational inequalities and the class of BVPs leading to hemivariational inequalities. The first class is related to convex energy functions and has being studied over the last forty years and the second class is related to nonconvex energy functions and has a shorter research "life" beginning with the works of the second author of the present book in the year 1981. Nevertheless a variety of important results have been produced within the framework of the theory of hemivariational inequalities and their numerical treatment, both in Mathematics and in Applied Sciences, especially in Engineering.
It is worth noting that inequality problems, i.e. BVPs leading to variational or to hemivariational inequalities, have within a very short time had a remarkable and precipitate development in both Pure and Applied Mathematics, as well as in Mechanics and the Engineering Sciences, largely because of the possibility of applying and further developing new and efficient mathematical methods in this field, taken generally from convex and/or nonconvex Nonsmooth Analysis. The evolution of these areas of Mathematics has facilitated the solution of many open questions in Applied Sciences generally, and also allowed the formulation and the definitive mathematical and numerical study of new classes of interesting problems. Most of these problems have as a common characteristic the development of free boundaries between areas of different behaviour. These developments have given birth during the last two decades to a new area of Mechanics, called Nonsmooth Mechanics, which has already had beneficial effects on the related Engineering Sciences, e.g. in Structural Mechanics, from the point of view of more elegant, more economical and more correct constructional solutions.
Here we should also point out that the relation of the inequality problems to Mechanics and Engineering Sciences is more inherent and fundamental. Indeed the variational and the hemivariational inequalities are expressions of the "'principle" of virtual work or power in inequality form. Therefore they are mathematical expressions of equilibrium problems in the static case and of time evolution problems in the dynamic case. Accordingly the study of the corresponding eigenvalue problems is of paramount importance in order to investigate the qualitative properties of the solutions. The mathematical theory of eigenvalue problems for hemivariational inequalities and the qualitative behaviour of their

solutions, as well as their possible applications, are the subjects of the present book.
The book is a result of seven years of cooperation between the two authors and the outgrowth of five years of seminars and courses they have delivered to a variety of audiences in the University of Bucharest and Iasi, the RWTH Aachen, the University of Namur and the Aristotle University of Thessaloniki, as well as in numerous Conferences. The book which is addressed to people working in Pure and Applied Mathematics and to research workers in Mechanics, Engineering and Applied Sciences, is to a large extent self-contained, at least concerning the critical point theory in Nonsmooth Analysis. The aim of the book is to introduce innovative mathematical methods, which are applied to the study for, as yet, unsolved or partially solved problems.
We would like to acknowledge the great assistance we received from Dipl. Ing. Aimilios Kalyvas who diligently prepared the final text with the LATEXprogram and Dipl. Ing. Thanassis Makrodimopoulos for the preparation of the final figures. We also wish to acknowledge the very helpful comments received from Prof. Daniel Goeleven concerning the corrections of some errors in the applications.
Many thanks are also due to our editors at Kluwer Academic Publishers for their friendly assistance and cooperation during this project. We would like to apologize to those whose work was inadvertently neglected in compiling the literature for this book. We shall welcome all comments and corrections from readers.

D. Motreanu

P.D. Panagiotopoulos

Table of Contents

Introduction

The scope of the present book is the study of the qualitative properties of the solutions of hemivariational inequalities. To this end we deal with the mathematical theory of eigenvalue problems for hemivariational inequalities. This new type of eigenvalue problem arises, e.g. in the variational formulation of mechanical problems, whenever nonconvex nonsmooth energy functionals are involved. The basic forms of the problems are the following. Suppose that V is a Hilbert space, $a(\cdot,\cdot)$ is a bilinear form on $V \times V$, $J(\cdot)$ a locally Lipschitz functional on V and B is an operator mapping V into V^*, the dual space of V. Denoting by $\langle \cdot,\cdot \rangle$ the duality pairing we seek a $u \in V$ and a $\lambda \in \mathbf{R}$ such as to satisfy

$$a(u, v-u) + J^0(u, v-u) \geq \langle \lambda Bu, v-u \rangle \quad \forall v \in V.$$

Another form results if we replace in the previous problem $\langle \lambda Bu, v-u \rangle$ by $(\lambda \bar{B}u, v-u)$, where $(\cdot,\cdot)$ denotes the inner product in V and $\bar{B}$ maps V into V. These types of eigenvalue problems are generalizations of the classical eigenvalue problems for variational inequalities, because $J(\cdot)$ is nonconvex and $J^0(\cdot,\cdot)$ is the directional differential in the sense of F.H. Clarke, i.e. $J(\cdot)$ is in general nonsmooth. More general forms of eigenvalue problems studied in this book are related to variational-hemivariational inequalities whose solution is sought in convex and closed admissible sets. The theory of eigenvalue problems for hemivariational inequalities is a new field of Nonsmooth Analysis.

In Chapter 1 notions and propositions of Nonsmooth Analysis which are used in the next chapters are presented. Also, some elements of the theory of critical points for locally Lipschitz functions are given and their relation to the eigenvalue problems of hemivariational inequalities is illustrated. The chapter closes with a section concerning the formulation and study of hemivariational inequalities. The material of this chapter is sufficient for the understanding of the forthcoming chapters.

Chapter 2 deals with the presentation of general results, many of which have been proved by the authors, that can be applied to locally Lipschitz functionals possibly invariant under a compact Lie group of linear isometries, concerning the nonsmooth critical point theory in the locally Lipschitz case. Here the results of Chang are deduced from a general principle that also incorporates the results of Du. The minimax principles presented are based on a deformation theorem that unifies different classical deformation results. Certain linking properties and elements of the equivariance theory are also presented. Finally certain applications of the abstract critical point theory are given, related to nonsmooth elliptic BVPs.

In Chapter 3 we study the critical point theory for functionals that are not locally Lipschitz as was the case in Chapter 2. This chapter presents an extension of Szulkin's minimax principles for functions of the form $I = \Phi + \Psi$, with $\Phi \in C^1(X, \mathbb{R})$, where X denotes a real Banach space, Ψ a convex proper and lower semicontinuous functional and Φ is locally Lipschitz. Thus it extends also Chang's minimax principle. The development of the new critical point theory in this chapter is necessary for the treatment of the eigenvalue problems for variational-hemivariational inequalities. This chapter ends with some applications concerning variational-hemivariational inequalities and semicoercive unilateral problems.

In Chapter 4 two types of eigenvalue problems for hemivariational inequalities are formulated and studied. The first is the most simple eigenvalue problem which can be formulated, and the second is a generalization of the first involving a term with a compact nonlinear operator. Existence results are given and certain applications from Mechanics illustrate the theory.

Chapter 5 deals with multiplicity results for the eigenvalue problems of hemivariational inequalities. A general minimax approach is developed, which permits the use of the corresponding linear eigenvalue problem for the determination of eigenvectors of the hemivariational inequality. Then the case of even nonconvex superpotential $j(x, \cdot)$ is studied by means of the theory of genus of Krasnosel'skii. The chapter concludes with some applications from Mechanics and Engineering Sciences.

In Chapter 6 we study a new type of eigenvalue problems for hemivariational inequalities whose solutions must belong to a sphere. Focusing here on problems in Mechanics and in Economics (network flow problems) leading to hemivariational inequalities, for which, either the solution remains a priori bounded, or the cost of the structure is prescribed, we are led to the aforementioned type of eigenvalue problem on a sphere. The existence of a solution as well as the multiplicity of solutions are discussed. Applications illustrate the theory.

Chapter 7 deals with the mathematical study of the resonant case of eigenvalue problems for hemivariational inequalities. We obtain new Landesman-Lazer type results allowing unbounded nonlinearities and resonance at multiple eigenvalues. A model of buckling in Mechanics illustrates the practical motivation and interest in these topics. We use two different methods: a nonsmooth version of the Saddle-Point Thorem and the Leray- Schauder topological degree. The results of this chapter improve and extend all the known results related to resonant problems.

In Chapter 8 we study a new type of eigenvalue problem, called "double eigenvalue problems", which arise in the theory of hemivariational inequalities. We give existence results, as well as some qualitative properties for the solutions of these eigenvalue problems. Three different approaches to the double eigenvalue problems for general hemivariational inequalities are developed: minimization, minimax methods and (sub) critical point theory on a sphere. Certain multiplicity and perturbation results are also proved. The chapter concludes with various applications concerning problems in Mechanics.

Chapter 9 is devoted to the study of some evolutive or periodic hemivariational inequalities. The following three types of problems are studied: i) hyperbolic hemivariational inequalities modelling nonlinear wave equations with nonlinearities, ii) homoclinic solutions for second order Hamiltonian systems with discontinuous nonlinearities and iii) multiplicity results for hemivariational inequalities containing periodic energy functionals. For the nonsmooth hyperbolic problems both the nonresonant and the resonant cases are studied separately.

Guidelines for the Reader. Abbreviations

The material of Chapter 1 has been appropriately chosen in order to comply with the requirements of the other chapters of the book. All propositions of Nonsmooth Analysis (Convex and Nonconvex) needed in this book are given in this chapter. We expect the reader to have some knowledge of Convex Analysis and Nonsmooth Analysis, but this is not absolutely necessary to understand the book. Indeed the definitions of all new mathematical notions are included in the text. Thus, only a basic knowledge of Functional Analysis is necessary.

Certain important notations and abbreviations used in the text and not defined there, are listed here. All other notations are defined in the text. Throughout the book the summation convention with respect to a repeated index is employed, unless otherwise stated. We use the same symbol ∂f for both the subdifferential and the generalized gradient.

$\mathbb{N}$	the set of natural numbers
$\mathbb{R}$	the set of real numbers
$\mathbb{R}^+$	the set of nonnegative numbers
$\overline{\mathbb{R}}$	the set of real numbers including $\pm\infty$
$\mathbb{R}^n$	Euclidean n-dimensional space
$\|x\| = (\sum_{i=1}^n x_i^2)^{1/2}$	length of $x \in \mathbb{R}$
$C_0^\infty(\Omega)$	space of infinitely differentiable functions with compact support in Ω
a.e.	almost every (almost everywhere)
μ – a.e.	almost every (almost everywhere) with respect to a measure μ
ker$T$	kernel of T
epi$f$	epigraph of a functional f
l.s.c., u.s.c.	lower semicontinuous, upper semicontinuous
$\partial f(x)$	subdifferential of f at x, generalized gradient of f at x
$w(x) \in \partial_y j(x, u(x))$	generalized gradient with respect to the second variable $y \to j(\cdot, y)$
$\overset{d}{=}$, :=	equal by definition (if no ambiguity occurs d and : may be omitted)

supp	support of a function
sgn	signum
$\Omega - \omega$ or Ω/ω	difference of two sets
$x = \{x_1, \ldots, x_n\} = \{x_i\}$, $i = 1, \ldots, n$	a point $x \in R^n$
$\{a_1, \ldots a_n \ldots\}$	set with elements $a_1, \ldots a_n, \ldots$ also written as (a_i) or $\{a_i\}$
$\bar{A}$	closure of a set A
co$A$	convex hull of a set A
$X, X^\star$	a topological space X an its dual $X^\star$ (denoted also as X')
$u_n \overset{w}{\rightarrow} u$, $u_n \rightharpoonup u$	
$u_n \rightarrow u$ weakly:	three notations for the weak convergence
BVP(BVPs):	Boundary Value Problem(Problems)
$\{a_i\}$ $i = 1, 2, 3$	vector in $\mathbb{R}^3$
$\{a_{ij}\}$ $i, j = 1, 2, 3$	tensor in $\mathbb{R}^3$
$\{u_n\}$	sequence $\{u_n : n \in \mathbb{N}\}$
imbedding or embedding	=continuous injection
$\{u \in V \| ....\} = \{u \in V : ....\} = \{u \in V; ....\}$	three equivalent notations
$V(\mathbb{R}/TZ, \mathbb{R})$	space V of T-periodic functions ($\mathbb{R}/TZ$ denotes the interval $[0, T]$ with the identification of T with the zero of the next interval $[0, T]$)
$\|\Omega\| = \text{mes } \Omega$	=measure of $\Omega \subset \mathbb{R}^n$
$\partial\Omega$	the boundary Γ of Ω
$\text{id}_V = I$	=identity operator in V
$\beta(t\pm)$	notation for $\beta(t \pm 0)$
pr	=projection operator

1. Elements of Nonsmooth Analysis. Hemivariational Inequalities

This chapter is devoted to the calculus developed by Clarke [9]-[14] for locally Lipschitz functions on Banach spaces, and some elements of the theory of hemivariational inequalities. We restrict ourselves to those basic elements of Clarke's theory that are needed in the sequel. Precisely, we discuss generalized directional derivative and generalized gradient, Lebourg's mean value theorem, chain rule, generalized gradient of integral functions and of restrictions to submanifolds. Finally, the definition of critical point in the sense of Chang [8] for a locally Lipschitz function is given and the relationship between the hemivariational inequalities and the generalized critical point problem is stressed. We close this chapter with some elements from the mathematical theory of hemivariational inequalities [33],[36],[37].

1.1 Generalized Directional Derivative and the Generalized Gradient of Clarke

A function $f : U \to \mathbb{R}$ on an open subset of a real Banach space X is called locally Lipschitz if each point $u \in U$ possesses a neighborhood $N_u \subset U$ such that

$$| f(u_1) - f(u_2) | \leq K\|u_1 - u_2\|, \quad \forall\, u_1, u_2 \in N_u,$$

for a constant $K > 0$ depending on N_u.

Definition 1.1 (Clarke [9]-[12]). The generalized directional derivative of the locally Lipschitz function $f : U \to \mathbb{R}$ at the point $u \in U$ in the direction $v \in X$ is defined by

$$f^0(u; v) := \limsup_{\substack{w \to u \\ t \downarrow 0}} \frac{1}{t}(f(w + tv) - f(w)).$$

The next two results establish in some important situations the relationship between the generalized directional derivative $f^0(u; v)$ and the classical (one-sided) directional derivative

$$f'(u; v) := \lim_{t \downarrow 0} \frac{1}{t}(f(u + tv) - f(u)),$$

when this exists.

Proposition 1.1 If $f : U \to \mathbb{R}$ is continuously differentiable, then the equality below holds

$$f^0(u;v) = f'(u;v), \quad \forall\, u \in U, \quad \forall\, v \in X. \tag{1.1}$$

Proof. Fix $u \in U$, $v \in X$ and $w \in U$. The function $g(t) = f(w+tv)$, for all $t \geq 0$ sufficiently small, is continuously differentiable with the derivative

$$g'(t) = f'(w+tv;v).$$

Given any admissible $t > 0$ the mean value theorem yields some $s \in (0,t)$ such that

$$\frac{1}{t}(f(w+tv) - f(w)) = \frac{1}{t}(g(t) - g(0)) = g'(s) = f'(w+sv;v).$$

In view of the continuity of the differential of f, letting $w \to u$ in X and $t \to 0$ in $\mathbb{R}$, implies the desired result. □

Proposition 1.2 Let the open set U be convex in X and suppose that $f : U \to \mathbb{R}$ is a convex and locally Lipschitz function (for instance, f is convex and bounded from above near some point in U). Then (1.1) is valid.

Proof. The convexity of $f : U \to \mathbb{R}$ ensures the existence of the one-sided directional derivative $f'(u;v)$. Let us fix $u \in U$, $v \in X$ and an arbitrary small number $\delta > 0$ such that the Lipschitz condition for f holds on an open ball centered at u with radius δ and the Lipschitz constant $K > 0$. Definition 1.1 enables us to write

$$\begin{aligned} f^0(u;v) &= \lim_{\varepsilon \downarrow 0} \sup_{\|w-u\|<\varepsilon\delta} \sup_{0<t<\varepsilon} \frac{1}{t}(f(w+tv) - f(w)) \\ &= \lim_{\varepsilon \downarrow 0} \sup_{\|w-u\|<\varepsilon\delta} \frac{1}{\varepsilon}(f(w+\varepsilon v) - f(w)), \end{aligned}$$

the second equality being the consequence of the convexity of f. The fact that f is local Lipschitz implies that

$$\left| \frac{1}{\varepsilon}(f(w+\varepsilon v) - f(w)) - \frac{1}{\varepsilon}(f(u+\varepsilon v) - f(u)) \right| \leq 2\delta K$$

for $\|w-u\| < \varepsilon\delta$ with $\varepsilon \in (0,1)$. We derive that

$$f^0(u;v) \leq \lim_{\varepsilon \downarrow 0} \frac{1}{\varepsilon}(f(u+\varepsilon v) - f(u)) + 2\delta K = f'(u;v) + 2\delta K.$$

Letting $\delta \to 0$ we arrive at the relation (1.1). □

Remarks 1.1 i) A function $f : U \to \mathbb{R}$ on an open set U of a Banach space X is said to be regular at $u \in U$ if the one-sided directional derivative $f'(u;v)$

exists for $v \in X$ and the relation (1.1) is true. Therefore Propositions 1.1 and 1.2 provide relevant examples of regular functions.

The following result collects essential properties of the generalized directional derivative.

Proposition 1.3 Let $f : U \to \mathbb{R}$ be a locally Lipschitz function on an open set U of a Banach space X. Then the following facts are true:

(a) For every $u \in U$ the function $f^0(u, \cdot\,) : X \to \mathbb{R}$ is positively homogeneous and subadditive (thus convex) and satisfies

$$|f^0(u; v)| \leq K\|v\|, \quad \forall\, v \in X. \tag{1.2}$$

Moreover, it is Lipschitz continuous on X with the Lipschitz constant K, where $K > 0$ is a Lipschitz constant of f near u.

(b) $f^0(\,\cdot\,;\,\cdot\,) : U \times X \to \mathbb{R}$ is upper semicontinuous.

(c) $f^0(u; -v) = (-f)^0(u; v), \quad \forall\, u \in U, \quad \forall\, v \in X.$

Proof. (a) The positive homogeneity of $f^0(u; \cdot)$ and inequality (1.2) follow readily from Definition 1.1. To check the subadditivity of $f^0(u; \cdot)$ let $v_1, v_2 \in X$ be fixed. Then we see that

$$\begin{aligned} f^0(u; v_1 + v_2) &= \limsup_{\substack{w \to u \\ t \downarrow 0}} \frac{1}{t}(f(w + t(v_1 + v_2)) - f(w)) \\ &\leq \limsup_{\substack{w \to u \\ t \downarrow 0}} \frac{1}{t}(f(w + tv_1 + tv_2) - f(w + tv_2)) \\ &\quad + \limsup_{\substack{w \to u \\ t \downarrow 0}} \frac{1}{t}(f(w + tv_2) - f(w)) \leq f^0(u; v_1) + f^0(u; v_2). \end{aligned}$$

For arbitrary $v_1, v_2 \in X$, using the Lipschitz constant K on a neighborhood of u, we obtain

$$\begin{aligned} & f(w + tv_1) - f(w) = \\ & \quad f(w + tv_1) - f(w + tv_2) + f(w + tv_2) - f(w) \\ & \quad \leq K\|v_1 - v_2\|t + f(w + tv_2) - f(w) \end{aligned}$$

whenever w is close to u and $t > 0$ is small enough. Then we have

$$f^0(u; v_1) \leq K\|v_1 - v_2\| + f^0(u; v_2).$$

Assertion (a) is thus verified.

(b) In order to justify the upper semicontinuity of f^0 let $\{u_n\} \subset U$ and $\{v_n\} \subset X$ be sequences with $u_n \to u \in U$ and $v_n \to v \in X$ as $n \to \infty$. Take any $x_n \in U$ and $t_n > 0$ with

$$\|x_n - u_n\| + t_n < \frac{1}{n}$$

and

$$f^0(u_n; v_n) \le t_n^{-1}(f(x_n + t_n v_n) - f(x_n)) + \frac{1}{n}.$$

It turns out that

$$\begin{aligned} f^0(u_n; v_n) - \frac{1}{n} &\le t_n^{-1}(f(x_n + t_n v) - f(x_n)) + t_n^{-1}(f(x_n + t_n v_n) \\ &\quad - f(x_n + t_n v)) \le t_n^{-1}(f(x_n + t_n v) - f(x_n)) + K\|v_n - v\|, \end{aligned}$$

where $K > 0$ is the Lipschitz constant of f around u. Passing to the limit for $n \to \infty$ it results

$$\limsup_{n\to\infty} f^0(u_n; v_n) \le \limsup_{n\to\infty} t_n^{-1}(f(x_n + t_n v) - f(x_n)) \le f^0(u; v)$$

which ends the proof of (b).

(c) For given $u \in U$, $v \in X$, we have by Definition 1.1

$$\begin{aligned} f^0(u; -v) &= \limsup_{\substack{w\to u \\ t\downarrow 0}} \frac{1}{t}(f(w - tv) - f(w)) \\ &= \limsup_{\substack{w\to u \\ t\downarrow 0}} \frac{1}{t}(-f(w - tv + tv) + f(w - tv)) = (-f)^0(u; v). \end{aligned}$$

This completes the proof. □

The definition below introduces the main concept in this Section.

Definition 1.2 (Clarke [9]-[12]). Let $f : U \to \mathbb{R}$ be a locally Lipschitz function on the open set U of a Banach space X. The generalized gradient $\partial f(u)$ of f at a point $u \in U$ is the subset of the dual space X^* defined as follows

$$\partial f(u) = \{z \in X^* : \langle z, v\rangle \le f^0(u; v),\ \forall\, v \in X\}.$$

Useful properties of the generalized gradient are given below.

Proposition 1.4 Let $f : U \to \mathbb{R}$ be as in Definition 1.2. Then the following assertions hold:

(i) For every $u \in U$, $\partial f(u)$ is a nonempty, convex and weakly*-compact subset of X^* which is bounded by the Lipschitz constant $K > 0$ of f near u.

(ii) For every $u \in U$, $f^0(u; \cdot)$ is the support function of $\partial f(u)$, i.e.,

$$f^0(u; v) = \max\{\langle z, v\rangle : z \in \partial f(u)\},\quad \forall\, v \in X.$$

(iii) The set valued map ∂f from U to X^* is weakly*-closed, that is, if $\{u_n\} \subset U$ and $\{z_n\} \subset X^*$ are sequences such that $u_n \to u$ strongly in X, $z_n \in$

$\partial f(u_n)$ and $z_n \to z$ weakly* in X^* for $u \in U$, $z \in X^*$, then $z \in \partial f(u)$. In particular, if X is finite dimensional, ∂f is upper semicontinuous.

(iv) The set-valued map ∂f is weakly*-upper semicontinuous from U to X^* in the following sense: for any $u \in U$, $\varepsilon > 0$, $v \in X$ there is a $\delta > 0$ such that for each $z \in \partial f(x)$ with $\|x - u\| < \delta$ there is some $\zeta \in \partial f(u)$ satisfying $|\langle z - \zeta, v\rangle| < \varepsilon$.

Proof. (i) Applying (a) in Proposition 1.3 and the Hahn-Banach Theorem there exists $z \in X^*$ satisfying

$$\langle z, v\rangle \leq f^0(u; v), \quad \forall\, v \in X.$$

Hence $\partial f(u)$ is nonempty. The convexity of $\partial f(u)$ is clear from Definition 1.2. For $z \in \partial f(u)$ and $v \in X$, by Definition 1.2 relation (1.2) and (c) in Proposition 1.3 we see that

$$-K\|v\| \leq -(-f)^0(u; v) \leq \langle z, v\rangle \leq f^0(u; v) \leq K\|v\|.$$

Thus

$$|\langle z, v\rangle| \leq K\|v\|.$$

It follows

$$\|z\|_{X^*} \leq K, \quad \forall\, z \in \partial f(u). \tag{1.3}$$

Since $\partial f(u)$ is weakly*-closed, the boundedness in (1.3) and Alaoglu's Theorem (see, e.g., Brézis [7], p.42) ensures that $\partial f(u)$ is weakly*-compact in X^*.

(ii) Suppose by condradiction that there exists $v \in X$ with

$$f^0(u; v) > \max\{\langle z, v\rangle : z \in \partial f(u)\}. \tag{1.4}$$

The Hahn-Banach Theorem enables us to find $\zeta \in X^*$ that fulfills the properties $\langle \zeta, v\rangle = f^0(u; v)$ and

$$\langle \zeta, x\rangle \leq f^0(u; x), \quad \forall\, x \in X.$$

Here we used (a) in Proposition 1.3. Therefore $\zeta \in \partial f(u)$ contradicts (1.4) which establishes the stated result.

(iii) Suppose that the sequences $\{u_n\} \subset U$ and $\{z_n\} \subset X^*$ satisfy the requirements in (iii). Fix some $v \in X$. In view of (b) in Proposition 1.3, by passing to the limit in the inequality

$$\langle z_n, v\rangle \leq f^0(u_n; v)$$

we obtain

$$\langle z, v\rangle \leq \limsup_{n\to\infty} f^0(u_n; v) \leq f^0(u; v).$$

Since $v \in X$ is arbitrary we conclude $z \in \partial f(u)$.

(iv) Arguing by contradiction we assume that the data $u \in U$, $\varepsilon > 0$ and $v \in X$ can be found such that there exist sequences $\{x_n\} \subset U$ and $\{z_n\} \subset X^*$ with $x_n \to u$ strongly in X, $z_n \in \partial f(x_n)$ and

$$|\langle z_n - \zeta, v\rangle| \geq \varepsilon, \quad \forall\, \zeta \in \partial f(u). \tag{1.5}$$

From (i) we deduce the boundedness $\|z_n\|_{X^*} \leq K$ for n sufficiently large. Thus, along a subsequence, we have for some $z \in X^*$ that

$$z_n \to z \quad \text{weakly}^* \text{ in } X^*.$$

Then assertion (iii) implies $z \in \partial f(u)$. This contradicts (1.5). The proof is complete. □

We determine the generalized gradient for the situations described in Propositions 1.1 and 1.2.

Proposition 1.5 Let $f : U \to \mathbb{R}$ be a locally Lipschitz function on an open set U of a Banach space X. The following properties are valid:

(i) If f is Gâteaux differentiable at $u \in U$, then its Gâteaux derivative $f'(u)$ belongs to $\partial f(u)$.

(ii) If f is continuously differentiable at $u \in U$, then $\partial f(u) = \{f'(u)\}$. More generally, f is strictly differentiable at $u \in U$ if and only if f is locally Lipschitz near u and $\partial f(u)$ reduces to a singleton which is necessarily the strict derivative of f at u (see Clarke [12], p. 30-33). In particular, if X is finite dimensional, $\partial f(u)$ is a singleton set for all $u \in U$, if and only if f is continuously differentiable on U.

(iii) If U is a convex open set and $f : U \to \mathbb{R}$ is a convex function, then $\partial f(u)$, at any $u \in U$, coincides with the subdifferential of f at u in the sense of convex analysis.

(iv) If X is finite dimensional, then $\partial f(u)$ at any $u \in U$ is the convex hull in $X^* \equiv X$ of all points

$$z = \lim_{n \to \infty} f'(u_n) \tag{1.6}$$

with $u_n \to u$ as $n \to \infty$, where $\{u_n\}$ is contained in the complement of a Lebesgue measure zero set and f is differentiable at each u_n.

Proof. (i) The Gâteaux derivative $f'(u)$ satisfies

$$\langle f'(u), v\rangle = f'(u; v) = \lim_{t \downarrow 0} \frac{1}{t}(f(u + tv) - f(u)) \leq f^0(u; v), \quad \forall\, v \in V.$$

By Definition 1.2 this means that $f'(u) \in \partial f(u)$.

(ii) Proposition 1.1, specifically its proof, shows that if f is continuously differentiable at $u \in U$ then formula (1.1) holds. The linearity of the map

$v \to \langle f'(u), v\rangle = f'(u;v)$ and Definition 1.2 imply $\partial f(u) = \{f'(u)\}$. For the second assertion of (ii) we refer to the careful analysis in Clarke [12]. Finally, assuming that X is finite dimensional and $\partial f(u)$ is a singleton for each $u \in U$, implies that f is differentiable on U. Then, by the final remark in (iii) of Proposition 1.4, we deduce the continuity of the derivative f', and thus f is continuously differentiable.

(iii) This is the direct consequence of Proposition 1.1, of assertion (ii) in Proposition 1.4 and of the convexity of f (see, e.g., Ekeland and Temam [14]).

(iv) By Rademacher's Theorem f is known to be differentiable a.e., and so the involved sequences $\{u_n\}$ exist. In view of (iii) in Proposition 1.4 we see that each limit z in (1.6) belongs to $\partial f(u)$. Hence the convex hull of the foregoing limits z forms a compact subset of $\partial f(u)$. The converse inclusion will be obtained from

$$f^0(u;v) \le \limsup_{w\to u}\langle f'(w), v\rangle, \quad \forall\, v \in X, \tag{1.7}$$

where w in (1.7) runs in the admissible subset of U. Denote by α the right-hand side of (1.7) and fix a nonzero $v \in X$. There is $\delta > 0$ with

$$\langle f'(w), v\rangle \le \alpha + \varepsilon, \quad \text{for a.e. } \|w-u\| < \delta.$$

Then, making use of Fubini's Theorem, we get

$$f(w+tv) = f(w) + \int_0^t \langle f'(w+sv, v)\rangle ds$$

for all $\|w-u\| < \delta/2$ and $0 \le t < \delta/2\|v\|$. Clearly, this leads to

$$f^0(u;v) \le \alpha + \varepsilon\,.$$

Letting $\varepsilon \to 0$ we arrive at (1.7). This completes the proof. □

Remark 1.2 It is worth to point out that for a locally Lipschitz function $f : U \to \mathbb{R}$ on an open subset U of a Banach space X, which is (Gâteaux) differentiable at $u \in U$ without being continuously differentiable at u, the generalized gradient $\partial f(u)$ does not generally reduce to $\{f'(u)\}$ (for comparison see (i) of Proposition 1.5). For instance, in the case of the locally Lipschitz function $f : \mathbb{R} \to \mathbb{R}$ given by $f(0) = 0$ and $f(x) = x^2 \sin(1/x)$ otherwise, one has $f'(0) = 0$ and $\partial f(0) = [-1, 1]$. In this respect we mention a notion of generalized derivative due to Michel and Penot [23] which is equal to the usual one-sided directional derivative under a weak condition. Namely, if $f : U \to \mathbb{R}$ is a function on an open subset of a Banach space X, the radial strict derivative of f at $u \in U$ in the direction $v \in X$ is

$$f^{\square}(u;v) = \sup_{w\in X}\limsup_{t\downarrow 0}\frac{1}{t}(f(u+tw+tv) - f(u+tw)).$$

The radial strict subdifferential of f at $u \in U$ is

$$\partial^{\square} f(u) = \{z \in X^* : \langle z, v \rangle \leq f^{\square}(u; v), \ \forall \ v \in X\}.$$

It is proved by Michel and Penot [23] that if f has a convex one-sided directional derivative $f'(u; \cdot)$ at $u \in U$ then $f^{\square}(u; \cdot) = f'(u; \cdot)$. Moreover, if f is Gâteaux differentiable at $u \in U$ then $\partial^{\square} f(u) = \{f'(u)\}$.

Employing the notion of generalized gradient in Definition 1.2 the corresponding calculus can be developed.

Proposition 1.6 (i) For each $\lambda \in \mathbb{R}$ and a locally Lipschitz function $f : U \to \mathbb{R}$ one has

$$\partial(\lambda f)(u) = \lambda \partial f(u), \quad \forall \ u \in U.$$

(ii) Given the locally Lipschitz functions $f, g : U \to \mathbb{R}$ the relation below holds

$$\partial(f + g)(u) \subset \partial f(u) + \partial g(u), \quad \forall \ u \in U.$$

Proof. (i) is obvious for $\lambda \geq 0$. Thus it suffices to justify it for $\lambda = -1$. This follows from (c) of Proposition 1.3.

(ii) It is equivalent to show that

$$(f + g)^0(u; v) \leq f^0(u; v) + g^0(u; v), \quad \forall \ v \in X$$

which is a direct consequence of Definition 1.1. □

Corollary 1.1 If $u \in U$ is a local minimum or maximum of the locally Lipschitz function $f : U \to \mathbb{R}$ on an open set of a Banach space X, then $0 \in \partial f(u)$.

Proof. Applying (i) of Proposition 1.6 for $\lambda = -1$ we see that it is sufficient to consider the case where $u \in U$ is a local minimum. Then we note that

$$f^0(u; v) \geq \limsup_{t \downarrow 0} \frac{1}{t}(f(u + tv) - f(u)) \geq 0$$

for all $v \in X$. By Definition 1.2 the result follows. □

The result below will be a basic tool in the following.

Theorem 1.1 (Lebourg's Mean Value Theorem [21]). Let U be an open subset of a Banach space X, let x, y be two points of U such that the line segment

$$[x, y] = \{(1 - t)x + ty : 0 \leq t \leq 1\}$$

is contained in U and let $f : U \to \mathbb{R}$ be a locally Lipschitz function. Then there exists $u \in [x, y] \setminus \{x, y\}$ satisfying

$$f(y) - f(x) = \langle z, y - x \rangle$$

for some $z \in \partial f(u)$.

Proof. The function $g : [0,1] \to \mathbb{R}$ given by

$$g(t) = f((1-t)x + ty), \quad \forall\, t \in [0,1]$$

is locally Lipschitz. It verifies

$$\begin{aligned} g^0(t;s) &= \limsup_{\substack{\tau \to t \\ \lambda \downarrow 0}} \frac{1}{\lambda}(g(\tau + \lambda s) - g(\tau)) \\ &= \limsup_{\substack{\tau \to t \\ \lambda \downarrow 0}} \frac{1}{\lambda}(f(x + (\tau + \lambda s)(y-x)) - f(x + \tau(y-x))) \\ &\leq \limsup_{\substack{w \to (1-t)x+ty \\ \lambda \downarrow 0}} \frac{1}{\lambda}(f(w + \lambda s(y-x)) - f(w)) \\ &= f^0((1-t)x + ty; s(y-x)), \qquad \forall\, t \in (0,1), \ \forall\, s \in \mathbb{R}. \end{aligned}$$

We introduce the function $h : [0,1] \to \mathbb{R}$, where

$$h(t) = g(t) + t(f(x) - f(y)), \quad \forall\, t \in [0,1].$$

Notice that $h(0) = h(1) = f(x)$, so h admits a local minimum or maximum point some $t \in (0,1)$. By Corollary 1.1 this implies

$$0 \in \partial h(t) = \partial g(t) + f(x) - f(y),$$

so, by the computation above and (ii) of Proposition 1.4

$$f(y) - f(x) \in \partial g(t) \subset \partial f((1-t)x + ty)(y-x).$$

The proof is complete. □

1.2 Subdifferentiation of Composite Mappings and Restrictions

The next objective is to establish the chain rule for the generalized gradients. To this end we recall that a map $F : U \to Y$ from an open set of a Banach space X to a Banach space Y is strictly differentiable at a point $u \in U$ if there is a continuous linear map $F'(u) : X \to Y$ provided

$$\lim_{\substack{w \to u \\ t \downarrow 0}} \frac{1}{t}(F(w + tv) - F(w)) = F'(u)v, \quad \forall\, v \in X,$$

and the convergence is uniform for v in compact sets.

Theorem 1.2 (Chain Rules in Clarke [9]-[12]).

(i) Let $F : U \to Y$ be a continuously differentiable map from an open set U of X with X and Y Banach spaces and let $g : Y \to \mathbb{R}$ be a locally Lipschitz function. Then one has

$$\partial(g \circ F)(u) \subset \partial g(F(u)) \circ F'(u), \quad \forall\, u \in U. \tag{1.8}$$

More generally, (1.8) is valid if F is strictly differentiable. Equality holds in (1.8) if, for instance, either g is regular at $F(u)$, or if $F'(u)$ is surjective.

(ii) Let $f : U \to \mathbb{R}$ be a locally Lipschitz function on an open set U of a Banach space X and let $h : \mathbb{R} \to \mathbb{R}$ be a locally Lipschitz function. Then one has

$$\partial(h \circ f)(u) \subset \overline{\text{co}}(\partial h(f(u)) \cdot \partial f(u)), \quad \forall\, u \in U, \tag{1.9}$$

where the notation $\overline{\text{co}}$ stands for the weakly*-closed convex hull. Furthermore, if either h is strictly differentiable (in particular, continuously differentiable) at $f(u)$ or if h is regular at $f(u)$ and f is strictly differentiable at u, in (1.9) the equality holds and the symbol $\overline{\text{co}}$ is superfluous.

Proof. (i) For simplicity we suppose that F is continuously differentiable.

The case of strict differentiability for F is treated in Clarke [12], Theorem 2.3.10. Clearly, by (ii) of Proposition 1.4 it is sufficient to show that

$$(g \circ F)^0(u; v) \leq \max\{\langle z, F'(u)v\rangle : z \in \partial g(F(u))\}, \quad \forall\, v \in X. \tag{1.10}$$

Accordingly, Theorem 1.1 allows us to write

$$g \circ F(w + tv) - g \circ F(w) = \langle \zeta, F(w + tv) - F(w)\rangle$$

for some $\zeta \in \partial g(y)$ and y belonging to the line segment that joins $F(w)$ and $F(w + tv)$ in Y. Then, the classical Mean Value Theorem ensures that

$$F(w + tv) - F(w) = tF'(x)v$$

for a point x in the interval between w and $w + tv$. Then we get

$$\begin{aligned}
(g \circ F)^0(u; v) &= \limsup_{\substack{w \to u \\ t \downarrow 0}} \frac{1}{t}(g \circ F(w + tv) - g \circ F(w)) \\
&= \limsup_{\substack{w \to u \\ t \downarrow 0}} \langle \zeta, F'(x)v\rangle = \langle \zeta, F'(u)v\rangle \\
&\leq \max\{\langle z, F'(u)v\rangle, \ \forall\, z \in \partial g(F(u))\}.
\end{aligned}$$

Suppose now that g is regular at $F(u)$ (see Remark 1.1). By (ii) of Proposition 1.4 and the regularity assumption we obtain for any $v \in X$ that

$$\begin{aligned}
&\max\{\langle z, F'(u)v\rangle : z \in \partial g(F(u))\} = \\
&= g^0(F(u); F'(u)v) = g'(F(u); F'(u)v) \\
&= \lim_{t \downarrow 0} \frac{1}{t}(g(F(u) + tF'(u)v) - g(F(u))) \\
&= \lim_{t \downarrow 0} \frac{1}{t}(g \circ F(u + tv) - g \circ F(u)) \leq (g \circ F)^0(u; v).
\end{aligned}$$

This yields the equality in (1.8).

The remaining situation is when $F'(u) : X \to Y$ is surjective. In view of Graves' Theorem [19] the mapping F is locally open at $u \in U$, so it is permitted to write

$$\begin{aligned} &\max\{\langle z, F'(u)v\rangle : z \in \partial g(F(u))\} = \\ &= g^0(F(u)); F'(u)v) = \limsup_{\substack{y \to F(u) \\ t \downarrow 0}} \frac{1}{t}(g(y + tF'(u)v) - g(y)) \\ &= \limsup_{\substack{x \to u \\ t \downarrow 0}} \frac{1}{t}(g(F(x) + tF'(u)v) - g(F(x))) \\ &= \limsup_{\substack{x \to u \\ t \downarrow 0}} \frac{1}{t}(g \circ F(x + tv) - g \circ F(x)) = (g \circ F)^0(u; v) \end{aligned}$$

for all $v \in X$. It follows that (1.8) is verified as an equality, as required.

(ii) By a double application of Theorem 1.1 we derive, with a fixed $v \in X$, that

$$h \circ f(w + tv) - h \circ f(w) = \alpha(f(w + tv) - f(w)) = \alpha\langle z, tv\rangle \tag{1.11}$$

for all w near u in U and all sufficiently small $t > 0$, where $\alpha \in \mathbb{R}$ belongs to $\partial h(s)$ for some s in the interval $[f(w), f(w + tv)]$ and $z \in \partial f(x)$, with $x \in U$ in the segment $[w, w + tv]$. Then, making use of property (iii) in Proposition 1.4 it is possible to show that (1.11) implies

$$(h \circ f)^0(u; v) \le \max\{a\langle \zeta, v\rangle : \zeta \in \partial f(u),\ a \in \partial h(f(u))\}.$$

Then formula (1.9) holds true. The proof of the assertions regarding the equality cases in (1.9) follows the same lines as in statement (i) and will be omitted. In the mentioned cases the symbol $\overline{co}$ is not necessary in (1.9), due to (ii) of Proposition 1.5, under the additional assumptions of strict differentiability. This completes the proof. □

For a later use we point out a special case of (i) in Theorem 1.2.

Corollary 1.2 (Chang [8], Theorem 2.2, or Clarke [12], p.47). Let X and Y be Banach spaces such that X is continuously imbedded in Y and is dense in Y. Let $g : Y \to \mathbb{R}$ be a locally Lipschitz function. Then, for every point $u \in X$ one has for the restriction $g\mid_X$ the formula

$$\partial(g\mid_X)(u) = \{z\mid_X : z \in \partial g(u)\} \equiv \partial g(u),$$

in the sense that every element z of $\partial(g|_X)$ admits a unique extension to an element of $\partial g(x)$.

Proof. From the proof of the last assertion of (i) in Theorem 1.2 we see that the same conclusion is achieved if the surjectivity assumption for $F'(u)$ is relaxed to the condition that F maps every neighborhood of u to a set which is dense in a neighborhood of $F(u)$. We observe that this is true in the case where F is the inclusion map of X into Y because it is supposed that X is dense in Y and the embedding $X \subset Y$ is continuous. Then (1.8) with equality yields the stated result. □

The theory of generalized gradients can be extended for locally Lipschitz functions on Banach manifolds. This can be done in a straightforward way by using local charts as shown for the first time in Motreanu and Pavel [32]. According to our strict goals we limit ourselves to the case of submanifolds of Banach spaces.

Definition 1.3 Let S be a C^1 submanifold of a Banach space X and let $f : X \to \mathbb{R}$ be a locally Lipschitz function. The generalized gradient of the restriction $f\mid_S$ of f to S at a point $u \in S$, denoted $\partial(f\mid_S)(u)$, is defined by

$$\partial(f\mid_S)(u) = \partial(f\mid_{T_uS})(0),$$

and S is identified (locally) with T_uS, thus $u \in S$ with $0 \in T_uS$.

All the preceding results admit appropriate versions in the setting of Definition 1.3. As an illustration of Definition 1.3 we determine the generalized gradient in the case where S is a sphere in a Hilbert space (see also Chang [8]). This is relevant in the nonlinear eigenvalue problems as shown, e.g., in Motreanu and Panagiotopoulos [27]-[31].

Example 1.1 Let X be a Hilbert space with the scalar product $(\cdot,\cdot)$ and the associated norm $\|\cdot\|$ and let $f : X \to \mathbb{R}$ be a locally Lipschitz function. For any number $r > 0$ consider the sphere

$$S_r = \{v \in X : \|v\| = r\}$$

which is known to be a smooth submanifold of X. Then $\partial(f\mid_{S_r})(u)$, at every $u \in S_r$, can be (isometrically) identified with the next subset of X^* described by

$$\begin{aligned}\partial(f\mid_{S_r})(u) &= \partial f(u) - \frac{1}{r^2}\langle \partial f(u), u\rangle (u,\cdot) \\ &= \{z - \frac{1}{r^2}\langle z, u\rangle (u,\cdot) : z \in \partial f(u)\}.\end{aligned}$$

Indeed, in view of Definition 1.3 and the characterization of the tangent space T_uS_r as

$$T_uS_r = \{v \in X : (v,u) = 0\},$$

to each $\zeta \in \partial(f\mid_{S_r})(u)$ we can associate a unique $z \in \partial f(u)$ such that

$$\zeta = z - \frac{1}{r^2}\langle z, u\rangle (u,\cdot) \quad \text{on } T_uS_r$$

and

$$\|\zeta\|_{T_u^* S_r} = \|z - \frac{1}{r^2}\langle z, u\rangle (u, \cdot)\|_{X^*}.$$

The situation presented in Example 1.1 can be extended as follows.

Example 1.2 On the Hilbert space H let $g : H \to \mathbb{R}$ be a C^1 function for which $0 \in \mathbb{R}$ is a regular value, i.e., $g'(x) \neq 0$ whenever $g(x) = 0$. It is well-known that

$$S = g^{-1}(0)$$

is a C^1-submanifold of H, whose tangent space T_uS at any $u \in S$ is expressed by

$$T_uS = g'(u)^{-1}(0) = \{x \in H : \langle g'(u), x\rangle = 0\}$$

(see, e.g., Abraham and Robbin [1]). Given a locally Lipschitz functional $f : H \to \mathbb{R}$ a reasoning similar to the one in Example 1.1 shows that, in the sense of Definition 1.3, the generalized gradient $\partial(f\mid_S)(u)$ at some $u \in S$ admits the following characterization

$$\partial(f\mid_S)(u) \equiv \partial f(u) - \frac{\langle \partial f(u), \nabla g(u)\rangle}{\|\nabla g(u)\|^2} g'(u) = \{z - \frac{\langle z, \nabla g(u)\rangle}{\|\nabla g(u)\|^2} g'(u) : z \in \partial f(u)\},$$

where $\nabla g(u)$ means the gradient of g at u, that is the element $\nabla g(u) \in H$ satisfying

$$\langle g'(u), v\rangle = (\nabla g(u), v), \quad \forall\, v \in H.$$

1.3 Subdifferentiation of Integral Functionals

First of all we discuss the subdifferentiation of the real-valued function in a form of a primitive $j : \mathbb{R} \to \mathbb{R}$,

$$j(t) = \int_0^t \beta(\tau) d\tau, \quad \forall\, t \in \mathbb{R}, \tag{1.12}$$

for $\beta \in L^\infty_{loc}(\mathbb{R})$. It is clear that j in (1.12) is locally Lipschitz. Its generalized gradient is described below. For other related results we indicate Chang [8] and Clarke [12].

Proposition 1.7 Under the foregoing conditions the next relation holds

$$\partial j(t) = [\underline{\beta}(t), \overline{\beta}(t)], \quad \forall\, t \in \mathbb{R},$$

where we use the notations

$$\underline{\beta}(t) = \lim_{\delta \to 0} \operatorname*{essinf}_{|\tau - t| < \delta} \beta(\tau)$$

and

$$\overline{\beta}(t) = \lim_{\delta\to 0} \operatorname*{esssup}_{|\tau-t|<\delta} \beta(\tau).$$

In particular, if there exist $\beta(t-0)$ and $\beta(t+0)$ at some $t \in \mathbb{R}$, then it is true that

$$\partial j(t) = [\min\{\beta(t-0), \beta(t+0)\}, \max\{\beta(t-0), \beta(t+0)\}].$$

Proof. It is straightforward to check that

$$\begin{aligned} j^0(t;1) &= \limsup_{\substack{\tau\to t\\ \lambda\downarrow 0}} \frac{1}{\lambda}(j(\tau+\lambda) - j(\tau)) \\ &= \limsup_{\substack{\tau\to t\\ \lambda\downarrow 0}} \frac{1}{\lambda}\int_\tau^{\tau+\lambda} \beta(s)ds \le \overline{\beta}(t). \end{aligned}$$

This implies that

$$\max \partial j(t) \le \overline{\beta}(t).$$

Analogously one shows that

$$\underline{\beta}(t) \le \min \partial j(t).$$

From the definition of $\overline{\beta}$ it is seen that there is a sequence $t_n \to t$ with $\beta(t_n) \to \overline{\beta}(t)$. Since j in (1.13) is differentiable a.e., we can choose $\{t_n\}$ such that $j'(t_n) = \beta(t_n)$. By (i) of Proposition 1.5 we know that $\beta(t_n) \in \partial j(t_n)$, thus by virtue of (iii) in Proposition 1.4 we deduce that $\overline{\beta}(t) \in \partial j(t)$. In a similar way we get $\underline{\beta}(t) \in \partial j(t)$. The assertions above and the fact that $\partial j(t)$ is an interval yield the stated result. □

In the following we deal with integral functionals on the spaces $L^p(T, \mathbb{R}^m)$, where $m \ge 1$, $p \ge 1$ and T is a positive complete measure space with mes $(T) < \infty$. The cases of interest for us will be $T = \Omega$ and $T = \partial\Omega$ for a bounded and open subset Ω of an Euclidean space $\mathbb{R}^n$ with the corresponding Lebesgue measure $\mathcal{L}_n$ of dimension n and Hausdorff measure H_{n-1} of dimension $n-1$. Other related information can be found in Aubin and Clarke [3], Chang [8] and Clarke [11], [12].

Let $j : T \times \mathbb{R}^m \to \mathbb{R}$ be a function such that $j(\cdot, y) : T \to \mathbb{R}$ is measurable whenever $y \in \mathbb{R}^m$ and satisfies either

$$|j(x, y_1) - j(x, y_2)| \le k(x)|y_1 - y_2|, \quad \forall\, x \in T,\ y_1, y_2 \in \mathbb{R}^m, \tag{1.13}$$

for a function $k \in L^q(T)$, with $1/p + 1/q = 1$, or, $j(x, \cdot) : \mathbb{R}^m \to \mathbb{R}$ is locally Lipschitz for all $x \in T$ and there is a constant $c > 0$ such that

$$|z| \le c(1 + |y|^{p-1}), \quad \forall\, x \in T,\ y \in \mathbb{R}^m,\ z \in \partial_y j(x, y). \tag{1.14}$$

The notation $|\cdot|$ utilized in (1.13),(1.14) stands for the Euclidean norm in $\mathbb{R}^m$, while $\partial_y j(x, y)$ in (1.14) means the generalized gradient of j with respect to the second variable $y \in \mathbb{R}^m$, i.e., $\partial j(x,\cdot)(y)$. We are now in a position to handle the functional $J : L^p(T,\mathbb{R}^m) \to \mathbb{R}$ introduced by

$$J(v) = \int_T j(x, v(x))dx, \quad \forall\, v \in L^p(T,\mathbb{R}^m), \tag{1.15}$$

where the integral is with respect to the complete measure on T. The main subdifferentiation result regarding J in (1.15) is formulated below.

Theorem 1.3 Under either (1.13) or (1.14) the functional $J : L^p(T,\mathbb{R}^m) \to \mathbb{R}$ of (1.15) is Lipschitz continuous on bounded sets and satisfies

$$\partial J(u) \subset \int_T \partial_y j(x, u(x))dx, \quad \forall\, u \in L^p(T,\mathbb{R}^m) \tag{1.16}$$

in the sense that for every $z \in \partial J(u)$ and $x \in T$ there is $w(x) \in \mathbb{R}^m$ such that

$$w(t) \in \partial_y j(t, u(t)) \;\text{ for a.e. } t \in T,$$

$w(\cdot)\xi \in L^1(T)$ whenever $\xi \in \mathbb{R}^m$ and

$$\langle z, v\rangle = \int_T w(x)v(x)dx, \quad \forall\, v \in L^p(T,\mathbb{R}^m).$$

Moreover, if $j(x,\cdot)$ is regular at $u(x)$ for all $x \in T$, then J is regular at u and (1.16) holds with equality.

Proof. The first step in the proof is to check that J is Lipschitz continuous on bounded subsets of $L^p(T,\mathbb{R}^m)$. Suppose that (1.13) is verified. Then it is straightforward to establish that J is just Lipschitz continuous on $L^p(T,\mathbb{R}^m)$. Assume now that (1.14) holds. For a fixed number $r > 0$ and arbitrary elements $u, v \in L^p(T,\mathbb{R}^m)$ with $\|u\|_{L^p} \le r,\ \|v\|_{L^p} \le r$ we can write the estimate

$$\begin{aligned} |J(u) - J(v)| &\le \int_T |j(x,u(x)) - j(x,v(x))|dx \\ &\le c_1 \int_T (1 + |u(x)|^{p-1} + |v(x)|^{p-1})|u(x) - v(x)|dx \\ &\le c_1 \left(\int_T (1 + |u(x)|^{p-1} + |v(x)|^{p-1})^{p/(p-1)}dx\right)^{(p-1)/p} \|u - v\|_{L^p} \\ &\le c_2 \left(\int_T (1 + |u(x)|^{p} + |v(x)|^{p})dx\right)^{(p-1)/p} \|u - v\|_{L^p} \le c_3\|u - v\|_{L^p}, \end{aligned}$$

with constants $c_1, c_2, c_3 > 0$ where c_3 depends on r. The inequalities above have been derived by using Theorem 1.1, assumption (1.14) and Hölder's inequality. The Lipschitz property on bounded sets for J is thus verified.

The map $x \to j_y^0(x, u(x); v(x))$ is measurable on T. Since $j(x,\cdot)$ is continuous, we may express $j_y^0(x, u(x); v(x))$ as the upper limit of

$$\frac{j(x, y + \lambda v(x)) - j(x, y)}{\lambda}$$

where $\lambda \downarrow 0$ taking rational values and $y \to u(x)$ taking values in a countable dense subset of $\mathbb{R}^m$. Thus $j_y^0(x, u(x); v(x))$ is measurable as the "countable limsup" of measurable functions of x.

The next task is to prove (1.16). To this end we are firstly concerned with the proof of the inequality

$$J^0(u;v) \leq \int_T j_y^0(x, u(x); v(x))dx, \quad \forall\, u, v \in L^p(T, \mathbb{R}^m), \tag{1.17}$$

where the subscript y indicates that the generalized directional derivative j^0 is taken with respect to $y \in \mathbb{R}^m$.

Assuming (1.13), it is permitted to apply Fatou's Lemma that leads directly to (1.17). Suppose now that assumption (1.14) is satisfied. Then using Theorem 1.1, we obtain

$$\frac{j(x, u(x) + \lambda v(x)) - j(x, u(x))}{\lambda} = \langle \zeta_x, v(x) \rangle$$

for some $\zeta_x \in \partial j(x, u^*(x))$ and for $u^*(x)$ lying in the interval $[u(x), u(x)+\lambda v(x)]$. We can now also use Fatou's lemma to obtain (1.17).

The final step, that we only sketch, is to pass from (1.17) to (1.16). Here the essential thing is to observe that, in view of (1.17), any $z \in \partial J(u)$ belongs to the subdifferential at $0 \in L^p(T, \mathbb{R}^m)$ (in the sense of convex analysis) of the convex function on $L^p(T, \mathbb{R}^m)$ mapping $v \in L^p(T, \mathbb{R}^m)$ to

$$\int_T j_y^0(x, u(x); v(x))dx \in \mathbb{R}. \tag{1.18}$$

These properties and the subdifferentiation results in § 2 of Ioffe and Levin [20] applied to (1.18) yield (1.16). Finally, we are dealing with the regularity assertion in the statement. Under either of hypotheses (1.13) or (1.14) we may apply Fatou's lemma to get, if the regularity of $j(x, \cdot)$ at $u(x)$ is imposed,

$$\begin{aligned} \liminf_{\lambda \downarrow 0} \frac{1}{\lambda}(J(u + \lambda v) - J(u)) \;&\geq\; \int_T j_y'(x, u(x); v(x))dx \\ &=\; \int_T j_y^0(x, u(x); v(x))dx, \qquad \forall\, v \in L^p(T; \mathbb{R}^m). \end{aligned}$$

Combining with (1.17) it is seen that there exists $J'(u;v)$ and $J'(u;v) = J^0(u;v)$, whenever $v \in L^p(T; \mathbb{R}^m)$, which means the regularity of J at u. Moreover we deduced the equality

$$J^0(u;v) = \int_T j_y'(x, u(x); v(x))dx, \quad \forall\, v \in L^p(T; \mathbb{R}^m).$$

Taking z in the right-hand side of (1.16), the regularity assumption for $j(x, \cdot)$ implies through the formula above that

$$\langle z, v \rangle = \int_T \langle z(x), v(x) \rangle dx \leq J^0(u;v), \quad \forall\, v \in L^p(T; \mathbb{R}^m),$$

so $v \in \partial J(u)$. This completes the proof. □

In the study of nonlinear boundary value problem an essential role is played by the potential constructed by means of Nemytski operator (see, e.g., Ambrosetti and Rabinowitz [2], Kavian [22], Rabinowitz [39]). In our nonsmooth setting this represents functionals J of the form (1.15) restricted to appropriate Sobolev spaces. As a consequence of Theorem 1.3 we point out a result due to Chang [8] that is very useful in such problems.

Corollary 1.3 Let $\beta \in L^{\infty}_{loc}(\mathbf{R})$ verify the growth condition

$$|\beta(t)| \leq c(1 + |t|^{p-1}), \quad \forall\, t \in \mathbf{R}, \tag{1.14$'$}$$

for constants $c > 0$ and $p \geq 1$. Then the functional $J : L^p(\Omega) \to \mathbf{R}$, where Ω is an open and bounded subset of $\mathbf{R}^N$, described by the Lebesgue integral

$$J(v) = \int_{\Omega} \int_0^{v(x)} \beta(t)dtdx, \quad \forall\, v \in L^p(\Omega), \tag{1.15$'$}$$

is Lipschitz continuous on bounded sets in $L^p(\Omega)$ and satisfies at any $u \in L^p(\Omega)$ the relation

$$\partial J(u) \subset [\underline{\beta}(u(x)), \overline{\beta}(u(x))] \text{ for a.e. } x \in \Omega, \tag{1.19}$$

where the notations in Proposition 1.7 have been used.

Proof. Notice that with the notation j in (1.12) the functional J of (1.15)′ can be expressed in the form of (1.15). Since, by making use of Proposition 1.7, condition (1.14) is valid from (1.14)′, we may invoke Theorem 1.3. The fact that J is Lipschitz on bounded subsets of $L^p(\Omega)$ is established. Now, using (1.16), Proposition 1.18 and (ii) of Proposition 1.7, we obtain for each $u \in L^p(\Omega)$ and $z \in \partial J(u)$ that

$$\langle z, v \rangle \leq \int_{v(x)<0} v(x)\underline{\beta}(u(x))dx + \int_{v(x)>0} v(x)\overline{\beta}(u(x))dx, \quad \forall\, v \in L^p(\Omega). \tag{1.20}$$

We justify (1.19) arguing by contradiction. For instance, let us admit that on a set A in Ω of positive Lebesgue measure one has

$$z(x) < \underline{\beta}(u(x)), \quad \forall\, x \in A.$$

Since (1.20) yields for $v = -\chi_A$ (the characteristic function of A) the inequality

$$\int_A (z(x) - \underline{\beta}(u(x))dx \geq 0,$$

we arrive at a contradiction. A similar reasoning shows that one cannot have the inequality

$$z(x) > \overline{\beta}(u(x))$$

on a set in Ω of positive measure. In this argument we employed implicitly the fact that $z \in L^{p/(p-1)}(\Omega)$. The proof of (1.19) is achieved. □

Remark 1.3 The conclusion of Corollary 1.3 remains true for functional $J : L^p(\Omega) \to \mathbb{R}$ in (1.15), with $T = \Omega$ and $m = 1$, if in addition to (1.13) or (1.14) a measurability condition for $\partial_y j$ holds. Namely, denoting for $j : \Omega \times \mathbb{R} \to \mathbb{R}$,

$$\partial_y j(x,y) = [\underline{j}(x,y), \overline{j}(x,y)]$$

for a.e. $x \in \Omega$ and $y \in \mathbb{R}$, we require that the functions $\underline{j}(x,y)$ and $\overline{j}(x,y)$ are Baire measurable on $\Omega \times \mathbb{R}$. Then the following extension of formula (1.19) holds

$$\partial J(u) \subset [\underline{j}(x,y), \overline{j}(x,y)] \quad \text{for a.e.} \quad x \in \Omega$$

and any $u \in L^p(\Omega)$. In the case where j does not depend on $x \in \Omega$ the functions $\underline{j}$ and $\overline{j}$ are Baire measurable, so the assumption above is satisfied. In particular, we obtain relation (1.19), if j is given by (1.12). For results of this type we refer to Chang [8]. These remarks hold also for other possible choices for T in (1.16), for example when $T = \partial\Omega$ for a smooth bounded domain Ω in $\mathbb{R}^n$.

1.4 Hemivariational Inequalities and Generalized Critical Point Problem

The critical point theory is nowdays one of the most important and fascinating tools in the modern nonlinear analysis. For our goals this tool is used in solving equations or differential inclusions whose structure is variational, i.e., those equations that are obtained by vanishing the differential (possibly in a generalized sense) of a functional on a Banach space or a Banach manifold. Among the various generalizations of the classical notion of critical (or stationary) point in smooth analysis we recall two significant concepts of critical points due to Chang [8] and Szulkin [44] that are very useful in nonsmooth analysis. Later in Chapter 3 we shall encounter and study more general new critical point notions.

Definition 1.4 (Chang [8]). An element $u \in X$ is said to be a critical point of a locally Lipschitz functional $I : X \to \mathbb{R}$ on a Banach space or, generally, a C^1 Banach manifold X, if

$$0 \in \partial I(u).$$

An equivalent formulation of the foregoing condition is that

$$I^0(u; v) \geq 0, \quad \forall\, v \in X,$$

if X is a Banach space.

Examples 1.3 With X as above and I continuously differentiable on X, Definition 1.4 coincides with the usual concept of critical point for I, i.e., $I'(u) = 0$. A local minimum (or maximum) of a locally Lipschitz function $I : X \to \mathbb{R}$ is a critical point of I in the sense of Definition 1.4 as shown in Corollary 1.1. If X is a (convex open subset of a) Banach space and $I : X \to \mathbb{R}$ is locally Lipschitz and convex, then the converse is true, and thus every critical point of I is a global minimum of I because $\partial I(u)$ is equal to the subdifferential of I at $u \in X$ (cf. (iii) of Proposition 1.5). □

We indicate at this point a second notion of critical point in the nonsmooth setting related to possibly discontinuous functions.

Definition 1.5 (Szulkin [44]). Let $I : X \to \mathbb{R} \cup \{+\infty\}$ be a functional on the Banach space X satisfying the hypothesis $I = E + F$ with $E \in C^1(X, \mathbb{R})$ and $F : X \to \mathbb{R} \cup \{+\infty\}$ convex, proper (in the sense that is not $\equiv +\infty$) and lower semicontinuous. A point $u \in X$ is called a critical point of I if $-E'(u) \in \partial F(u)$, where as usually $E'(u)$ and $\partial F(u)$ stand for the Fréchet differential of E at u and the subdifferential of F at u, respectively. Consequently, this means that

$$\langle E'(u), v - u\rangle + F(v) - F(u) \geq 0, \quad \forall\, v \in X. \tag{1.21}$$

Examples 1.4 Given $I : X \to \mathbb{R} \cup \{+\infty\}$ as in Definition 1.5, a local minimum $u \in X$ is a critical point of I in the sense of Definition 1.5. Indeed, if $u \in X$ minimizes locally I, for all small $t > 0$ the differentiability of E and the convexity of F yield

$$\begin{aligned} 0 &\leq I((1-t)u + tv) - I(u) \\ &\leq E(u + t(v-u)) - E(u) + t(F(v) - F(u)), \quad \forall\, v \in X, \end{aligned}$$

and thus (1.21) dividing by t and then letting $t \to 0$. A very useful situation in Definition 1.5, which is relevant especially for solving variational inequalities is when F is equal to the indicator function of a nonempty closed convex set K in X, i.e., $F = 0$ on K and $F = +\infty$ otherwise. Then (1.21) reduces to

$$\langle E'(u), v - u\rangle \geq 0, \quad \forall\, v \in K.$$

We briefly discuss now the notion of hemivariational inequality that has been introduced in 1981 by P. D. Panagiotopoulos [33]-[37].

Definition 1.6 (Panagiotopoulos [33]-[37]). Let V be a Banach space that is continuously embedded in $L^p(\Omega; \mathbb{R}^m)$, with $p \geq 1$ and an integer $m \geq 1$, for a bounded domain Ω in $\mathbb{R}^N$. Let $a : V \times V \to \mathbb{R}$ be a continuous, symmetric, bilinear form on V, let $f \in V^*$ and let $j : \Omega \times \mathbb{R}^m \to \mathbb{R}$ be a function measurable in the first variable satisfying the requirement (1.13) or (1.14). By a hemivariational inequality we mean the following problem: find $u \in V$ such that

$$a(u, v) + \langle f, v\rangle + \int_\Omega j_y^0(x, u(x); v(x))dx \geq 0, \quad \forall\, v \in V. \tag{1.22}$$

Here, as previously, the notation j_y^0 denotes the generalized directional derivative of j with respect to the second variable $y \in \mathbb{R}^m$.

The main result of this Section reveals the relationship between the concept of critical point introduced in Definition 1.4 and the solution of the hemivariational inequality in (1.22).

Theorem 1.4 Assume that the data a, f and j fulfill the conditions required in Definition 1.6. Define the locally Lipschitz functional $I : V \to \mathbb{R}$ by

$$I(v) = \frac{1}{2}a(v,v) + \langle f, v\rangle + \int_\Omega j(x, v(x))dx, \quad \forall\, v \in V. \tag{1.23}$$

Then any critical point $u \in V$ of I is a solution of (1.22). If $j(x, \cdot)$ is regular for all $x \in \Omega$, then conversely every solution $u \in V$ of (1.22) is a critical point of the associated functional I in (1.23).

Proof. Notice that the functional I of (1.23) can be expressed as

$$I(v) = \frac{1}{2}a(v,v) + \langle f, v\rangle + J(v), \quad \forall\, v \in V,$$

with $J : L^p(\Omega; \mathbb{R}^m) \to \mathbb{R}$ given by (1.15). So I is locally Lipschitz and it makes sense to consider the critical points of I in the framework of Definition 1.4.

Assume that $u \in V$ is a critical point of I. Then Definition 1.4 and assertion (i) of Theorem 1.3 imply

$$0 \in a(u, \cdot) + f + \partial J(u)\mid_V .$$

Using relation (1.16) in Theorem 1.3 we infer that there exists $z \in L^{p/(p-1)}(\Omega; \mathbb{R})$ such that

$$a(u,v) + \langle f, v\rangle + \int_\Omega z(x)\cdot v(x)dx = 0, \quad \forall\, v \in V,$$

and

$$z(x) \in \partial_y j(x, u(x)) \ \text{ for a.e. } \ x \in \Omega.$$

It is now clear that by means of (ii) in Proposition 1.4 it follows that u is a solution to (1.22).

Conversely, if $u \in V$ solves (1.22) and $j(x, \cdot)$ is regular for all $x \in \Omega$ then the last assertion of Theorem 1.4 ensures that (1.22) can be written as $0 \in \partial I(u)$ which ends the proof. □

Corollary 1.4 In addition to the assumptions of Theorem 1.4 in the case $m = 1$ we admit that $f \in L^{p/(p-1)}(\Omega)$ and the measurability condition in Remark 1.4 for $\underline{j}$ and $\overline{j}$ defined therein. Then any critical point $u \in V$ of I in (1.23) satisfies

$$-a(u, \cdot) - f \in [\underline{j}(x, u(x)), \overline{j}(x, u(x))] \ \text{ for a.e. } \ x \in \Omega.$$

Proof. It suffices to apply Theorem 1.4 and Corollary 1.3 in the form stated in Remark 1.3. □

We close this Section with a brief discussion concerning the eigenvalue problems that can be considered in the setting of hemivariational inequalities.

Throughout the rest of the Section we assume that the conditions in Definition 1.6 for the data a and j are met. Then we formulate the following eigenvalue problem: find $u \in V\backslash\{0\}$ and $\lambda \in \mathbb{R}\backslash\{0\}$ such that

$$(P_1) \qquad a(u,v) + \lambda \int_\Omega j^0(x, u(x); v(x))dx \geq 0, \quad \forall\, v \in V.$$

Using the critical points on manifolds we get the result below.

Proposition 1.8 Assume in addition that a is coercive in the sense that there is a constant $\alpha > 0$ provided

$$a(v,v) \geq \alpha \|v\|^2, \quad \forall\, v \in V.$$

For some $r > 0$ let us consider the smooth submanifold S of V given by

$$S = \{v \in V : \; a(v,v) = r\}.$$

If $u \in S$ is a critical point of $J\mid_S$ (in the sense of Definition 1.3) with $J : L^p(\Omega, \mathbb{R}^m) \to \mathbb{R}$ defined by (1.15) such that

$$\int_\Omega z(x) \cdot u(x)dx < 0 \tag{1.24}$$

for all $z \in L^{p/(p-1)}(\Omega, \mathbb{R}^m)$ satisfying

$$z(x) \in \partial_y j(x, u(x)) \text{ for a.e. } x \in \Omega, \tag{1.25}$$

then u is an eigenfunction of (P_1) whose corresponding eigenvalue $\lambda > 0$ is expressed as

$$\lambda = -\left(\int_\Omega z(x) \cdot u(x)dx\right)^{-1} r$$

for some $z \in L^{p/(p-1)}(\Omega, \mathbb{R}^m)$ which verifies (1.25).

Proof. Since $u \in S$ is a critical point of $J\mid_S$, in view of Example 1.2 we know that there is $z \in L^{p/(p-1)}(\Omega)$ such that (1.25) holds and

$$z - r^{-1} \int_\Omega z(x) \cdot u(x)dx \; a(u, \cdot) = 0 \text{ in } V^*.$$

Here we have also used Theorem 1.3 and V was considered as a Hilbert space endowed with the scalar product $a(\cdot,\cdot)$. By (1.24) we derive

$$a(u, \cdot) + \lambda z = 0 \text{ in } V^*$$

where λ has the expression in the statement because $a(u,u) = r$. Taking into account that $\lambda > 0$, property (ii) of Proposition 1.4 yields that $u \in S$ solves the eigenvalue problem (P_1). □

Remark 1.4 Various results concerning the behaviour of eigensolutions (u, λ) to problem (P_1) in the smooth setting are established in Rabinowitz [38],[39],

Schechter and Tintarev [42],[43], Motreanu [24],[25]. In the locally Lipschitz case such problems are considered in Chang [8].

Another eigenvalue problem is formulated as follows: find $u \in V$ and $\lambda \in \mathbb{R}$ such that

$$(P_2) \qquad a(u,v) + \langle f, v\rangle + \textstyle\int_\Omega j^0(x, u(x); v(x))dx \geq \lambda(u,v)_V, \quad \forall\, v \in V,$$

where V is a Hilbert space with the scalar product $(\cdot,\cdot)_V$. The data a, f and j have the meaning and satisfy the assumptions in Definition 1.6. The eigenvalue problem (P_2) can be studied via the critical point methods in two different ways. For specific applications we indicate [27]-[31].

Proposition 1.9 Assume that the assumptions in Definition 1.6 are satisfied. Given any $\lambda \in \mathbb{R}$, each critical point $u \in V$ of the locally Lipschitz functional $I_\lambda : V \to \mathbb{R}$ expressed by

$$I_\lambda(v) = \frac{1}{2}(a(v,v) - \lambda\|v\|_V^2) + \langle f, v\rangle + \int_\Omega j(x, v(x))dx, \quad \forall\, v \in V,$$

is an eigenfunction in (P_2) whose corresponding eigenvalue is λ. □

Proof. Clearly, $0 \in \partial I_\lambda(u)$ is equivalent to

$$-a(u,\cdot) + \lambda(u,\cdot)_V - f \in \partial\left(\int_\Omega j(x,\cdot)dx\right)(u).$$

It suffices now to apply Theorem 1.3.

Proposition 1.10 Suppose that the assumptions in Definition 1.6 are satisfied with V a Hilbert space. Then for every number $r > 0$, a critical point $u \in S_r$ of $I\,|_{S_r}$, where $I : V \to \mathbb{R}$ is the locally Lipschitz functional in (1.23) and

$$S_r = \{v \in V : \ \|v\|_V = r\},$$

is an eigenfunction of (P_2), whose eigenvalue $\lambda \in \mathbb{R}$ satisfies

$$\lambda = r^{-2}(a(u,u) + \langle f, u\rangle + \int_\Omega z(x)u(x)dx)$$

with some $z \in L^{p/(p-1)}(\Omega; \mathbb{R}^m)$ verifying (1.25).

Proof. It is enough to combine Definition 1.4, Example 1.1 and Theorem 1.3. □

A third eigenvalue problem of hemivariational type which is very important in applications (cf. Goeleven, Motreanu and Panagiotopoulos [16]-[18]) is considered below. Assume that the Banach space V is densely and compactly embedded in $L^2(\Omega; \mathbb{R}^m)$. We state the next eigenvalue problem: find $u \in V$ and $\lambda \in \mathbb{R}$ such that

$$(P_3) \qquad a(u,v) + \langle f, u\rangle + \textstyle\int_\Omega j^0(x,u(x);v(x))dx \geq$$

$$\geq \lambda \int_\Omega u(x)v(x)dx, \quad \forall\, v \in V,$$

where the data a, f, j have the meaning above.

Proposition 1.11 Assume that the hypotheses in Definition 1.6 are fulfilled. Then, for every $\lambda \in \mathbf{R}$, a critical point $u \in V$ of the locally Lipschitz functional $I : V \to \mathbf{R}$ given by

$$\begin{aligned} I(v) &= \frac{1}{2}(a(v,v) - \lambda \int_\Omega v(x)^2 dx) + \langle f, u\rangle \\ &+ \int_\Omega j(x, v(x))dx, \quad \forall\, v \in V, \end{aligned}$$

is an eigenfunction of (P_3) corresponding to λ.

Proof. The result follows from Definition 1.4 and Theorem 1.3.

Remark 1.5 Eigenvalue problem (P_3) can be treated as a critical point problem on the submanifold S of V described by

$$S = \{v \in V : \; \|v\|_{L^2(\Omega;\mathbf{R}^m)} = r\}$$

with some $r > 0$ (following the pattern of Proposition 1.10). Many other versions of (P_3) are of interest in applications. For instance, assuming that Ω has a C^1 boundary $\partial\Omega$, we state the eigenvalue problem:

$$(P_3') \qquad a(u,v) + \langle f, u\rangle + \textstyle\int_\Omega j^0(x,u(x);v(x))dx \geq$$

$$\geq \lambda \int_{\partial\Omega} \hat{u}(\sigma)\hat{v}(\sigma)d\sigma, \quad \forall\, v \in V,$$

where $(u, \lambda) \in V \times \mathbf{R}$ denotes the eigensolution and $\hat{v} := Lv$ for some, "trace" operator $L : V \to L^2(\partial\Omega)$. □

We illustrate the eigenvalue problem (P_3) with the Dirichlet problem for semilinear elliptic equations containing discontinuous nonlinearities (see Rauch [40], Chang [8], Goeleven, Motreanu and Panagiotopoulos [16]-[18]).

Example 1.5 Given $\beta \in L^\infty_{loc}(\mathbf{R})$, on a bounded domain Ω in $\mathbf{R}^N$ with a C^1 boundary $\partial\Omega$ we consider the Dirichlet problem

$$\Delta u + \lambda u \in [\underline{\beta}(u(x)), \overline{\beta}(u(x))] \;\text{ for a.e. } x \in \Omega,$$

$$u = 0 \;\text{ on }\; \partial\Omega,$$

with as unknowns the function $u : \overline{\Omega} \to \mathbf{R}$ and the number $\lambda \in \mathbf{R}$, where the notations in Proposition 1.7 are used. Then Corollary 1.3 shows that every eigenfunction $u \in V$ of (P_3) with $f = 0$ and

$$a(v,w) = \frac{1}{2}\int_\Omega \nabla v \cdot \nabla w \, dx, \quad \forall\, v, w \in V,$$

for $V = H_0^1(\Omega)$, is an eigenfunction of our problem. □

Finally, we mention the following double eigenvalue problem for hemivariational inequalities: find $u_1, u_2 \in V$ and $\lambda_1, \lambda_2 \in \mathbb{R}$ such that

$$(P_4) \qquad a_1(u_1,v_1) + a_2(u_2,v_2) + \textstyle\int_\Omega j_y^0(x,(u_1-u_2)(x);(v_1-v_2)(x))dx \geq$$

$$\geq \lambda_1(B_1u_1,v_1)_V + \lambda_2(B_2u_2,v_2)_V\ , \quad \forall\, v_1, v_2 \in V,$$

where V stands for a Hilbert space with the scalar product $(\cdot,\cdot)_V$ which is densely and compactly embedded in $L^2(\Omega;\mathbb{R}^N)$, $a_1, a_2 : V \times V \to \mathbb{R}$ are continuous symmetric bilinear forms and $B_1, B_2 : V \to V$ are self-adjoint bounded linear operators. The foregoing critical point approach can be extended to study the eigenvalue problem (P_4).

1.5 Elements of the Theory of Hemivariational Inequalities

Several types of hemivariational inequalites can be formulated and studied. The most simple type first studied in [35],[36],[37] is the following. Let $u \in X$ (Hilbert space), $l \in X^*$ (the dual space) and $a(\cdot,\cdot) : X \times X \in \mathbb{R}$ a bilinear form. Moreover let $j : \mathbb{R} \times \mathbb{R}$ be a locally Lipschitz function. The problem reads:Find

$u \in V$ such that

$$a(u, v-u) + \int_\Omega j^0(u, v-u)d\Omega \geq (l, v-u) \qquad \forall\, v \in X.$$

An analogous problem arises for $J : V \to \mathbb{R}$ locally Lipschitz and $\Phi : V \to (-\infty, +\infty]$, $\Phi \not\equiv \infty$, convex and l.s.c.:Find $u \in X$ such that

$$a(u, v-u) + J^0(u, v-u) + \Phi(v) - \Phi(u) \geq (l, v-u) \qquad \forall\, v \in X.$$

which in the case of $\Phi = I_K$ (indicator of K convex and closed in V) takes the form:Find $u \in K$ such that

$$a(u, v-u) + J^0(u, v-u) \geq (l, v-u) \qquad \forall\, v \in K.$$

This inequality is called also a variational – hemivariational inequality. Another interesting type of hemivariational inequality is the following. Let C be closed subset of X and let $T_C(u)$ be the tangent cone in the sense of Clarke at the point u (cf. e.g. [33] p.12). Then we look for $u \in C$ such that

$$a(u,v) \geq (l,v) \qquad \forall\, v \in T_C(u).$$

All the above types of hemivariational inequalities and their generalizations (e.g. for $a(u,v)$ replaced by (Au,v), where $A : X \to X^*$ is a pseudomonotone mapping etc.) are studied in [33] with several methods. Here we shall treat as a pilot problem a general variational hemivariational inequality using a method different from the methods applied in [33]. Indeed here we apply a critical point method, whereas in [33] methods based on compactness and/or multivalued pseudomonotonicity and quasipseudomonotonicity arguments are used.

We treat further the following variational-hemivariational inequality on a real Hilbert space X

$$\langle Tu, v-u\rangle + C^0(u; v-u) + \Psi(v) - \Psi(u) \geq 0, \quad \forall\, v \in X. \tag{1.26}$$

Precisely, we look for the nontrivial solutions of (1.26) in the sense that $u \notin \text{Ker } T$. The mapping $T : X \to X^*$ is a symmetric bounded linear operator whose kernel is finite dimensional

$$\dim \text{Ker } T < \infty \tag{1.27}$$

and which is semicoercive in the sense

$$\langle Tx, x\rangle \geq c\|Px\|^2, \quad \forall\, x \in X. \tag{1.28}$$

Here $c > 0$ is a constant, and $P : X \to (\text{Ker } T)^\perp$ denotes the orthogonal projection onto the orthogonal complement $(\text{Ker } T)^\perp$ of Ker T. The function $C : X \to \mathbb{R}$ is locally Lipschitz, weakly l.s.c., β-positively homogeneous for a given number β, i.e.

$$C(tx) = t^\beta C(x), \quad \forall\, x \in X, \quad t > 0. \tag{1.29}$$

The notation $C^0(\,\cdot\;\cdot\,)$ means the generalized directional derivative of C in the sense of Clarke that is

$$C^0(u;v) = \limsup_{\substack{w \to u \\ t \downarrow 0}} \frac{1}{t}(C(w+tv) - C(w)), \quad \forall\, u, v \in X. \tag{1.30}$$

Finally, we have that Ψ in (1.26) is a weakly lower semicontinuous function $\Psi : X \to (-\infty, +\infty]$.

We treat the variational inequality (1.26) by means of critical point methods. The closed unit ball in X is denoted by B_1. Let us put

$$\text{dom}\, \Psi := \{x \in X : \; \Psi(x) < +\infty\}.$$

and let us consider the functional $I : X \to (-\infty, +\infty]$

$$I(v) = \frac{1}{2}\langle Tv, v\rangle + C(v) + \Psi(v), \quad \forall\, v \in X. \tag{1.31}$$

We will make use of some new critical point results which will be given here and will be discussed and proved in detail in Chapter 3.

Definition 1.7 A critical point of I represents in the sequel any solution of the inequality problem

$$\Phi^0(u; v-u) + \Psi(v) - \Psi(u) \geq 0, \quad \forall\, v \in X.$$

where

$$\Phi(x) = \frac{1}{2}\langle Tx, x\rangle + C(x), \quad \forall\, x \in X.$$

and obviously $I = \Phi + \Psi$.

Moreover the functional $I : X \to (-\infty, +\infty]$ is said to satisfy the Palais-Smale condition at the level $c \in \mathbb{R}$ (in short, $(\mathrm{PS})_c$) if every sequence $\{u_n\} \subset X$ for which one has $I(u_n) \to c$ and

$$\Phi^0(u_n; v-u_n) + \Psi(v) - \Psi(u_n) \geq -\varepsilon_n \|v-u_n\|, \quad \forall\, v \in X \tag{1.32}$$

with $\varepsilon_n \downarrow 0$ in $\mathbb{R}$, contains a convergent subsequence. If this occurs for all $c \in \mathbb{R}$, I is said to satisfy the Palais-Smale condition (in short, (PS)).

We will apply the following result due to the authors, which is also proved in Corollary 3.3.

Theorem 1.5 Let $X = X_1 \oplus X_2$ with $\dim X_1 < \infty$ and let the functional $I : X \to (-\infty, +\infty]$ of (1.31) satisfy the (PS), and

$$I(v) \leq a, \quad \forall\, v \in X_1 \text{ with } \|v\| = \rho, \tag{1.33}$$

for constants $\rho > 0$ and a $a \in \mathbb{R}$, and

$$I(v) \geq b, \quad \forall\, v \in X_2, \tag{1.34}$$

for a constant $b > a$ satisfying

$$b > a. \tag{1.35}$$

Then the number

$$c = \inf_{f \in \Gamma} \max_{x \in Q} I(f(x)), \tag{1.36}$$

where

$$\Gamma = \{f \in C(Q, X) : f|_{\partial Q} = id_{\partial Q}\} \tag{1.37}$$

for

$$Q = \{x \in X_1 : \|x\| \leq 1\} \quad \text{and} \quad \partial Q = \{x \in X_1 : \|x\| = 1\},$$

is a critical value of I, i.e., there is an element $u \in X$ satisfying

$$\Phi^0(u; v-u) + \Psi(v) - \Psi(u) \geq 0, \quad \forall\, v \in X,$$

and

$$I(u) = c,$$

and in addition

$$c \geq b \tag{1.38}$$

□

Now we prove the following result.

Proposition 1.12 Assume that the symmetric operator $T \in L(X, X^*)$, the locally Lipschitz, weakly l.s.c. function $C : X \to \mathbb{R}$ and the weakly l.s.c. function $\Psi : X \to (-\infty, +\infty]$ satisfy (1.27),(1.28),(1.29)

$$\Psi(tx) = t^\alpha \Psi(x), \quad \forall \; \|x\| = 1, \quad t \geq 2, \tag{1.39}$$

$$\liminf_{n \to \infty} C^0(u_n; \overline{u} - u_n) \leq 0 \;\; \text{whenever } u_n \to \overline{u} \text{ weakly in } X, \tag{1.40}$$

$$\sup_{\partial B_1 \cap \text{Ker } T} C < 0, \qquad \inf_{\partial B_1 \cap (\text{Ker } T)^\perp} C \geq 0, \tag{1.41}$$

$$\Psi \text{ is bounded above on } \partial B_1 \cap Ker\ T, \tag{1.42}$$

$$\Psi \geq 0, \text{ and } \Psi > 0 \text{ on } \partial B_1 \cap Ker\ T, \tag{1.43}$$

$$\alpha > 1, \quad \beta > \max\{2, 2^\alpha - 1, \alpha\}. \tag{1.44}$$

The problem (1.26) admits a solution $u \in X$. If in addition there is no solution in $Ker\ T$ of the hemivariational inequality

$$C^0(u; v - u) + \Psi(v) - \Psi(u) \geq 0, \quad \forall\; v \in X,$$

then u is nontrivial in the sense that $u \notin Ker\ T$.

Proof. Our goal is to apply Theorem 1.5. To this end we start by checking that I verifies condition $(\text{PS})_c$ for any $c \in \mathbb{R}$. Let $\{u_n\}$ be a sequence in X satisfying $I(u_n) \to c$ and (1.32). If we set $v = 2u_n$ in (1.32) and we use (1.31), for all n sufficiently large the inequality below holds

$$\begin{aligned}\beta(c+1) + \|u_n\| \;\; &\geq \;\; (\frac{\beta}{2} - 1)\langle Tu_n, u_n\rangle + \beta C(u_n) \\ &\quad -C^0(u_n; u_n) + (\beta+1)\Psi(u_n) - \Psi(2u_n).\end{aligned}$$

Using the homogeneity assumption (1.29) we obtain

$$\beta(c+1) + \|u_n\| \geq (\frac{\beta}{2} - 1)\langle Tu_n, u_n\rangle - \Psi(2u_n) + (\beta+1)\Psi(u_n). \tag{1.45}$$

Firstly, we want to deduce from (1.45) that $\{u_n\}$ is bounded. On the contrary we assume that $\|u_n\| \to \infty$ as $n \to \infty$. If we set $x_n = u_n / \|u_n\|$, relations (1.45),(1.39),(1.43),(1.44) imply that

$$(\frac{\beta}{2} - 1)\langle Tu_n, u_n\rangle \leq \beta(c+1)\frac{1}{\|u_n\|^2} + \frac{1}{\|u_n\|} \tag{1.46}$$

for a sufficiently large n.

Along a subsequence we may suppose that $x_n \to x$ weakly in X. From (1.46) it turns out that $\langle Tx, x\rangle = 0$, hence due to (1.28) that $Px = 0$. Thus we have that

$$x \in \text{Ker } T \text{ and } \|x\| = 1.$$

Again from (1.45), together with (1.39) we derive that

$$(\beta + 1 - 2^\alpha)\Psi(x_n) \leq \beta(c+1)\frac{1}{\|u_n\|^\alpha} + \frac{1}{\|u_n\|^{\alpha-1}}. \tag{1.47}$$

The fact that Ψ is weakly l.s.c. and (1.44) ensure that $\Psi(x) \leq 0$. A simple comparison with (1.43),(1.47) yields a contradiction. This establishes the boundedness of the sequence $\{u_n\}$. Therefore we may assume that, up to a subsequence, $u_n \to \overline{u}$ weakly in X for some $\overline{u} \in X$. Then from (1.32) with $v = \overline{u}$ we get

$$\langle Tu_n, \overline{u} - u_n\rangle + C^0(u_n; \overline{u} - u_n) + \Psi(\overline{u}) - \Psi(u_n) \geq -\varepsilon_n \|\overline{u} - u_n\|.$$

If we pass to the limit as $n \to \infty$ and we use that $\varepsilon_n \to 0$, we obtain that

$$\liminf_{n\to\infty}(\langle Tu_n, u_n - \overline{u}\rangle + \Psi(u_n) - \Psi(\overline{u})) \leq \liminf_{n\to\infty} C^0(u_n; \overline{u} - u_n).$$

In this way we find

$$\liminf_{n\to\infty} \|P(u_n - \overline{u})\|^2 \leq \liminf_{n\to\infty}\langle T(u_n - \overline{u}), u_n - \overline{u}\rangle \leq 0.$$

It results that for a subsequence again denoted by u_n, $Pu_n \to P\overline{u}$ strongly in X. Due to (1.27) we can conclude that $\{u_n\}$ possesses a strongly convergent subsequence. This proves the (PS).

Take now a point $u \in$ Ker T with $\|u\| \geq 2$. Then (1.29) and (1.39) yield the estimate

$$\begin{aligned} I(u) &= \|u\|^\beta C(\frac{u}{\|u\|}) + \|u\|^\alpha \Psi(\frac{u}{\|u\|}) \\ &\leq \|u\|^\beta \sup_{\partial B_1 \cap \text{Ker } T} C + \|u\|^\alpha \sup_{\partial B_1 \cap \text{Ker } T} \Psi. \end{aligned}$$

Using the first property in (1.41) and assumptions (1.42),(1.44) we obtain that

$$\lim_{\substack{\|u\| \to \infty \\ u \in \text{Ker } T}} I(u) = -\infty. \tag{1.48}$$

For any $u \in (\text{Ker } T)^\perp \setminus \{0\}$ we have that

$$I(u) = \langle T(u), u\rangle + \|u\|^\beta C(\frac{u}{\|u\|}) + \Psi(u) \geq c\|u\|^2. \tag{1.49}$$

Here (1.28), the second property in (1.44) and (1.43) have been applied. Thus it is true from (1.49) that

$$I \mid_{(\text{Ker } T)^\perp} \geq 0. \tag{1.50}$$

Notice that (1.50) implies (1.34) with $b = 0$ and $X_2 = (\text{Ker } T)^{\perp}$. Relation (1.48) allows to obtain (1.33) for some $\alpha < 0$, $\rho > 0$ sufficiently large and $X_1 = \text{Ker } T$. Hence (1.35) holds true. Consequently, all the requirements of Theorem 1.5 are verified for the locally Lipschitz functional in (1.31). Thus a critical point $u \in X$ of I is obtained in the sense of Definition 1.7. Due to the inequality in this Definition the $u \in X$ obtained is a solution of problem (1.26). The last assuption of the Proposition implies that the solution $u \in X$ of problem (1.26) does not belong to Ker T. This completes the proof. □

In the next Chapters of the book certain multiplicity results for the solutions of (1.26) will be proved.

1.6 Historical and Bibliographical Notes

The calculus with generalized gradients for locally Lipschitz functions has been introduced and developed by F. H. Clarke [9]-[12]. The notion of generalized gradient of a locally Lipschitz function plays the same crucial role as the concept of derivative or differential in the case of a smooth function. It extends and unifies the fundamental notions of Fréchét differential for a continuously differentiable function and of subdifferential for convex function. Different geometric aspects in optimization are related to the theory of generalized gradients (see, e.g., Aubin and Frankovska [4], Ekeland [13], Rockafellar [41]). Among the basic calculus results discovered for generalized gradients there are the chain rules (Clarke [11]), the mean-value theorem (Lebourg [21]) and the subdifferentiation of integral functionals (Aubin and Clarke [3]).

The critical point approach with generalized gradients has been initiated by Chang [8]. This works also on Banach manifolds. The possibility to define the generalized gradients on Banach manifolds was first observed by Motreanu and Pavel [32]. Another critical point theory for nonsmooth functionals which is different from Chang's approach is due to Szulkin [44]. The difference lies in the fact that Chang studies locally Lipschitz functionals, while Szulkin deals with sums of a continuously differentiable function and a lower semicontinuous convex function. Further in the book these two fundamental approaches are unified (cf. Chapter 3).

The application of nonsmooth critical point theory to differential equations with discontinuities is due to Chang [8]. The concept of solution in an appropriate weak sense was considered before Chang [8] by Rauch [40]. This goes back to Filippov [15]. Related concepts and results are presented in Barbu [5],[6], Motreanu [24] and Motreanu and Naniewicz [26]. The minimax methods in the critical point theory have been developed by many authors starting with the celebrated paper of Ambrosetti and Rabinowitz [2]. A rich field of application for these variational methods was the study of nonlinear elliptic boundary value problems (see the surveys of Rabinowitz [38],[39]). In the setting of differential equations with discontinuous nonlinearities such methods were applied by Chang [8].

In order to treat various phenomena in Mechanics under the presence of nonconvex nonsmooth energy functions (nonconvex superpotentials) P. D. Panagiotopoulos introduced the notion of hemivariational inequalities [33]-[37]. There is a close relationship between the theory of hemivariational inequalities and the study of semilinear elliptic equations. The variational approach in the study of hemivariational inequality permitts a progress in the study of the qualitative behaviour of the solutions of hemivariational inequalities. The eigenvalue problems for hemivariational inequalities were first discussed by the authors (cf. [27]-[31],[16]-[18]).

References

[1]R.Abraham and J. Robbin, *Transversal Mappings and Flows*, Benjamin, New York, 1967.

[2]A. Ambrosetti and P. H. Rabinowitz, Dual Variational Methods in Critical Point Theory, *J. Funct. Anal.* **14** (1973), 349-381.

[3]J. P. Aubin, and F.H. Clarke, Shadow Prices and Duality for a Class of Optimal Control Problems, *SIAM J. Control Optimization* **17** (1979), 567-586.

[4]J. P. Aubin, and H. Frankovska, *Set-Valued Analysis,* Birkhäuser Verlag, Basel, Boston 1990.

[5]V. Barbu, Necessary Conditions for Multiple Integral Problem in the Calculus of Variations, *Math. Ann.* **260** (1982), 175-189.

[6]V. Barbu, *Mathematical Methods in Optimization of Differential Systems* (in Romanian), Ed. Acad., Bucuresti, 1989.

[7]H. Brézis, *Analyse Fonctionnelle. Théorie et Applications*, Masson, Paris, 1992.

[8]K. C. Chang, Variational Methods for Non-Differentiable Functionals and their Applications to Partial Differential Equations, *J. Math. Anal. Appl.* **80** (1981), 102-129.

[9]F. H. Clarke, Generalized Gradients and Applications, *Trans. A.M.S.* **205** (1975), 247-262.

[10]F. H. Clarke, A New Approach to Lagrange Multipliers, *Math. Oper. Res.* **1** (1976), 165-174.

[11]F. H. Clarke, Generalized Gradients of Lipschitz Functionals, *Adv. Math.* **40** (1981), 52-67.

[12]F. H. Clarke, *Optimization and Nonsmooth Analysis,* John Wiley & Sons, New York, 1983.

[13]I. Ekeland, Nonconvex Minimization Problems, *Bull. (New Series) Amer. Math. Soc.* **1** (1979), 443-474.

[14]I. Ekeland and R. Temam, *Convex Analysis and Variational Problems,* North Holland, Amsterdam, 1976.

[15]A. F. Filippov, Classical Solutions of Differential Equations with Multivalued Right-Hand Side, *SIAM J. Control* **5** (1967), 609-621.

[16]D. Goeleven, D. Motreanu and P. D. Panagiotopoulos, Multiple Solutions for a Class of Eigenvalue Problems in Hemivariational Inequalities, *Nonlinear Anal. TMA* **29** (1997), 9-26.

[17]D. Goeleven, D. Motreanu and P. D. Panagiotopoulos, Multiple Solutions for a Class of Hemivariational Inequalities Involving Periodic Energy Functionals, *Math. Methods Appl. Sciences* **20** (1997), 547-568.

[18]D. Goeleven, D. Motreanu and P. D. Panagiotopoulos, Eigenvalue Problems for Variational - Hemivariational Inequalities at Resonance, *Nonlin. Anal. TMA* **33** (1998), 161-180.

[19]L. M. Graves, Some Mappings Theorems, *Duke Math. J.* **17** (1950), 111-114.

[20]A. D. Ioffe and V. L. Levin, Subdifferentials of Convex Functions, *Trans. Moscow Math. Soc.* **26** (1972), 1-72.

[21]G. Lebourg, Valeur moyenne pour gradient généralisé, *C. R. Acad. Sci.* Paris **281** (1975), 795-797.

[22]O. Kavian, *Introduction à la théorie des points critiques et applications aux problèmes elliptiques*, Springer-Verlag, Paris, 1993.

[23]Ph. Michel and J. P. Penot, A Generalized Derivative for Calm and Stable Functions, *Diff. Int. Equations* **5** (1992), 433-454.

[24]D. Motreanu, Existence of Critical Points in a General Setting, *Set-Valued Anal.* **3** (1995), 295-305.

[25]D. Motreanu, Nonlinear Eigenvalue Problems with Constraints, *Topological Meth. in Nonlin. Anal.*, (to appear).

[26]D. Motreanu and Z. Naniewicz, Discontinuous Semilinear Problems in Vector-Valued Function Spaces, *Differ. Int. Equations* **9** (1996), 581-598.

[27]D. Motreanu and P. D. Panagiotopoulos, Hysteresis: The Eigenvalue Problem for Hemivariational Inequalities, in: Models of Hysteresis , *Longman Scientific Publ.*, Harlow, 1993, pp. 102-117.

[28]D. Motreanu and P. D. Panagiotopoulos, A Minimax Approach to the Eigenvalue Problem of Hemivariational Inequalities, *Appl. Anal.* **58** (1995), 53-76.

[29]D. Motreanu and P. D. Panagiotopoulos, Nonconvex Energy Functions, Related Eigenvalue Hemivariational Inequalities on the Sphere and Applications, *J. Global Optimiz.* **6** (1995), 163-177.

[30]D. Motreanu and P. D. Panagiotopoulos, An Eigenvalue Problem for a Hemivariational Inequality Involving a Nonlinear Compact Operator, *Set-Valued Analysis* **3** (1995), 157-166.

[31]D. Motreanu and P. D. Panagiotopoulos, On the Eigenvalue Problem for Hemivariational Inequalities: Existence and Multiplicity of Solutions, *J. Math. Anal. Appl.* **197** (1996), 75-89.

[32]D. Motreanu and N. H. Pavel, Quasi-tangent Vectors in Flow-Invariance and Optimization Problems on Banach Manifolds, *J. Math. Anal. Appl.* **88** (1982), 116-132.

[33]Z. Naniewicz and P. D. Panagiotopoulos, *Mathematical Theory of Hemivariational Inequalities and Applications*, Marcel Dekker, Inc., New York, 1995.

[34]P. D. Panagiotopoulos, Nonconvex Superpotentials in the Sense of F. H. Clarke and Applications, *Mech. Res. Comm.* **8** (1981), 335-340.

[35]P. D. Panagiotopoulos, Coercive and Semicoercive Hemivariational Inequalities, *Nonlin. Anal. TMA* **16** (1991), 209-231.

[36]P. D. Panagiotopoulos, *Inequality Problems in Mechanics and Applications. Convex and Nonconvex Energy Functions.*, Birkhäuser Verlag, Basel, Boston, 1985.(Russian translation, MIR Publisher, Moscow, 1989).

[37]P. D. Panagiotopoulos, *Hemivariational Inequalities. Applications in Mechanics and Engineering*, Springer-Verlag, Berlin, 1993.

[38]P. H. Rabinowitz, Variational Methods for Nonlinear Eigenvalue Problems, in: Eigenvalues of Nonlinear Problems, C.I.M.E., *Edizioni Cremonese*, Roma, 1975, pp. 141-195.

[39]P. H. Rabinowitz, Minimax Methods in Critical Point Theory with Applications to Differential Equations, CBMS Reg. Conf. Ser. Math. 65, *Amer. Math. Soc.*, Providence, R. I., 1986.

[40]J. Rauch, Discontinuous Semilinear Differential Equations and Multiple Valued Maps, *Proc. Amer. Math. Soc.* **64** (1977), 277-282.

[41]R. T. Rockafellar, *The Theory of Subgradients and its Applications to Problems of Optimization. Convex and Nonconvex Functions*, Heldermann Verlag, Berlin 1979.

[42]M. Schechter and K. Tintarev, Spherical Maxima in Hilbert Space and Semilinear Elliptic Eigenvalue Problems, *Differ. Int. Equations* **3** (1990), 889-899.

[43]M. Schechter and K. Tintarev, Points of Spherical Maxima and Solvability of Semilinear Elliptic Equations, *Can. J. Math.* **43** (1991), 825-831.

[44]A. Szulkin, Minimax Principles for Lower Semicontinuous Functions and Applications to Nonlinear Boundary Value Problems, *Ann. Inst. Henri Poincaré, Anal. Non Linéaire* **3** (1986), 77-109.

2. Nonsmooth Critical Point Theory

The aim of this chapter is to present general results, many of them belonging to the authors, that can be applied to locally Lipschitz functionals, possibly invariant under a compact Lie group of linear isometries. The nonsmooth critical point theory in the locally Lipschitz case originates in the work of Chang [4]. Here the results of Chang [4] are deduced from a general principle that also incorporates the results of Du [7]. Our minimax principles are based on a deformation theorem that unifies different classical deformation results. Two main ideas are the bases of the mathematical approach in this chapter: the linking properties and the equivariance theory. A certain structure of the locally Lipschitz functionals that is particularly appropriate in the setting of our minimax methods is also pointed out. The applications of the abstract critical point results refer to nonsmooth elliptic boundary value problems.

2.1 Nonsmooth Deformation Results

The topic that we treat in this Section concerns the deformation for locally Lipschitz functionals. Let X be a real Banach space and let $I : X \to \mathbb{R}$ be a locally Lipschitz functional. We recall from Definition 1.4 in Chapter 1 that a critical point of I in the sense of Chang [4] means a point $u \in X$ satisfying $0 \in \partial I(u)$, where the notation $\partial I(u)$ stands for Clarke's generalized gradient of I at u. The following variant of Palais-Smale condition is an extension to the locally Lipschitz case of the one introduced by Ghoussoub and Preiss [8].

Definition 2.1 Given a subset B of X and a real number c it is said that the locally Lipschitz functional I satisfies the Palais-Smale condition around B at level c (shortly, $(PS)_{B,c}$) if every sequence $\{x_n\} \subset X$ with $I(x_n) \to c$, $\mathrm{dist}(x_n, B) \to 0$ and $\lambda(x_n) \to 0$ as $n \to \infty$, contains a convergent subsequence in X.

In the formulation above we used the notation in Chang [5]

$$\lambda(x) := \min\{\|w\|_{X^*} : \ w \in \partial I(x)\}, \quad \forall x \in X. \tag{2.1}$$

In particular, we put $(PS)_c = (PS)_{X,c}$ and simply (PS) if $(PS)_c$ holds for every c.

In the following we use the notations below corresponding to a fixed number $c \in \mathbb{R}$:

$$I_c = \{x \in X : I(x) \leq c\},$$

$$I^c = \{x \in X : I(x) \geq c\},$$

$$K_c(I) = \{x \in X : x \text{ is a critical point of } I \text{ and } I(x) = c\}.$$

For a fixed subset $B \subseteq X$ and a fixed number $\delta > 0$, we denote the closed δ-neighborhood of B by $N_\delta(B)$, that is,

$$N_\delta(B) = \{x \in X : \operatorname{dist}(x, B) \leq \delta\}.$$

A basic tool in our approach will be an appropriate notion of pseudo-gradient vector field for a locally Lipschitz function. For the classical notion of pseudo-gradient vector field we refer, e.g. to Rabinowitz [22].

Definition 2.2 A generalized normalized pseudo-gradient vector field of the locally Lipschitz functional I on X with respect to a subset $B \subset X$ and a number $c \in \mathbb{R}$ is a locally Lipschitz mapping $v : N_\delta(B) \cap I^{-1}[c-\delta, c+\delta] \to X$, with some $\delta > 0$, such that $\|v(x)\| \leq 1$ and

$$\langle z, v(x) \rangle > \frac{1}{2} \inf_{x \in \mathrm{dom} v} \lambda(x) > 0$$

for all $z \in \partial I(x)$ and $x \in \mathrm{dom} v := N_\delta(B) \cap I^{-1}[c-\delta, c+\delta]$.

The existence of a generalized normalized pseudo-gradient vector field in the sense of Definition 2.2. is given by the result below.

Lemma 2.1 Let $I : X \to \mathbb{R}$ be a locally Lipschitz function, $c \in \mathbb{R}$ and a closed subset B of X such that $(PS)_{B,c}$ is satisfied together with $B \cap K_c(I) = \emptyset$ and $B \subset I^c$. Then there exist a number $\delta > 0$ and a generalized normalized pseudo-gradient vector field $v : N_\delta(B) \cap I^{-1}[c-\delta, c+\delta] \to X$ of I with respect to B and c.

Proof. Let us show that there exists a number $\delta > 0$ such that

$$\lambda(x) \geq \sigma, \quad \forall x \in N_\delta(B) \cap I^{-1}[c-\delta, c+\delta], \tag{2.2}$$

with

$$\sigma := \inf\{\lambda(x) : x \in N_\delta(B) \cap I^{-1}[c-\delta, c+\delta]\} > 0.$$

Indeed, arguing by contradiction we assume that there exists a sequence $\{x_n\}$ in X with $\lambda(x_n) \to 0$, $\operatorname{dist}(x_n, B) \to 0$ and $I(x_n) \to c$. By $(PS)_{B,c}$ we derive the existence of a convergent subsequence of $\{x_n\}$ denoted again by $\{x_n\}$, say $x_n \to x$ in X as $n \to \infty$. The lower semicontinuity of the function λ in (2.1) (see Chang [4]) yields $\lambda(x) \leq \liminf_{n\to\infty} \lambda(x_n) = 0$. We deduce that $x \in K_c(I)$ which contradicts the condition $B \cap K_c(I) = \emptyset$. The claim in (2.2) is verified.

Along the lines of the proof of Lemma 3.3 in Chang [4], now with the use of separation theorem in the weak*-topology, $(X^*, \sigma(X^*, X))^* = X$ (see, e.g.,

Kantorovitch and Akilov [12], p. 109) and the property (2.2), we construct a locally Lipschitz map

$$v : N_\delta(B) \cap I^{c-\delta} \cap I_{c+\delta} \to X$$

such that

$$\|v(x)\| \leq 1 \tag{2.3}$$

and

$$\langle z, v(x)\rangle > \frac{1}{2}\sigma, \ \forall z \in \partial I(x), \ x \in N_\delta(B) \cap I^{c-\delta} \cap I_{c+\delta}. \tag{2.4}$$

The relations (2.3) and (2.4), together with a usual partition of unity argument, complete the proof of Lemma 2.2. □

Our main deformation result is the following one stated in Motreanu and Varga [24].

Theorem 2.1 Let $I : X \to \mathbb{R}$ be a locally Lipschitz functional, $c \in \mathbb{R}$ and a closed subset B of X provided one has $(PS)_{B,c}$, $B \cap K_c(I) = \emptyset$ and $B \subset I_c$. Let v be a generalized normalized pseudo-gradient vector field of I with respect to B and c. Then for every $\bar{\varepsilon} > 0$ there exist an $\varepsilon \in (0, \bar{\varepsilon})$ and a $\delta < c$ such that for each closed subset A of X with $A \cap B = \emptyset$ and $A \subset I^{c-\varepsilon_A}$, where

$$\varepsilon_A := \min(\varepsilon, \varepsilon d(A, B)), \tag{2.5}$$

and $d(A, B) := \inf\{\|x - y\| : \ x \in A, \ y \in B\}$, there is a continuous mapping $\eta_A : \mathbb{R} \times X \to X$ with the properties below

(i) $\eta_A(\cdot, x)$ is the solution of the vector field $V_A = -\varphi_A v$ with the initial condition $x \in X$ for some locally Lipschitz function $\varphi_A : X \to [0, 1]$ whose support is contained in $(X \setminus A)$;

(ii) $\eta_A(t, x) = x$ for all $t \in \mathbb{R}$ and $x \in A \cup I_{c-\bar{\varepsilon}} \cup I^{c+\bar{\varepsilon}}$;

(iii) for every $\delta \leq d \leq c$ one has $\eta_A(1, B \cap I_d) \subset I_{d-\varepsilon}$.

Proof. Let us note that the existence of a normalized generalized pseudo-gradient vector field $v : N_{3\delta_1}(B) \cap I^{-1}[c - 3\varepsilon_1, c + 3\varepsilon_1] \to X$ of I with respect to B and c is assured by Lemma 2.1, for some constants $\delta_1 > 0$ and $\varepsilon_1 > 0$. Consequently, a constant $\sigma_1 > 0$ can be found such that

$$\langle z, v(x)\rangle > \frac{1}{2}\sigma_1, \ \forall z \in \partial I(x), \ x \in N_{3\delta_1}(B) \cap I^{c-3\varepsilon_1} \cap I_{c+3\varepsilon_1}. \tag{2.6}$$

We claim that the result of Theorem 2.1 holds for every $\varepsilon > 0$ with

$$\varepsilon < \min\{\bar{\varepsilon}, \varepsilon_1, \frac{1}{2}\sigma_1, \frac{1}{2}\sigma_1\delta_1\}. \tag{2.7}$$

In order to check the claim in (2.7) let us fix two locally Lipschitz functions $\varphi,\ \psi : X \to [0,1]$ satisfying

$$\varphi = 1 \text{ on } N_{\delta_1}(B) \cap I_{c+\varepsilon_1} \cap I^{c-\varepsilon_1};$$

$$\varphi = 0 \quad \text{on} \quad X \setminus (N_{2\delta_1}(B) \cap I_{c+2\varepsilon_1} \cap I^{c-2\varepsilon_1});$$

$$\psi = 0 \quad \text{on} \quad I_{c-\bar{\varepsilon}} \cup I^{c+\bar{\varepsilon}};$$

$$\psi = 1 \quad \text{on} \quad I_{c+\varepsilon_0} \cap I^{c-\varepsilon_0},$$

for some number ε_0 with

$$\varepsilon < \varepsilon_0 < \min(\bar{\varepsilon}, \varepsilon_1). \tag{2.8}$$

Then we are able to construct the locally Lipschitz vector field $V : X \to X$ by setting

$$V(x) = \begin{cases} -\delta_1 \varphi(x)\psi(x)v(x), & \forall x \in N_{3\delta_1}(B) \cap I^{c-3\varepsilon_1} \cap I_{c+3\varepsilon_1}, \\ 0, & \text{otherwise.} \end{cases} \tag{2.9}$$

Using (2.9) we see that the vector field V is locally Lipschitz and bounded, namely

$$\|V(x)\| \le \delta_1, \quad x \in X. \tag{2.10}$$

From (2.6), (2.9) and (2.10) we derive

$$-\langle z, V(x)\rangle = \delta_1 \langle z, v(x)\rangle \ge \frac{1}{2}\delta_1\sigma_1,$$

$$\forall x \in N_{\delta_1}(B) \cap I^{c-\varepsilon_0} \cap I_{c+\varepsilon_0},\ z \in \partial I(x). \tag{2.11}$$

In view of (2.10) we may consider the global flow $\gamma : \mathbb{R} \times X \to X$ of V defined by (2.9), i.e.,

$$\frac{d\gamma}{dt}(t,x) = V(\gamma(t,x)),\ \forall (t,x) \in \mathbb{R} \times X,$$

$$\gamma(0,x) = x,\ \forall x \in X.$$

Inspired by an idea in Du [7] we set

$$B_1 := \gamma([0,1] \times B). \tag{2.12}$$

We notice that B_1 in (2.12) is a closed subset of X. To see this let $y_n = \gamma(t_n, x_n) \in B_1$ be a sequence with $t_n \in [0,1]$, $x_n \in B$ and $y_n \to y$ in X. Passing to a subsequence we can suppose $t_n \to t \in [0,1]$ in $\mathbb{R}$. Putting $u_n = \gamma(t, x_n)$ we get

$$\|u_n - y_n\| = \|\gamma(t,x_n) - \gamma(t_n,x_n)\| = \|\int_{t_n}^{t} \frac{d}{dt}\gamma(\tau,x_n)d\tau\| \le \delta_1|t_n - t|,$$

where (2.10) has been used. Since $u_n \to y$ in X, it turns out that $x_n \to \gamma(-t,y) \in B$. Finally, we obtain $y = \gamma(t,\gamma(-t,y)) \in B_1$ which establishes the closedness of B_1.

The next step is to justify that $I(\gamma(t,x))$ is a decreasing function of $t \in \mathbb{R}$, for each $x \in X$. Towards this, by applying Lebourg's mean value theorem and the chain rule for generalized gradients (see Theorems 1.1 and 1.2 in Chapter 1) we infer for arbitrary real numbers $t > t_0$ the following inclusions

$$I(\gamma(t,x)) - I(\gamma(t_0,x)) \in \partial_t(I(\gamma(t,x)))\Big|_{t=\tau}(t-t_0)$$

$$\subset \partial I(\gamma(\tau,x))\frac{d\gamma}{dt}(\tau,x)(t-t_0) = \partial I(\gamma(\tau,x))V(\gamma(\tau,x))(t-t_0)$$

with some $\tau \in (t_0,t)$, where the notation ∂_t means the generalized gradient with respect to t. By (2.6) and (2.9) we derive that $I(t,x) \le I(t_0,x)$.

Now we prove the relation

$$A \cap B_1 = \emptyset. \tag{2.13}$$

To check (2.13), we admit by contradiction that there exist $x_0 \in B$ and $t_0 \in [0,1]$ provided $\gamma(t_0,x_0) \in A$. Since A and B are disjoint we have necesarily that $t_0 > 0$. From the relations $A \subset I^{c-\varepsilon_A}$ and $B \subset I_c$ we deduce

$$c - \varepsilon_A \le I(\gamma(t_0,x_0)) \le I(\gamma(t,x_0)) \le I(x_0) \le c, \ \forall t \in [0,t_0]. \tag{2.14}$$

It turns out that

$$\gamma(t,x_0) \in N_{\delta_1}(B) \cap I_c \cap I^{c-\varepsilon_A}, \quad \forall t \in [0,t_0].$$

On the other hand from (2.10) we infer the estimate

$$d(A,B) \le \|\gamma(t_0,x_0) - x_0\| = \|\int_0^{t_0} V(\gamma(s,x_0))ds\| \le \delta_1 t_0.$$

If we denote $h(t) = I(\gamma(t,x_0))$, then h is a locally Lipschitz function, and (2.9), (2.11) allow to write

$$\begin{aligned} h'(s) &\le \max\{\langle z, \frac{d\gamma}{ds}(s,x_0)\rangle : z \in \partial I(\gamma(s,x_0))\} = \\ &= \max\{\langle z, V(\gamma(s,x_0))\rangle : z \in \partial I(\gamma(s,x_0))\} \le -\frac{1}{2}\delta_1\sigma_1 \end{aligned}$$

for a.e. $s \in [0,t_0]$.

Therefore, by virtue of (2.7), we have the following estimate

$$\begin{aligned} I(\gamma(t_0,x_0)) - I(x_0) &= h(t_0) - h(0) = \\ &= \int_0^{t_0} h'(s)ds \le -\frac{1}{2}\delta_1\sigma_1 t_0 < \\ &< -\delta_1\varepsilon t_0 \le -\varepsilon d(A,B) \le -\varepsilon_A. \end{aligned} \tag{2.15}$$

The contradiction between (2.14) and (2.15) shows that the property (2.13) is true.

Taking into account (2.13) there is a locally Lipschitz function $\psi_A : X \to \mathbb{R}$ verifying $\psi_A = 0$ on a neighborhood of A and $\psi_A = 1$ on B_1. Then we define the homotopy $\eta_A : \mathbb{R} \times X \to X$ as being the global flow of the vector field $V_A = \psi_A V$. The assertion (i) is clear from the construction of η_A because one can take $\varphi_A = -\delta_1 \psi_A \varphi \psi$.

The assertion (ii) follows easily because $V_A = 0$ on $A \cup I_{c-\bar{\varepsilon}} \cup I^{c+\bar{\varepsilon}}$.

We show that (iii) is valid for $\delta = c + \varepsilon - \varepsilon_0$ with ε described in (2.7) and ε_0 in (2.8). To this end we argue by contradiction. Suppose that for some $d \in [\delta, c]$ there exists $x \in B \cap I_d$ such that

$$I(\eta_A(1,x)) > d - \varepsilon. \tag{2.16}$$

It is straightforward to deduce

$$\eta_A(t,x) = \gamma(t,x) \in N_{\delta_1}(B) \cap I_d \cap I^{d-\varepsilon}, \quad \forall t \in [0,1].$$

Here we employed essentially the fact that $\psi_A = 1$ on B_1. Then a reasoning similar to the one in (2.15) can be carried out to write

$$I(\eta_A(1,x)) - I(x) \le -\frac{1}{2}\delta_1\sigma_1 < -\varepsilon.$$

This contradicts the relation (2.16) because $I(x) \le d$. The proof of the assertion (iii) is complete. □

Remark 2.1 Theorem 2.1 unifies different deformation results as for example those in Chang [2], Du [7], Motreanu [13], Motreanu and Varga [15], Pucci and Serrin [21].

We illustrate this aspect here only by the following result showing that Theorem 2.1 contains Chang's deformation theorem [4] (and a fortiori the classic deformation lemma for continuously differentiable functionals, see Proposition 2.1 in Rabinowitz [22]). It achives in fact a unification between the deformation results in Chang [4] and Motreanu and Varga [15]. In particular, one refinds the unification between the classic deformation lemma (see Rabinowitz [22]) and the one of Du [7], a fact discussed in Motreanu [13]. For the sake of simplicity we point out only the part in Chang's deformation result that is used in the sequel to establish the existence of critical points.

Corollary 2.1 (Chang [4]) Let $I : X \to \mathbb{R}$ be a locally Lipschitz function which satisfies the (PS) condition. If c is not a critical value of I in the sense of Chang [4], i.e. $K_c(I) = \emptyset$, then given any $\bar{\varepsilon} > 0$ there exist an $\varepsilon \in (0, \bar{\varepsilon})$ and a homeomophism $\eta : X \to X$ such that

1° $\eta(x) = x, \quad \forall x \in I_{c-\bar{\varepsilon}} \cup I^{c+\bar{\varepsilon}}$;

2° $\eta(I_{c+\varepsilon}) \subset I_{c-\varepsilon}$.

Proof. Let us fix a positive number $\bar{a} < \bar{\varepsilon}$ such that the interval $[c - \bar{a}, c + \bar{a}]$ be without critical values of I in the sense of Chang [4]. We apply Theorem

2.1 for $B_a = I_{c+a} \cap I^{c-a}$ and $c+a$ in place of B and c, respectively, for each $a \in (0, \bar{a}]$. Theorem 2.1 provides $\varepsilon_a > 0$, $\delta_a < c+a$ and, with $A = I^{c+\bar{\varepsilon}}$, the homotopy $\eta_a \in C(\mathbb{R} \times X, X)$ satisfying the requirements $(i) - (iii)$ for ε, δ, η_A replaced by ε_a, δ_a, η_a, respectively. Note that this can be done because $I^{c+\bar{\varepsilon}} \subset I^{c-\varepsilon_{a,A}}$, where $\varepsilon_{a,A} = \min\{\varepsilon_a, \varepsilon_a d(A, B_a)\}$. Then 1° follows from (ii) of Theorem 2.1. The relations (2.6) and (2.7) show that $\varepsilon_{a,A}$ is bounded away from zero, say $\varepsilon_{a,A} \geq \tilde{\varepsilon} > 0$ for $a \in (0, \tilde{a}]$. Set $d = c + \min\{a, \bar{\varepsilon}\}/2$. We observe that if $a > 0$ is small enough, d can be used in (iii) of Theorem 2.1 relative to η_a, that is $\delta_a \leq d \leq c+a$, because ε_0 in (2.8) can be chosen independently of $a \in (0, \bar{a}]$. Then 2° is checked with $\eta(x) := \eta_a(x, 1)$, $\forall x \in X$ and $\varepsilon = \min\{a, \tilde{\varepsilon}\}/2$ by means of property (iii) in Theorem 2.1 for B_a and $c+a$ in place B and c, respectively, with $a > 0$ sufficiently small. This occurs in view of the relations $c + \varepsilon = d$ and $d - \varepsilon_a \leq c - \varepsilon$, so one can conclude. □

The following result extends Lemma 1.1 in Du [7] to the case of locally Lipschitz functions (see Motreanu and Varga [15]).

Corollary 2.2 Let $I : X \to \mathbb{R})$ be a locally Lipschitz function, let A and B be two closed disjoint subsets of X and let $c \in \mathbb{R}$ such that $B \cap K_c(I) = \emptyset$, $B \subset I_c$, $A \subset I^c$ and I satisfies the $(PS)_{B,c}$ condition. Then there exist a number $\varepsilon > 0$ and a homeomorphism η of X such that

(i) $I(\eta(x)) \leq I(x), \quad \forall\, x \in X$;

(ii) $\eta(x) = x, \quad \forall\, x \in A$;

(iii) $\eta(B) \subset I_{c-\varepsilon}$.

Proof. Apply Theorem 2.1 for the set B and the number c. One obtains an $\varepsilon > 0$ and $\eta := \eta_A(1, \cdot) \in C(X, X)$ corresponding to $A \subset I^c \subset I^{c-\varepsilon_A}$. It is obvious that the conclusion of Corollary 2.2 follows from Theorem 2.1, where (iii) is deduced for $c = d$. □

2.2 Equivariant Version of Deformation Result

This Section is devoted to the equivariant version of Theorem 2.1. Let G be a compact Lie group which acts linear isometrically on the real Banach space X, i.e. the action $G \times X \to X$ is differentiable and for every $g \in G$, $g : X \to X$ is a linear isometry. The action on X induces an action of the same type on the dual space X^* defined by $(gx^*)(x) = x^*(gx)$, for all $g \in G$, $x \in X$ and $x^* \in X^*$. Since

$$\|gx^*\| = \sup_{\|x\|=1} |(gx^*)(x)| = \sup_{\|x\|=1} |x^*(gx)|,$$

the isometry assumption for the action of G implies

$$\|gx^*\| = \sup_{\|x\|=1} |x^*(x)| = \|x^*\|, \ \forall\, x^* \in X^*, \ g \in G.$$

We suppose that $I : X \to \mathbb{R}$ is a locally Lipschitz and G-invariant function, i.e., $I(gx) = I(x)$ for every $g \in G$ and $x \in X$. It is known the relation

$$g\partial I(x) = \partial I(gx) = \partial I(x), \text{ for every } g \in G \text{ and } x \in X$$

(see Krawcewicz-Marzantowicz [11]). Therefore the subset $\partial f(x) \subset X^*$ is G-invariant, so the function $\lambda(x) = \inf_{w \in \partial I(x)} \|w\|$, $x \in X$, is G-invariant.

We are now in a position to state the equivariant version of Theorem 2.1.

Theorem 2.2 Let $I : X \to \mathbb{R}$ be a locally Lipschitz G-invariant functional, $B \subset X$ a G-invariant closed subset and $c \in \mathbb{R}$ a real number provided one has $(PS)_{B,c}$, $B \cap K_c(I) = \emptyset$ and $B \subset I_c$. Then for every $\bar{\varepsilon} > 0$ there exist an $\varepsilon \in (0, \bar{\varepsilon})$ and a $\delta < c$ such that for each closed G-invariant subset A of X with $A \subseteq I^{c-\varepsilon_A}$ and $A \cap B = \emptyset$ one finds an (equivariant) homotopy $\eta_A \in \mathcal{C}(\mathbb{R} \times X, X)$ with the properties below

(i) $\eta_A(t, x) = x$ for all $t \in \mathbb{R}$ and $x \in A \cup I_{c-\bar{\varepsilon}} \cup I^{c+\bar{\varepsilon}}$;

(ii) $\eta_A(t, \cdot)$ is a G-homeomorphism of X onto X for each $t \in [0, 1]$ such that $\eta_A(t, \cdot)(M^H) = M^H$ for every subgroup H of G, where

$$M^H = \{x \in X : \ x = hx, \ \forall\, h \in H\};$$

(iii) for every $\delta \le d \le c$ one has $\eta_A(1, B \cap I_d) \subset I_{d-\varepsilon}$.

Proof. Let $v : N_{3\delta_1}(B) \cap I^{c-3\varepsilon_1} \cap I_{c+3\varepsilon_1} \to X$ be a generalized normalized pseudo-gradient vector field of I with respect to B and c satisfying (2.6) (see Lemma 2.1). Using a standard method (cf. Chang [5]) we define the mapping $\widehat{v} : N_{3\delta_1}(B) \cap I^{c-3\varepsilon_1} \cap I_{c+3\varepsilon_1} \to X$ by

$$\widehat{v}(x) = \int_G g^{-1} v(gx)\, d\mu, \tag{2.17}$$

where $d\mu$ is the right Haar measure on G with $\int_G d\mu = 1$. The mapping $\widehat{v}$ is well defined because $N_{3\delta_1}(B) \cap I^{c-3\varepsilon_1} \cap I_{c+3\varepsilon_1}$ is a G-equivariant subset of X. For any $g' \in G$ we can write

$$\widehat{v}(g'x) = \int_G g^{-1} v(gg'x)\, d\mu = g' \int_G (gg')^{-1} v(gg'x)\, d\mu = g'\widehat{v}(x),$$

$$\forall x \in N_{3\delta_1}(B) \cap I^{c-3\varepsilon_1} \cap I_{c+3\varepsilon_1},$$

so $\widehat{v}$ is G-equivariant. In addition, the mapping $\widehat{v}$ in (2.17) is locally Lipschitz, because v has the same property and the orbits are compact (see Chang [5], p. 68-69). The condition $\|\widehat{v}(x)\| \le 1$, $x \in X$, required in Definition 2.2, is fulfilled because G acts linear isometrically and v is normalized. Therefore $\widehat{v}$ (defined by (2.17)) is a generalized normalized pseudo-gradient vector field which is also G-equivariant. We may now apply Theorem 2.1 with $\widehat{v}$ in place of v. We point out that the functions φ, ψ and ψ_A used in the proof of Theorem 2.1 can be chosen to be G-invariant. Taking into account that the deformation η obtained by Theorem 2.1 is the global flow of an equivariant vector field, it results that $\eta(t,\cdot)$ is an equivariant homeomorphism of X for every t. Assertions (i) and (iii) follow directly from Theorem 2.1.

In order to check the property (ii) it suffices to show that the inclusion $\eta_A(t,\cdot)(M^H) \subset M^H$ holds true. The equivariance of the deformation $\eta_A(t,\cdot)$ implies that

$$h\eta_A(t,x) = \eta_A(t,hx) = \eta_A(t,x) = \eta_A(t,hx)$$

for all $h \in H$ and $x \in M^H$. This completes the proof. □

The following corollary presents a version of Theorem 3.5 in Goeleven, Motreanu and Panagiotopoulos [9].

Corollary 2.3 Assume that $I : X \to \mathbb{R}$ is a locally Lipschitz, G-invariant function, for a finite subgroup G of the group of linear isometries of X, which satisfies the next compactness condition of Palais-Smale type: for every sequence $\{x_n\} \subset X$ with $I(x_n)$ bounded and $\lambda(x_n) \to 0$ there is a sequence $\{y_n\} \subset X$ containing a relabelled convergent subsequence provided x_n and y_n belong to the same orbit. If c is not a critical value of I in the sense of Chang [4], then given any $\bar{\varepsilon} > 0$ there exist an $\varepsilon \in (0, \bar{\varepsilon})$ and a homeomorphism $\eta : X \to X$ such that

1° $\eta(x) = x, \quad \forall x \in I_{c-\bar{\varepsilon}} \cup I^{c+\bar{\varepsilon}}$;

2° $\eta(I_{c+\varepsilon}) \subset I_{c-\varepsilon}$;

3° $\eta(gx) = g\eta(x), \quad \forall x \in X, \ \forall g \in G$.

Proof. The Palais-Smale condition imposed permits to construct an equivariant generalized normalized pseudo-gradient vector field of I with respect to c (taking the set B in Definition 2.2 to be X). Then arguing as in the proof of Corollary 2.1 on the basis of Theorem 2.1 we deduce from Theorem 2.2 the conclusion of Corollary 2.3. □

2.3 A General Minimax Theorem

In this Section we suppose that G is a compact Lie group which acts linear isometrically on a real Banach space X. We state from Motreanu and Varga [15] the main result of the Section describing a general minimax principle.

Theorem 2.3 Let $I : X \to \mathbb{R}$ a G-invariant locally Lipschitz functional and $B \subseteq X$ a G-invariant closed set such that $c := \inf_B I > -\infty$ and I satisfies $(PS)_{B,c}$. Let $\mathcal{M}$ be a nonempty family of subsets M of X such that

$$c = \inf_{M \in \mathcal{M}} \sup_{x \in M} I(x). \tag{2.18}$$

Assume that for an equivariant generalized normalized pseudo-gradient vector field $\hat{v}$ of I with respect to B and c the following hypothesis holds

(H) for each set $M \in \mathcal{M}$ and each number $\varepsilon > 0$ with $I|_M < c + \varepsilon$ there exists a G-invariant closed subset A of X with $I|_A \leq c + \varepsilon_A$ (see (2.5)) and $A \cap B = \emptyset$ such that for each locally Lipschitz function $\varphi_A : X \to [0,1]$ with $\operatorname{supp} \varphi_A \subset (X \setminus A)$ the global flow ξ_A of $\varphi_A \widehat{v}$ satisfies $\xi_A(1, M) \cap B \neq \emptyset$.

Then the assertions below are true

(i) $c = \inf_B I$ is attained;

(ii) $K_c(I) \setminus A \neq \emptyset$ for each set A entering (H);

(iii) $K_c(I) \cap B \neq \emptyset$.

Proof. The assertions (i) and (ii) are direct consequences of the property (iii). The proof of (iii) is achieved arguing by contradiction. Accordingly, we suppose $K_{-c}(-I) \cap B = \emptyset$. By hypothesis we know that $B \subset (-I)_{-c}$, so Theorem 2.2 can be applied for $-I$ and $-c$ (in place of I and c, respectively). Thus Theorem 2.2 yields an $\varepsilon > 0$ with the properties there stated. Then from the minimax description of c, by means of $\mathcal{M}$, we obtain the existence of a set $M \in \mathcal{M}$ satisfying $I|_M < c + \varepsilon$. Corresponding to M, assumption (H) allows to find a closed G-invariant set $A \subset X \setminus B$ which satisfies $A \subset (-I)^{-c-\varepsilon_A}$ and the linking property formulated in (H). Theorem 2.2 gives rise to the equivariant deformation $\eta_A \in \mathcal{C}(\mathbb{R} \times X, X)$ which verifies $\eta_A(1, B \cap (-I)_{-c}) \subset (-I)_{-c-\varepsilon}$. This reads as

$$\eta_A(1, B) \subset I^{c+\varepsilon}. \tag{2.19}$$

From Theorem 2.2 and assumption (H) it is seen that

$$\xi_A(t, x) = \eta_A(-t, x), \tag{2.20}$$

for all $(t, x) \in \mathbb{R} \times X$. As shown in (H) one has the intersection property

$$\xi(1, M) \cap B \neq \emptyset.$$

Combining with (2.20) it turns out

$$\eta_A(1, B) \cap M \neq \emptyset.$$

Taking into account (2.19) we obtain the existence of some point $x_0 \in M$ with $I(x_0) \geq c + \varepsilon$. This contradicts the choice of the set M. □

Corollary 2.4 Let $I : X \to \mathbb{R}$ be a locally Lipschitz G-invariant functional satisfying (PS) and let a family $\mathcal{M}$ of subsets M of X be such that c defined by (2.18) is a real number. Assume that the hypothesis below holds

(H') for each $M \in \mathcal{M}$ there exists a closed set A in X with $I|_A < c$ such that for every homeomorphism h of X with $h|_A = id_A$ one has $h(M) \cap I^c \neq \emptyset$.

Then c in (2.18) is a critical value of I and $K_c(I) \cap A = \emptyset$ for every A in (H').

Proof. We consider the global flow ξ_A (see (H)) and we apply Theorem 2.3 with $B = I^c$. It is clear that (H') implies (H) because $A \subset M \setminus B$ and $\xi_A(1, \cdot)$ is a G-homeomorphism of X with $\xi_A(1,\cdot) = id$ on A. Then Theorem 2.3 concludes the proof. □

Remark 2.2 The minimax principle in Corollary 2.4 includes and extends to the locally Lipschitz and invariant functionals many classic minimax results, e.g. those in Ambrosetti and Rabinowitz [2], Chang [4], Du [7] , Ghoussoub and Preiss [8], Motreanu [13], Motreanu and Varga [15], Varga [24].

Theorem 2.3 is useful in locating the critical points (in fact the critical orbits). We illustrate this aspect by deriving from Theorem 2.3 an extension of a result due to Ghoussoub and Preiss [8].

Corollary 2.5 Let $I : X \to \mathbb{R}$ be a locally Lipschitz, G-invariant functional, with G as above, and for the points $u, v \in X$ let the number

$$c = \inf_{g\in\Gamma} \max_{0\leq t\leq 1} I(g(t)),$$

where Γ is the set of paths $g \in C([0,1], X)$ joining u and v. Suppose F is a closed, G-invariant subset of X such that $F \cap I^c$ separates u and v, i.e. u, v belong to disjoint connected components of $X \setminus F \cap I^c$, and condition $(PS)_{F,c}$ is verified. Then there exists a critical point of I in F with critical value c.

Proof. Set $\mathcal{M} = \{g([0,1]) : g \in \Gamma\}$, $B = F \cap I^c$ and $A = \{u, v\}$. Applying Theorem 2.3 we see that $\xi_A(1, M) \in \mathcal{M}$ whenever $M \in \mathcal{M}$. Thus hypothesis (H) is verified. Theorem 2.3 implies the conclusion of Corollary 2.5. □

Theorem 2.3 is suitable for applications to multiple linking problems. In this respect, the next concept of linking is an equivariant version of that in Du [7] for locally Lipschitz, G-invariant functionals.

Definition 2.3 Let Q, Q_0 be closed G-invariant subsets of X, with $Q_0 \neq \emptyset$, $Q_0 \subset Q$, and let S be a G-invariant subset of X such that $Q_0 \cap S = \emptyset$. We say that the pair (Q, Q_0) links with S if for each G-equivariant mapping $g \in C(Q, X)$ with $g|_{Q_0} = id|_{Q_0}$ one has $g(Q) \cap S \neq \emptyset$.

A common situation of linking is presented in the following result given in Motreanu and Varga [15] (it unifies the minimax principles in Chang [4] and Du [7]).

Corollary 2.6 Given the subsets Q, Q_0, S of the real Banach space X we assume that (Q, Q_0) links with S in X in the G-equivariant sense of Definition 2.3. Let $I : X \to \mathbb{R}$ be a locally Lipschitz, G-invariant functional such that $\sup_Q I < \infty$ and, for some number $\alpha \in \mathbb{R}_+$,

$$Q_0 \subset I_\alpha, \quad S \subset I^\alpha.$$

Then assuming that for the minimax value

$$c = \inf_{g \in \Gamma} \sup_{x \in Q} I(g(x)),$$

where

$$\Gamma = \{g \in \mathcal{C}(Q, X) : g|_{Q_0} = id|_{Q_0}\},$$

$(PS)_{S,c}$ is satisfied, the following properties hold

(i) $c \geq \alpha$;

(ii) $K_c(I) \setminus Q_0 \neq \emptyset$;

(iii) $K_c(I) \cap S \neq \emptyset$ if $c = \alpha$.

Proof. Since the case $\alpha < c$ follows immediately from Chang [5] we discuss only the situation where $\alpha = c$. The conclusion is readily obtained from Theorem 2.3 by choosing $\mathcal{M} = \{g(Q) : g \in \Gamma\}$ and $B = S$. □

The final part of this Section is devoted to a symmetric version of Mountain Pass Theorem for locally Lipschitz functionals that are invariant with respect to the action of a compact Lie group G. We follow closely the approach in Bartsch [3] developed for the smooth case. The classic symmetric version of Mountain Pass Theorem in the smooth case is obtained for $G = Z_2$ (see Rabinowitz [22]). The core of the argument in the next result is the construction of a multiple linking corresponding to a Galerkin basis of the space. In comparison with the classic Mountain Pass Theorem where Krasnoselski's genus is used, here the notion of $\mathcal{A}$-genus is employed (for the definition and various properties of $\mathcal{A}$-genus we refer to Bartsch [3]).

Corollary 2.7 Let G be a compact Lie group acting on a finite dimensional vector space V such that the admissibility condition in the sense of Bartsch [3] holds, i.e., there does not exist a G-invariant continuous map $\partial U \to V^{k-1} \setminus 0$, $k \geq 1$, where U is an open, bounded, G-invariant neighborhood of 0 in V. Let $X = \oplus_{j=1}^{\infty} E^j$ be a G-Hilbert space such that E^j is G-isomorphic to V for each j. Assume that a G-invariant functional $I : X \to \mathbb{R}$ satisfies the following conditions:

(i) the Palais-Smale condition in the sense of Chang [4];

(ii) for every $k \geq 1$ there exists $R_k > 0$ such that $I(u) \leq I(0)$ for every $u \in E_k = \oplus_{j=1}^{k} E^j$ with $\|u\| \geq R_k$;

(iii) there exist $k_0 \geq 1$, $b > I(0)$ and $\rho > 0$ such that $I(u) \geq b$ for every $u \in E_{k_0}^{\perp}$ with $\|u\| = \rho$.

Then I possesses an unbounded sequence of critical values in the sense of Chang [4].

Proof. Since I is G-invariant, we can construct a generalized normalized pseudo-gradient vector field of I which is G-equivariant (see Lemma 2.1). We introduce the sequence of minimax values

$$c_k = \inf_{B\in\Gamma_k} \sup_{x\in B} I(x),$$

where Γ_k denotes the set of all compact G-invariant subsets B of X of the form $B = h(\overline{B_m - Y})$, where $m \geq k$, $h : B_m := \{u \in E_m : \|u\| \leq R\} \to X$ is a G-map with $h = id$ on ∂B_m and Y is a closed G-invariant subset of B_m with $\mathcal{A}-$genus$(Y) \leq m-k$. Then, using the properties of $\mathcal{A}-$genus, it can be shown that the set $\mathcal{M} = \Gamma_k$ verifies the assumption (H) in Thorem 2.3 for every $k \geq 1$. Theorem 2.3 implies that each c_k is a critical value of I in the sense of Chang [4]. The fact that $c_k \to \infty$ as $k \to \infty$ is established as in the smooth case (see Bartsch [3] and Rabinowitz [22]). □

2.4 A General Critical Point Setting

The theory of hemivariational inequalities created by P. D. Panagiotopoulos [18]-[19] for studying nonconvex and nonsmooth energy functions under nonmonotone multivalued laws, as well as other important theories including semilinear Dirichlet problems (see Rabinowitz [22]), suggests a certain structure for the associated locally Lipschitz functionals. This framework is particularly suitable for treating the nonsmooth eigenvalue problems for which a typical example is provided by the analysis of bifurcation phenomena. Such specific situations can be illustrated for instance by the loading and unloading problems with hysteresis modes or unilateral bending problems in elasticity (see Panagiotopolos [18]-[19], Motreanu and Panagiotopoulos [16], Goeleven, Motreanu and Panagiotopoulos [9]). In the sequel we follow closely the presentation in Motreanu [14].

Given a Banach space V with the norm $\|\cdot\|$ and a bounded domain Ω in $\mathbb{R}^m$, $m \geq 1$, let $T : V \to L^s(\Omega;\mathbb{R}^N)$, $1 < s < \infty$, be a compact linear operator. Let $F : V \to \mathbb{R}$ be a locally Lipschitz function and let $G : \Omega \times \mathbb{R}^N \to \mathbb{R}$ be a (Carathéodory) function such that $G(x,y)$ is measurable in $x \in \Omega$, locally Lipschitz in $y \in \mathbb{R}^N$ and $G(x,0) = F(0) = 0$, $x \in \Omega$. The hypotheses below are imposed

$$(H_1) \qquad |w| \leq c(1+|y|^{s-1}),\ \forall w \in \partial_y G(x,y),\ x\in\Omega,\ y\in\mathbb{R}^N,$$

with a constant $c > 0$;

$$(H_2)(i) \qquad F(v) - r < z, v >_V \geq \alpha\|v\|_V^\sigma - \alpha_0,\ \forall v \in V,\ z \in \partial F(v);$$

$$(H_2)(ii) \qquad r < w, y > -G(x,y) \geq -b|y|^{\sigma_0} - b_0,$$

a.e. $x \in \Omega$, $y \in \mathbb{R}^N$, $w \in \partial_y G(x,y)$, with positive constants r, α, α_0, b, b_0, σ, σ_0, where $1 \leq \sigma_0 < \min\{\sigma, r^{-1}, s\}$;

(H_3) a bounded sequence$\{v_n\} \subset V$ for which there is a convergent sequence $\{z_n\}$ in V^* with $z_n \in \partial F(v_n)$ contains a convergent subsequence in V;

$(H_4)(i)$
$$\liminf_{v \to 0} F(v)\|v\|^{-p} > 0;$$

$(H_4)(ii)$
$$\limsup_{y \to 0} G(x,y)|y|^{-p} \le 0$$

uniformly with respect to x, $1 \le p < s$;

$(H_5)(i)$
$$\limsup_{n \to +\infty} F(nv_0)n^{-\frac{1}{r}} \le 0,$$

$(H_5)(ii)$
$$\limsup_{n \to +\infty} n^{-\sigma_0} \int_\Omega G(x, nTv_0)dx = +\infty, \text{for some } v_0 \in V.$$

The following statement is the main result in Motreanu [14].

Theorem 2.4 Assume that the hypotheses $(H_1)-(H_5)$ hold. Then there exists a nontrivial critical point $u \in V$ of the locally Lipschitz functional $I : V \to \mathbb{R}$ defined by
$$I(v) = F(v) - \int_\Omega G(x, (Tv)(x))dx, \ v \in V.$$

Moreover, there exist $z \in \partial F(u)$ and $w \in L^{\frac{s}{s-1}}(\Omega; \mathbb{R}^N)$ such that
$$w(x) \in \partial_y G(x, (Tu)(x)) \ a.e. \ x \in \Omega,$$
$$< z, v >_V - \int_\Omega < w(x), (Tv)(x) > dx = 0, \ v \in V.$$

Conversely, if $u \in V$ verifies the relations above, corresponding to some z and w, and the function $G(x, \cdot)$ is regular at $(Tu)(x)$ (in the sense of Clarke [6]) for each $x \in \Omega$, then u is a critical point of I.

Proof. Hypothesis (H_1) and Lebourg's mean value theorem (Theorem 1.1 in Chapter 1) show that
$$|G(x,y)| \le a_1 + a_2|y|^s \tag{2.21}$$

for all $x \in \Omega$ in $y \in R^N$, with constants $a_1, \ a_2 \ge 0$. It follows that the functional $J : L^s(\Omega; \mathbb{R}^N) \to \mathbb{R}$,
$$J(f) = \int_\Omega G(x, f(x))dx, \ \ \forall f \in L^s(\Omega; \mathbb{R}^N), \tag{2.22}$$

is well defined and Lipschitz continuous on the bounded subsets of $L^s(\Omega; \mathbb{R}^N)$. Since we have
$$I = F - JT, \tag{2.23}$$

it is seen that I is locally Lipschitz on V. Consequently, it makes sense to consider the critical points of I in the sense of Chang [4], i.e. the elements $u \in V$ verifying

$$0 \in \partial I(u) \tag{2.24}$$

(cf. Definition 1.4 in Chapter 1).

The foregoing locally Lipschitz functional I in (2.23) satisfies the Palais-Smale condition in the sense of Chang [4] (see Definition 2.1). Indeed, let $\{v_n\}$ be a sequence in V with $|I(v_n)| \le M$ and for which there exists a sequence $J_n \in \partial I(v_n)$ with $J_n \to 0$ in V^*. Then from (H_2) and taking into account that (2.21)-(2.23) imply

$$J_n = z_n - T^* w_n,\ z_n \in \partial F(v_n),\ w_n(x) \in \partial_y G(x, (Tv_n)(x))\ a.e.\ x \in \Omega,$$

we infer that

$$\begin{aligned} M + r\|v_n\|_V &\ge F(v_n) - r\langle z_n, v_n\rangle_V + \\ &+ \int_\Omega (-G(x, (Tv_n)(x)) + r\langle w_n(x), (Tv_n)(x)\rangle) dx \ge \\ &\ge \alpha\|v_n\|_V^\sigma + C_1\|v\|_V^{\sigma_0} + C_2, \end{aligned}$$

with real constants $C_1,\ C_2$, provided that n is large enough. It is clear that the estimate above implies that the sequence $\{v_n\}$ is bounded in V. Then a standard argument based on the assumption (H_3) allows to conclude that $\{v_n\}$ possesses a strongly convergent subsequence. Namely, the boundedness of (v_n) implies that $\{Tv_n\}$ is bounded in $L^s(\Omega; \mathbb{R}^N)$. Thus $\{w_n\}$ is bounded in $L^{\frac{s}{s-1}}(\Omega; \mathbb{R}^N)$ due essentially to the assumption (H_1). Since T^* is a compact operator and $J_n \to 0$ we derive that $\{z_n\}$ has a convergent subsequence in V^*. This fact combined with the boundedness of $\{v_n\}$ allows to use the hypothesis (H_3). The claim that the locally Lipschitz functional I verifies the Palais-Smale condition is proved.

Fix now an $\varepsilon > 0$. Then the assumption (H_4) insures the existence of some $\delta > 0$ such that

$$F(v) \ge A\|v\|_V^r,\ \ \|v\|_V \le \delta, \tag{2.25}$$

for a constant $A > 0$, and

$$G(x, y) \le \varepsilon|y|^p,\ \forall x \in \Omega,\ |y| \le \delta.$$

Combining the inequality above with (H_1) and using the Sobolev imbedding theorem one obtains that

$$J(Tv) \le \varepsilon B\|v\|_V^p,\ \ \|v\|_V \le \delta, \tag{2.26}$$

where $B > 0$ is a constant independent of ε. This is derived on the basis of (2.21), the boundedness above near 0 for G that we have obtained and the fact that $s > p$. By (2.25) and (2.26), and choosing $\varepsilon < AB^{-1}$, we arrive at the conclusion that there exist positive numbers $\rho,\ \eta$ such that

$$I(v) \geq \eta, \quad \|v\|_V = \rho. \tag{2.27}$$

The formula

$$\partial_t(t^{-\frac{1}{r}}G(x,ty)) = \frac{1}{r}t^{-1-\frac{1}{r}}[r < \partial_y G(x,ty), ty > -G(x,ty)],$$

Lebourg's mean value theorem and $(H_2)(ii)$ show that

$$(n+1)^{-\frac{1}{r}}G(x,(n+1)y) - n^{-\frac{1}{r}}G(x,ny)$$

$$\geq -\frac{1}{r}(bn^{\sigma_0-1-\frac{1}{r}}|y|^{\sigma_0} + b_0 n^{-1-\frac{1}{r}})$$

for a.e. $x \in \Omega$, $y \in \mathbf{R}^N$ and each positive integer n. The first $n-1$ inequalities above imply the estimate

$$G(x,ny) \geq n^{\frac{1}{r}}[G(x,y) - \frac{1}{r}(bK_1|y|^{\sigma_0} + b_0K_0)],$$

with a.e. $x \in \Omega$, $y \in \mathbf{R}^N$, $n \geq 1$, where K_0, K_1 are positive constants. This leads to

$$I(nv) \leq F(nv) - n^{\frac{1}{r}}[J(Tv) - \frac{1}{r}(bK_1\|Tv\|^{\sigma_0}_{L^{\sigma_0}} + b_0K_0|\Omega|)], \tag{2.28}$$

for $v \in V$ and $n \geq 1$. In view of $(H_5)(ii)$ we know that there exists $m_0 \geq 1$ provided

$$K := J(m_0Tv_0) - \frac{1}{r}(bK_1\|T(m_0v_0)\|^{\sigma_0}_{L^{\sigma_0}} + b_0K_0|\Omega|)) > 0.$$

Putting $v = m_0v_0$ in (2.28) it follows that

$$I(nm_0v_0) \leq F(nm_0v_0) - n^{\frac{1}{r}}K, \quad n \geq 1. \tag{2.29}$$

Letting $n \to \infty$ in (2.29), hypothesis $(H_5)(i)$ ensures that a sequence $t_n \to \infty$ satisfies

$$I(t_nv_0) \leq 0. \tag{2.30}$$

The Palais-Smale condition and the properties stated in (2.27) and (2.30) permit to apply the Mountain Pass Theorem in the nonsmooth version of Chang [4] (which is a particular case of Corollary 2.6). This yields the desired critical point u of I. The other assertions of the first part of Theorem 2.4 are direct consequences of what we have already proved and of Theorem 1.3 in Chapter 1.

The converse part of Theorem 2.4 follows from the next formula

$$\partial \int_\Omega G(x,u(x))dx = \int_\Omega \partial_y G(x,u(x))dx, \quad \forall u \in L^s(\Omega;\mathbf{R}^N),$$

which is valid under the growth condition in (H_1) and the regularity assumption for G (see Clarke [6]). The proof of Theorem 2.4 is thus complete. □.

2.5 Applications to Nonsmooth Boundary Value Problems

Among the various applications of the abstract variational methods in the nonsmooth critical point theory that have been developed in the previous sections a significant interest is directed to the nonlinear boundary value problems with discontinuities. This study has been initiated by Chang [4]. In the present Section we treat some examples that cover new situations or weaken some usual requirements.

Throughout the Section Ω denotes a bounded and smooth domain in $\mathbb{R}^m$, $m \geq 1$.

Example 2.1 Consider the nonsmooth Dirichlet problem containing both superlinear and sublinear terms

$$-\Delta u \in a(x)u + \lambda|u|^{s-2}u + \mu[\underline{\beta}(u(x)), \overline{\beta}(u(x))] \text{ a.e. } x \in \Omega,$$

$$u = 0 \text{ on } \partial\Omega.$$

The data entering the problem above are suppose to satisfy

$$a \in L^\infty(\Omega) \text{ with } \|a\|_{L^\infty} < \lambda_1,$$

where λ_1 denotes the first eigenvalue of $-\Delta$ on $H_0^1(\Omega)$, λ, μ, s are constants with $\lambda > 0$, $2 < s < 2m(m-2)^{-1}$ if $m \geq 3$,

$$\beta \in L^\infty_{loc}(\mathbb{R}), \quad |\beta(t)| \leq (1 + |t|^\gamma),$$

with constants $c > 0$, $0 \leq \gamma < 1$,

$$\limsup_{t \to 0} |t|^{-p} \int_0^t \beta(\sigma) d\sigma \leq 0$$

for a number p, $2 \leq p < s$. In stating the aforementioned problem we have used the notations $\underline{\beta}$ and $\overline{\beta}$ in Section 1.3 of Chapter 1. We apply Theorem 2.4 by choosing $V = H_0^1(\Omega)$, the compact linear operator $T : H_0^1(\Omega) \to L^s(\Omega)$ equal to the embedding of $H_0^1(\Omega)$ into $L^s(\Omega)$ with $2 < s < 2m(m-2)^{-1}$ if $m \geq 3$,

$$F(v) = \frac{1}{2}\int_\Omega (|\nabla v|^2 - a(x)v^2)dx, \quad \forall v \in H_0^1(\Omega),$$

$$G(t) = \lambda s^{-1}|t|^s + \mu \int_0^t \beta(\sigma)d\sigma, \ t \in \mathbb{R}.$$

Hypothesis (H_1) follows because, in view of Corollary 1.3 in Chapter 1, one has

$$\partial G(t) = \lambda|t|^{s-2}t + \mu[\underline{\beta}(t), \overline{\beta}(t)], \quad t \in \mathbb{R}.$$

Taking $\frac{1}{s} < r < \frac{1}{2}$ one obtains (H_2). The compactnesss of the embedding $H_0^1(\Omega) \subset L^s(\Omega)$ assures that (H_3) holds. Assertion (H_4) is deduced from the imposed conditions on a and β. Assumption (H_5) is checked for any $v_0 \in V\setminus\{0\}$ since $\sigma_0 = \gamma + 1 < 2 < \frac{1}{r} < s$. Then Theorem 2.4 and a standard regularity result ensure the existence of a nontrivial solution $u \in H_0^1(\Omega) \cap H^2(\Omega)$ of our nonlinear Dirichlet problem.

Example 2.2 The nonlinear problems involving the critical exponent $2^* = 2m(m-2)^{-1}$ with $m \geq 3$ may have no solutions (see, e.g., Passaseo [20]). However, Theorem 2.4 implies the solvability of the following problem

$$-\Delta u \in \lambda[K^*(|Ku|^{2^*-1}Ku) + \partial_t H(x,u(x))] \text{ a.e } x \in \Omega,$$

$$u = 0 \text{ on } \partial\Omega,$$

where λ is any positive number, $K : L^{2^*}(\Omega) \to L^{2^*}(\Omega)$ is a compact linear operator and $H : \Omega \times \mathbf{R} \to \mathbf{R}$ is the primitive

$$H(x,t) = \int_0^t h(x,\sigma)d\sigma, \quad \forall (x,t) \in \Omega \times \mathbf{R},$$

of a function $h : \Omega \times \mathbf{R} \to \mathbf{R}$, which is measurable with respect to the first variable and locally Lipschitz with respect to the second one, satisfying the conditions

$$|h(x,t)| \leq c(1 + |t|^{\gamma-1}), \quad \forall (x,t) \in \Omega \times \mathbf{R},$$

with constants $c > 0$ and $2 < \gamma < 2^*$ if $m \geq 3$,

$$\min(H(x, \pm q), rt\partial_t H(x,t) - H(x,t)) \geq 0$$

for a.e. $x \in \Omega$ and $t \in \mathbf{R}$, $|t| \geq q$, with constants $\frac{1}{2^*} < r < \frac{1}{2}$ and $q > 0$,

$$h(x,t) = o(|t|) \text{ as } t \to 0 \text{ uniformly with respect to } x \in \Omega.$$

The operator K can be for instance an integral operator in which case our problem becomes a nonlinear integro-diferential equation. Theorem 2.4 is applied with the following choices: $V = H_0^1(\Omega)$,

$$Tv = Kv, \ \forall v \in V,$$

$$F(v) = \int_\Omega (\frac{1}{2\lambda}|\nabla v|^2 - H(x,v(x)))dx, \ \forall v \in V,$$

$$G(t) = 2^{*-1}|t|^{2^*}, \quad t \in \mathbf{R}.$$

The local Lipschitziannes of F is a consequence of the growth condition for h. Hypothesis (H_1) in Theorem 2.4 is obviously satisfied, while (H_2) with $\sigma = 2$ is easily derived from the condition on H. Since the inclusion $V \subset L^\gamma(\Omega; \mathbf{R})$ is compact, it turns out that (H_3) holds. The condition concerning the behaviour of h near $0 \in \mathbf{R}$ yields that (H_4) is verified for $p = 2$. Clearly, (H_5), (ii) is valid for any $1 \leq \sigma_0 < 2$ and $Kv_0 \neq 0$. From the imposed condition on H we derive the estimate

$$H(x,t) \geq |t|^{1/r} q^{-1/r} \inf_{\Omega} H(\cdot, \pm q) + C \text{ a.e. } x \in \Omega \text{ and } t \in \mathbb{R},$$

with a constant C. Then it is straightforward to get $(H_5)(i)$. Then Theorem 4.2 provides a nontrivial solution of our boundary value problem.

Other related results and applications to eigenvalue problems in the form of hemivariational inequalities are given in the next chapters.

We focus now on applying the equivariant minimax principle for locally Lipschitz functionals formulated in Corollary 2.7 in the study of a nonsmooth Dirichlet problem. General information for the variational approach in solving nonlinear elliptic boundary value problems with symmetries can be found in Ambrosetti and Rabinowitz [2], Bartsch [3], Chang [4]-[5], Goeleven, Motreanu and Panagiotopoulos [9], Hulshoff and van der Vorst [10], Motreanu [13], Motreanu and Panagiotopulos [16], Motreanu and Varga [15], Rabinowitz [22], Struwe [23]. Precisely, our application here concerns the nonsmooth form of the equivariant Dirichlet problem in Bartsch [3], p. 27-29.

Let Ω denote a bounded domain in $\mathbb{R}^N$ whose boundary $\partial\Omega$ is regular of class C^1. Let the (Carathéodory) function $F : \Omega \times \mathbb{R}^m \to \mathbb{R}$ satisfy the conditions: $F(\cdot, u)$ is measurable for each $u \in \mathbb{R}^m$, $F(x, \cdot)$ is locally Lipschitz for each $x \in \Omega$ and $F(\cdot, 0) \in L^1(\Omega)$. We state the following nonsmooth Dirichlet problem

$$\text{(P)} \qquad \begin{aligned} -\Delta u &\in \partial_y F(x,u) \text{ a.e. } x \in \Omega, \\ u &= 0 \text{ on } \partial\Omega. \end{aligned}$$

We stress that Problem (P) is in fact a system of elliptic equations. The notation $\partial_y F$ stands for the partial generalized gradient of F with respect to the second variable $y \in \mathbb{R}^m$, while $u : \Omega \to \mathbb{R}^m$ denotes the unknown function in (P). Naturally, we seek the solutions u in $H_0^1(\Omega, \mathbb{R}^m)$. In this respect we associate to Problem (P) the functional $I : H_0^1(\Omega, \mathbb{R}^m) \to \mathbb{R}$ expressed as follows

$$I(u) = \int_{\Omega} (\frac{1}{2}|\nabla u|^2 - F(x,u))dx, \quad \forall u \in H_0^1(\Omega, \mathbb{R}^m). \tag{2.31}$$

We assume that the function $F : \Omega \times \mathbb{R}^m \to \mathbb{R}$ satisfies the following conditions:

(H_1) there are constants $a_1 \geq 0$, $a_2 \geq 0$ and $p \geq 0$ such that $p < (N+2)/(N-2)$ if $N \geq 3$, and

$$|z| \leq a_1 + a_2|y|^p, \quad \forall z \in \partial_y F(x,y), \ (x,y) \in \Omega \times \mathbb{R}^m;$$

(H_2) there are constants $b_1 \geq 0$, $b_2 \geq 0$, $\mu > 2$ and $1 \leq \sigma < 2$ such that the inequality below holds

$$F_y^0(x,y) - \mu^{-1}F(x,y) \geq -b_1|y|^\sigma - b_2, \quad \forall (x,y) \in \Omega \times \mathbb{R}^m;$$

(H_3) F is G-invariant with respect to the second variable, i.e.,

$$F(x, gy) = F(x, y), \quad \forall (x, y) \in \Omega \times \mathbb{R}^m, \ g \in G,$$

where G is a compact Lie group acting orthogonally on $\mathbb{R}^m$ such that the action is admissible in the sense of Bartsch [3], i.e. does not exist a continuous G-invariant map from a closed, bounded, G-invariant neighborhood of 0 in $(\mathbb{R}^m)^k$ into $(\mathbb{R}^m)^{k-1} \setminus \{0\}$, $k \geq 1$.

(H_4) there exist constants $r > 0$ and $\alpha > 0$ such that

$$F(x, y) \geq \alpha, \ \text{a.e. } x \in \Omega, \ |y| = r.$$

Theorem 2.5 Under the assumptions $(H_1) - (H_4)$, the nonsmooth Dirichlet problem (P) has an unbounded sequence of solutions in $H_0^1(\Omega, \mathbb{R}^m)$.

Proof. The approach is based on the search of critical points for the functional $I : H_0^1(\Omega, \mathbb{R}^m) \to \mathbb{R}$ given by (2.31). The assumption (H_1) entails that I is locally Lipschitz and any critical point $u \in H_0^1(\Omega, \mathbb{R}^m)$ of I in the sense of Chang [4] is a weak solution of our Dirichlet problem (P). Therefore it suffices to prove the existence of an unbounded sequence of critical points $\{u_n\} \subset H_0^1(\Omega, \mathbb{R}^m)$ of I.

The induced G-action on $H_0^1(\Omega, \mathbb{R}^m)$ is introduced by

$$(gu)(x) = g \cdot u(x), \quad \forall u \in H_0^1(\Omega, \mathbb{R}^m), \ x \in \Omega, \ g \in G. \tag{2.32}$$

The action of G on $H_0^1(\Omega, \mathbb{R}^m)$ given in (2.32) is orthogonal. By assumption (H_3) it is clear that the functional I is G-invariant with respect to the G-action in (2.32).

Consider the sequence $0 < \lambda_1 < \lambda_2 \leq \ldots$ of eigenvalues of $-\Delta$ on $H_0^1(\Omega)$ and let $\{\phi_n\}_{n \geq 1} \subset H_0^1(\Omega)$ be a corresponding sequence of eigenfunctions. For every $j \geq 1$ we set

$$E^j = \{\phi_j y : \ y \in \mathbb{R}^m\}$$

and then for every $k \geq 1$,

$$E_k = \oplus_{j=1}^k E^j.$$

It turns out that each space E^j is isomorphic to $\mathbb{R}^m$ by a G-equivariant isomorphism with respect to the G-action in (2.32). From the properties of the sequence $\{\phi_n\}_{n \geq 1}$ one obtains that

$$H_0^1(\Omega, \mathbb{R}^m) = \overline{\text{span} \cup_{k=1}^{\infty} E_k}.$$

With these preliminaries at our disposal we can apply Corollary 2.7 for the functional I, where it is chosen $V = \mathbb{R}^m$.

According to the requirement (i) we have to check the Palais-Smale condition in the sense of Chang [4] for the functional I in (2.31). To this end let $\{u_n\} \subset H_0^1(\Omega, \mathbb{R}^m)$ be a sequence such that $I(u_n)$ is bounded and $\{z_n\} \to 0$ in $[H_0^1(\Omega, \mathbb{R}^m)]^*$ as $n \to \infty$ for some $z_n \in \partial I(u_n)$. Then, taking $z_n = -\Delta u_n - w_n$, with $w_n \in L^{(p+1)/p}(\Omega, \mathbb{R}^m)$, the assumption (H_2) allows to write

$$
\begin{aligned}
I(u_n) - \mu^{-1}\langle z_n, u_n\rangle &= \\
&= (\frac{1}{2} - \frac{1}{\mu}) \int_\Omega |\nabla u_n|^2 dx + \int_\Omega (\mu^{-1}\langle w_n, u_n\rangle - F(x, u_n))dx \\
&\geq (\frac{1}{2} - \frac{1}{\mu})\|u_n\|^2_{H_0^1} - c_1\|u_n\|^\sigma_{H_0^1} - c_2, \qquad (2.33)
\end{aligned}
$$

for positive constants c_1 and c_2. The estimate (2.33) implies the boundedness of the sequence $\{u_n\}$ in $H_0^1(\Omega, \mathbb{R}^m)$. Then by a standard argument relying on the compactness embedding in the Sobolev embedding theorem (see Rabinowitz [22]) we infer that $\{u_n\}$ contains a strongly convergent subsequence. Therefore the functional I in (2.31) satisfies the Palais-Smale condition.

In order to verify the hypothesis (ii) in Corollary 2.7 let us fix a $k \geq 1$. The formula

$$\partial_t(t^{-\mu}F(x, ty)) = \mu t^{-1-\mu}[\frac{1}{\mu}\langle \partial_y F(x, ty), ty\rangle - F(x, ty)],$$

for all $t > 0$, $x \in \Omega$, $y \in \mathbb{R}$, yields by scalar integration over $[1, t]$ that

$$t^{-\mu}F(x, ty) - F(x, y) = \int_0^t \mu s^{-1-\mu}[\frac{1}{\mu}F_y^o(x, sy; sy) - F(x, sy)]ds$$

for all $t > 1$, $x \in \Omega$, $y \in \mathbb{R}^n$. Then the assumptions (H_2) and (H_4), using essentially that each E_k is finite dimensional, ensure that

$$I(u) \to -\infty \text{ in } E_k \text{ as } \|u\| \to \infty. \qquad (2.34)$$

The assertion (ii) in Corollary 2.7 is fulfilled.

We pass to the proof of (iii) in Corollary 2.7. For any fixed $k \geq 1$ let us consider the orthogonal complement $E_k^\perp$ of E_k in $H_0^1(\Omega, \mathbb{R}^m)$. By (H_1), the functional I can be estimated from below as follows

$$I(u) \geq \frac{1}{2}\|u\|^2_{H_0^1} - C_1\|u\|^{p+1}_{L^{p+1}} - C_2, \ \forall u \in H_0^1(\Omega, \mathbb{R}^m), \qquad (2.35)$$

where C_1, C_2 are positive constants. Then the argument in Rabinowitz [22], p. 60, based on (2.35), Gagliardo-Nirenberg interpolation inequality and the variational characterization of the eigenvalue λ_{k+1}, leads to the conclusion that a $k_0 \geq 1$ can be found to fulfill (iii) of Corollary 2.7.

The desired unboundedness of the sequence of critical points of I is a consequence of the unboundedness of the corresponding sequence of critical values that is provided by Corollary 2.7.

References

[1]S. Adly and D. Goeleven, Periodic solutions for a class of Hemivariational Inequality, *Comm. Appl. Nonlin. Anal.* **2** (1995), 45-57.

[2]A. Ambrosetti and P.H. Rabinowitz, Dual variational methods in critical point theory and applications, *J. Funct. Anal.* **14** (1973), 349-381.

[3]T. Bartsch, *Topological Methods for Variational Problems with Symmetries*, Springer-Verlag, Berlin, Heidelberg, New York, 1993.

[4]K.C. Chang, Variational methods for non-differentiable functionals and their applications to partial differential equations, *J. Math. Anal. Appl.* **80** (1981), 102-129.

[5]K.C. Chang, *Infinite dimensional Morse theory and Multiple Solutions Problems*, Birchäuser Verlag, Basel, Boston, 1993.

[6]F.H. Clarke, *Nonsmooth analysis and Optimization*, Wiley, New York, 1983.

[7]Y. Du, A deformation lemma and some critical point theorems, *Bull. Austral. Math. Soc.* **43** (1991), 161-168.

[8]N. Ghoussoub and D. Preiss, A general mountain pass principle for locating and classifying critical points, *Ann. Inst. H. Poicaré. Anal. Nonlinéaire* **6** (1989), 321-330.

[9]D. Goeleven, D. Motreanu and P. D. Panagiotopoulos, Multiple solutions for a class of hemivariational inequalities involving periodic energy functionals, *Math. Methods Appl. Sci.* **20** (1997), 547-568.

[10]J. Hulshoff and R. van der Vorst, *Differential systems with strongly indefinite variational structure, J. Funct. Anal.* **114** (1993), 97-105.

[11]W. Krawcewicz and W. Marzantowicz, Some remarks on the Lusternik-Schnirelmann methods for non-differentiable functionals invariant with respect to a finite group action, *Rocky Mountain J. Math.* **20** (1990), 1041-1049.

[12]L. Kantorovitch and G. Akilov, *Analyse Fonctionnelle*, Tome 1, Editions Mir, Moscow, 1981.

[13]D. Motreanu, A multiple linking minimax principle, *Bull. Austral. Math. Soc.* **53** (1996), 39-40.

[14]D. Motreanu, Existence of critical points in a general setting, *Set-Valued Anal.* **3** (1995), 295-305.

[15]D. Motreanu and Cs. Varga, A nonsmooth equivariant minimax principle, *Comm. Appl. Anal.*, in print.

[16]D. Motreanu and P.D. Panagiotopulos, Double eigenvalue problems for hemivariational inequalities, *Arch. Rational Mech. Anal.* **140** (1997), 225-251.

[17]Z. Naniewicz, On some nonconvex variational problems related to hemivariational inequalities, *Nonlinear Anal.* **13** (1989), 87-100.

[18]P.D. Panagiotopulos, *Inequality Problems in Mechanics and Applications. Convex and Nonconvex Energy Functions*, Birkhäuser, Boston, Basel, Stuttgart, 1985 (Russian Translation MIR Publ. Moscow, 1989).

[19]P.D. Panagiotopulos, *Hemivariational inequalities. Applications in Mechanics and Engineering*, Springer-Verlag, Berlin, Heidelberg, New-York, 1993.

[20]D. Passaseo, Nonexistence results for elliptic problems with supercritical nonlinearity in nontrivial domains, *J. Funct. Anal.* **114**(1993), 97-105.

[21]P. Pucci and J. Serrin, Extensions of the mountain pass theorem, *J. Funct. Anal.* **59** (1984), 185-210.

[22]P.H. Rabinowitz, Minimax methods in critical point theory with applications to differential equations, CBMS Reg. Conf. Ser. Math. No. 65, *Amer. Math. Soc.* Providence, R.I., 1986.

[23]M. Struwe, *Variational Methods*, Springer-Verlag, Berlin, 1990.

[24]Cs. Varga, *Topological Methods in Calculus of Variations*, Ph. D. Thesis, University of Cluj-Napoca, 1996.

3. Minimax Methods for Variational-Hemivariational Inequalities

The topic of this chapter is the critical point theory for the functionals that are not locally Lipschitz as was the case in Chapter 2. The setting is more general than in Chapter 2, and the results contain those in Chang [2]. In fact, this chapter presents an extension of Szulkin's minimax principles [32] for functions of the form $I = \Phi + \Psi$ with $\Phi \in C^1(X, \mathbf{R})$ (X denotes a real Banach space and Ψ a convex, proper and lower semicontinuous functional) to the case where Φ is locally Lipschitz. In this way it also extends Chang's minimax principles [2] by taking $\Psi = 0$. In this general context a point $u \in X$ is said to be a critical point if $\Phi^0(u; v-u) + \Psi(v) - \Psi(u) \geq 0,\ \forall\ v \in X$ (compare with the definitions of Section 1.4 of Chapter 1). Such a mathematical model - called a variational - hemivariational inequality - can be used to formulate various problems in Mechanics involving both monotone and non - monotone possibly multivalued boundary conditions, whose complete description can be found in the books of Panagiotopoulos [27] and Naniewicz and Panagiotopoulos [25]. The chapter is based on the paper of Motreanu and Panagiotopoulos [10]. It develops a deformation theorem and a general linking theorem for functionals of the above mentioned type. Then we deduce directly the generalization of the results of Szulkin and Chang. We end this chapter with some applications concerning variational - hemivariational inequalities and semicoercive unilateral problems.

3.1 Motivation and Introduction

Several problems in Mechanics leading to variational - hemivariational inequalities are presented in [25][27]. The corresponding stability problems (cf. [22]) lead to eigenvalue problems for variational - hemivariational inequalities. For their treatment one has to generalize the deformation lemma, the linking theorem and the corresponding results of Szulkin and Chang. This is the aim of the present chapter. Let us begin with a concrete mechanical problem.

Consider a linear elastic body which in its underformed state is identified with an open bounded subset $\Omega \subset \mathbf{R}^3$. The boundary Γ of the body Ω is assumed to consist of the three open disjoints parts Γ_1, Γ_2 and Γ_3, i.e. $\Gamma = \overline{\Gamma_1} \cup \overline{\Gamma_2} \cup \overline{\Gamma_3}$. Let us denote by $u = \{u_i\}, \sigma = \{\sigma_{ij}\}, \varepsilon = \{\varepsilon_{ij}\}, S = \{S_i\}$ the displacement vector, the stress tensor, the strain tensor and the stress vector, respectively.

We assume that on the part Γ_1 of the boundary of the body the constraint

$$u(x) \in Q(x), \quad \forall x \in \Gamma_1, \tag{A}$$

is satisfied, where $Q(x)$ is defined as follows

$$Q(x) := \{y \in \mathbf{R}^3 : f(x,y) \leq 0\}.$$

Here $f : \Gamma_1 \times \mathbf{R}^3 \to \mathbf{R}$ is continuous and convex in the second variable. Moreover, we assume that $f(x,0) \leq 0, \forall x \in \Gamma_1$. Thus $Q(x)$ is a nonempty, closed and convex subset for all $x \in \Gamma_1$. Now, the formulation of these unilateral constraints has to encompass the associated forces of reactions $r(x)$. We assume a normal reaction law of the form (i.e. without frictional effects)

$$-r(x) \in N_{Q(x)}(u(x)), \quad \forall x \in \Gamma_1,$$

where for a vector $v \in \mathbf{R}^3$, $N_{Q(x)}(v)$ denotes the normal cone to $Q(x)$ at $v \in \mathbf{R}^3$. Note that the previous relation is equivalent to

$$u(x) \in Q(x) : r(x)^T(v - u(x)) \geq 0, \quad \forall v \in Q(x), \ \forall x \in \Gamma_1. \tag{B}$$

The part Γ_2 (it is assumed that mes$\Gamma_2 > 0$) of the boundary is assumed to be fixed, that is

$$u(x) = 0, \quad \forall x \in \Gamma_2. \tag{C}$$

On Γ_3, we consider conditions of adhesive contact, i.e. a unilateral contact relation between the body and a Winkler-type support which may sustain only limited values of traction, and any compression. These conditions take the following form (we consider the usual decompositions $v = v_N n + v_T, S = S_N n + S_T, v_N, S_N \in \mathbf{R}, v_T, S_T \in \mathbf{R}^3$, where n denotes the unit outward normal vector on Γ, defining the positive direction)

$$\begin{array}{llll} \text{if} & -\varepsilon < u_N, & \text{then} & S_N + k u_N = 0, \\ \text{if} & u_N = -\varepsilon, & \text{then} & 0 \leq S_N \leq k\,\varepsilon \\ \text{if} & u_N < -\varepsilon, & \text{then} & S_N = 0 \end{array}$$

where ε is a positive real number, k is a positive function. Moreover, in the tangential direction the forces are given, i.e.

$$S_T = C_T$$

where

$$C_T : \Gamma_3 \times \mathbf{R}^3 \to \mathbf{R}^3,$$

$$\{x, u_{T_1}, u_{T_2}, u_{T_3}\} \to \{C_{T_1}(x, u_T), C_{T_2}(x, u_T), C_{T_3}(x, u_T)\}$$

is a vector-valued function. In order to describe the action of the adhesive material in the tangential direction we consider mappings $C_{T_i} (i = 1, 2, 3)$ depending of x and u_T as follows

$$C_{T_i}(x, u_T) = \alpha_i u_{T_i}^3 + \beta_i u_{T_i}^2 + \gamma_i u_{T_i} + \delta_i(x),$$

where α_i, β_i and $\gamma_i (i = 1,2,3)$ are real numbers and $\delta_i(i = 1,2,3)$ are given functions defined on Γ_3. Using the generalized gradient calculus rules defined by Clarke [3] (see Chapter 1), we can express the relations holding on Γ_3 in the normal direction by means of the following set-valued relation:

$$-S_N \in \partial j(x, u_N(x)), \forall x \in \Gamma_3,$$

where

$$j(x,y) := \begin{cases} \frac{1}{2}k(x)y^2 & \text{if } -\varepsilon \leq y \\ \\ \frac{1}{2}k(x)\varepsilon^2 & \text{if } \ y \leq -\varepsilon. \end{cases}$$

From the orthogonality between $S_N n$ and u_T and between S_T and $u_N n$, and the definition of the generalized gradient in the previous relation we obtain the hemivariational inequality

$$S^T v + j^o(x, u_N(x); v_N) - C_T^T v_T \geq 0, \quad \forall v \in \mathbb{R}^3, \forall x \in \Gamma_3. \qquad (D)$$

Here $\partial j(x,.)$ stands for the Clarke's subdifferential of j with respect to the second variable while $j^o(x, u; v)$ denotes the generalized directional derivative in the sense of Clarke of j at u in the direction v (cf. Definitions 1.1,1.2). Finally, we suppose that the body is subject to a body density force $f = \{f_i\}$. In the framework of a small deformation theory, we have that

$$\sigma_{ij,j} + f_i = 0,$$

$$\varepsilon_{ij}(u) = \frac{1}{2}(u_{i,j} + u_{j,i}),$$

$$\sigma_{ij} = C_{ijkl}\varepsilon_{kl}(u),$$

where C_{ijkl} denotes the elasticity tensor assumed to satisfy the usual symmetry and ellipticity properties:

$$C_{ijkl} = C_{jikl} = C_{klij}$$

and

$$C_{ijkl}\, \varepsilon_{ij}\varepsilon_{kl} \geq c\varepsilon_{ij}\varepsilon_{ij}, \quad \forall \varepsilon \in \mathbb{R}^6 \ \ (c = const > 0).$$

Let us now suppose for the present that all the data are sufficiently smooth. Using the Green-Gauss theorem and the boundary conditions we obtain

$$\int_\Omega C_{ijkl}\varepsilon_{ij}(u)\varepsilon_{kl}(v-u)dx = \int_\Omega f^T(v-u)dx + \int_{\Gamma_3} S_N(v_N - u_N)ds$$

$$+ \int_{\Gamma_3} C_T^T(v_T - u_T)ds + \int_{\Gamma_1} r^T(v-u)ds.$$

A combination of this last expression with the inequalities (D) and (B) implies the following variational-hemivariational inequality problem: Find u satisfying (A) and (C) such that

$$\int_{\Omega} C_{ijkl}\varepsilon_{ij}(u)\varepsilon_{kl}(v-u)dx + \int_{\Gamma_3} j^o(x, u_N; v_N - u_N)ds -$$

$$- \int_{\Omega} f^T(v-u)dx - \int_{\Gamma_3} C_T^T(v_T - u_T)ds \geq 0, \tag{E}$$

for all v satisfying (A) and (C).

Note that (E) is an expression of the virtual work "principle" for the present problem. Let us now define the functional framework. The boundary Γ of the body is assumed sufficiently regular ($C^{1,1}$ is sufficient) and we denote by $\gamma : H^1(\Omega, \mathbf{R}^3) \to H^{\frac{1}{2}}(\Gamma; \mathbf{R}^3), \gamma_N : H^1(\Omega; \mathbf{R}^3) \to H^{\frac{1}{2}}(\Gamma)$ and $\gamma_T : H^1(\Omega; \mathbf{R}^3) \to H_T$ (see [26] p. 32 for the notations) the usual trace mapping, normal trace mapping and tangential trace mapping, respectively. We set

$$X = \{u \in H^1(\Omega; \mathbf{R}^3) : \gamma(u(x)) = 0, a.e.\ x \in \Gamma_2\}$$

and

$$C = \{u \in X : \gamma(u(x)) \in Q(x), a.e.\ x \in \Gamma_1\}.$$

Let us now assume the following for the data: $C_{ijkl} \in L^\infty(\Omega)$, $f \in L^2(\Omega; \mathbf{R}^3)$, $\delta_i \in L^2(\Gamma_3)$ and $k \in L^2(\Gamma_3)$. We define the operators $A : X \to X^*, N : X \to X^*$, the element $h \in X^*$ and the functional $J : X \to \mathbf{R}$ by the formulae:

$$\begin{aligned}
\langle Au, v\rangle &:= \int_{\Omega} C_{ijkl}(x)\varepsilon_{ij}(u)\varepsilon_{kl}(v)dx, \quad \forall u, v \in X,\\
\langle N(u), v\rangle &:= \int_{\Gamma_3} (\alpha_i\gamma_{T_i}(u)^3 + \beta_i\gamma_{T_i}(u)^2 + \gamma_i\gamma_{T_i}(u) + \delta_i)\gamma_{T_i}(v)ds, \forall u, v \in X,\\
\langle h, v\rangle &:= \int_{\Omega} f_i v_i dx, \quad \forall v \in X,\\
J(v) &:= \int_{\Gamma_3} j(x, \gamma_N(v(x)))ds, \quad \forall v \in X.
\end{aligned}$$

The operator A is bounded, symmetric, linear and coercive, i.e. there exists $\alpha > 0$ such that

$$\langle Au, u\rangle \geq \alpha \parallel u \parallel^2, \quad \forall u \in X.$$

The function J defined previously is such that (i) $x \to j(x, u)$ is measurable for each $u \in \mathbf{R}$, (ii) $u \to j(x, u)$ is locally Lipschitz for each $x \in \Gamma_3$ and (iii) $x \to j(x, 0)$ is finitely integrable. Moreover, for a suare-integrable function k_o let

$$| j(x, y) - j(x, y') | \leq \varepsilon k_o(x) | y - y' |, \quad \forall y, y' \in \mathbf{R}, x \in \Gamma_3.$$

Thus using the rules concerning the subdifferentiation of integral functionals and composite mappings (see Chapter 1), we obtain

$$J^o(u, v) \leq \int_{\Gamma_3} j^o(x, \gamma_N(u(x)); \gamma_N(u(x))ds, \quad \forall u, v \in X.$$

If we set

$$F(u) = \frac{1}{2}\langle Au, u\rangle - \langle h, u\rangle - (\sum_{i=1}^{3} \int_{\Gamma_3} ((\alpha_i/4)\gamma_{T_i}(u)^4 + (\beta_i/3)\gamma_{T_i}(u)^3$$

$$+(\gamma_i/2)\gamma_{T_i}(u)^2 + \delta_i\gamma_{T_i}(u))ds)$$

then

$$\langle F'(u), v\rangle = \langle Au - N(u) - h, v\rangle, \quad \forall u, v \in X.$$

Using (E), it results that the positions of equilibrium of our system are solutions of the variational- hemivariational inequality:

$$u \in C : \langle F'(u), v-u\rangle + J^o(u; v-u) \geq 0, \quad \forall v \in C$$

or equivalently

$$u \in X : \langle F'(u), v-u\rangle + J^o(u; v-u) + \Psi_C(v) - \Psi_C(u) \geq 0, \quad \forall v \in X, \quad (F)$$

where Ψ_C denotes the indicator function of C. It is now natural to introduce a concept of generalized critical point of the energy functional $I = F + J + \Psi_C$ as a solution of the variational- hemivariational inequality (F).

That means that, if we intend to consider concrete problems in unilateral Mechanics which involve both monotone and nonmonotone boundary conditions, then we have in general to deal with a nonsmooth and nonconvex energy functional - expresssed as the sum of a locally Lipschitz function $\Phi : X \to \mathbb{R}$ and a proper, convex and lower semicontinuous function $\Psi : X \to \mathbb{R} \cup \{+\infty\}$ - whose critical points are defined as the solutions of the variational-hemivariational inequality

$$u \in X : \quad \Phi^0(u; v-u) + \Psi(v) - \Psi(u) \geq 0, \quad \forall\ v \in X. \quad (G)$$

Various examples in mechanics leading to a problem like (F) are described in the books of Panagiotopoulos [27] and the one of Naniewicz and Panagiotopoulos [25].

If $\Phi \in C^1(X; \mathbb{R})$ then problem (G) reduces to a classical variational inequality. This model is now well known and an efficient critical point theory has been developed by Szulkin [32]. If $\Psi = 0$ then problem (G) reduces to the one studied by Chang [2] and has been studied in an extended form in Chapter 2. The theory of Chang has now been used in several directions by Goeleven, Motreanu and Panagiotopoulos [9] to study the qualitative properties of the solutions of hemivariational inequalities arising in unilateral Mechanics. If $\Phi \in C^1(X; \mathbb{R})$ and $\Psi = 0$ then the problem reduces to the equation $\Phi'(u) = 0$ which has been the subject of intensive developments. See for example the work of Ambrosetti and Rabinowitz [1], Rabinowitz [29] and [30]. So, the theory of Szulkin and the one of Chang have been shown very efficient to deal respectively with classical variational inequalities and hemivariational inequalities. The aim of this chapter is to consider the more general problem (G) or (F).

The present approach combines the approach of Chang with the one of Szulkin in a nontrivial way. Indeed the two mentioned approaches are completely different and absolutely not "additive". Then we derive a general linking theorem, which is then used to generalize the well-known Mountain-Pass theorem, the Saddle-Point theorem and the main results for even functionals. Using the resulting theory, we study various unilateral problem which have not been considered before. Besides its practical interest in the theory of variational-hemivariational inequalities, our theory unifies the three basic ones, i.e. the one for smooth functionals developed by Rabinowitz [29], the one of Szulkin [32] for variational inequalities and the one of Chang [2] for locally Lipschitz functionals. Moreover, it applies to the sum of locally Lipschitz functionals and proper l.s.c. convex functionals and, to the best of our knowledge, it is the first theory in this direction.

3.2 The General Setting

The aim of this Section is to set up the framework of our critical point theory. Namely, we deal with lower semicontinuous functionals (in short, l.s.c.) $I : X \rightarrow (-\infty, +\infty]$ on a real Banach space X satisfying the following structure hypothesis

(H) $I = \Phi + \Psi$, with $\Phi : X \rightarrow \mathbb{R}$ locally Lipschitz and $\Psi : X \rightarrow (-\infty, +\infty]$ proper (i.e., $\not\equiv +\infty$), convex and l.s.c..

Therefore assumption (H) extends the corresponding statement (H) of Szulkin [32] (as well as the theory of Chang [2] exposed in a general framework in Chapter 2) in that the function Φ is demanded to be only locally Lipschitz in place of being continuously differentiable.

In order to develop a critical point theory for functionals of type (H), following the pattern in Chapters 1 and 2, we need to introduce the basic concepts of critical point and Palais-Smale condition.

Definition 3.1 An element $u \in X$ is called a critical point of functional I in (H) if

$$\Phi^0(u; v-u) + \Psi(v) - \Psi(u) \geq 0, \quad \forall\, v \in X. \tag{3.1}$$

Definition 3.2 The functional $I : X \rightarrow (-\infty, +\infty]$ in (H) is said to satisfy the Palais - Smale condition at the level $c \in \mathbb{R}$ (in short, $(\mathrm{PS})_c$) if every sequence $\{u_n\} \subset X$ satisfying $I(u_n) \rightarrow c$ and

$$\Phi^0(u_n; v-u_n) + \Psi(v) - \Psi(u_n) \geq -\varepsilon_n \|v - u_n\|, \quad \forall\, v \in X. \tag{3.2}$$

for a sequence $\{\varepsilon_n\} \subset \mathbb{R}^+$ with $\varepsilon_n \rightarrow 0$, contains a convergent subsequence. If $(\mathrm{PS})_c$ is verified for all $c \in \mathbb{R}$, I is said to satisfy the Palais - Smale condition (in short, (PS)).

The notation Φ^0 in (1), (2) stands for the generalized directional derivative of Clarke [3], namely

$$\Phi^0(u; v) = \limsup_{\substack{w \rightarrow u \\ t \downarrow 0}} \frac{1}{t}(\Phi(w + tv) - \Phi(w))$$

for all $u, v \in X$ (see Section 1.1 in Chapter 1).

Remark 3.1 Definitions 3.1 and 3.2 unify the approaches of nonsmooth critical point theory due to Chang [2] and Szulkin [32]. If $\Psi = 0$ Definitions 3.1 and 3.2 reduce to the corresponding definitions of Chang [2] . If $\Phi \in C^1(X, \mathbb{R})$, Definition 3.1 and 3.2 coincide with those of Szulkin [32].

For the functional $I : X \to (-\infty, +\infty]$ in (H) we denote by $K(I)$ the set of critical points of I in the sense of Definition 2.1. Given any number $c \in \mathbb{R}$ let us set

$$K_c(I) := K(I) \cap I^{-1}(c).$$

For a latter use we put

$$I_a := \{v \in X : \ I(v) \leq a\}, \quad \forall \, a \in \mathbb{R}.$$

These notations have been already used in Chapter 2. Following Szulkin [32] the result below points out some relevant aspects of Definition 3.2.

Proposition 3.1 (i) Any limit point u of a (Palais - Smale) sequence $\{u_n\}$ entering Definition 3.2 belongs to $K_c(I)$.

(ii) Inequality (3.2) is equivalent to

$$\Phi^0(u_n; v - u_n) + \Psi(v) - \Psi(u_n) \geq \langle z_n, v - u_n \rangle, \quad \forall \, v \in X, \tag{3.3}$$

for some sequence $\{z_n\} \subset X^*$ with $z_n \to 0$.

Proof. (i) Without loss of generality we may admit that $u_n \to u$ in X. Then, letting $n \to \infty$, the upper semicontinuity of Φ^0 (see Clarke [3] or Proposition 1.3) and the l.s.c. of Ψ allow us to derive (3.1) from (3.2), so $u \in K(I)$. Taking $v = u$ in (3.2) we obtain that

$$\begin{aligned} \Psi(u) &\leq \liminf_{n\to\infty} \Psi(u_n) \leq \limsup_{n\to\infty} \Psi(u_n) \\ &\leq \Phi^0(u; u - u) + \Psi(u) = \Psi(u). \end{aligned}$$

This ensures that $I(u) = \lim_{n\to\infty} I(u_n) = c$, so $u \in K_c(I)$.

(ii) The assertion (3.3)$\Longrightarrow$(3.2) is true because one can take $\varepsilon_n = \|z_n\|_{X^*}$. Conversely, assuming (3.2) we note that the left-hand side of (3.2) is, with respect to the variable $w := v - u_n$, convex, l.s.c. and vanishes at $w = 0$. Then Lemma 1.3 of Szulkin [31] (see Lemma 5.8) provides the existence of $z_n \in X^*$ for which $z_n \to 0$ in X^* and (3.3) is valid. □

3.3 A Deformation Result

The aim of this Section is to extend Szulkin's deformation result ([32], Proposition 2.3) to the functionals of type (H). To prove our deformation result two preliminary properties are necessary.

Lemma 3.1 Let the functional $I : X \to (-\infty, +\infty]$ satisfy (H) and $(\text{PS})_c$ for some $c \in \mathbb{R}$ and let U be a neighborhood of $K_c(I)$. Then, for each $\bar{\varepsilon} > 0$, there is an $\varepsilon \in (0, \bar{\varepsilon})$ such that, corresponding to every $u_0 \in I^{-1}([c-\varepsilon, c+\varepsilon]) \setminus U$, there exists an $v_0 \in X$ satisfying

$$\Phi^0(u_0; v_0 - u_0) + \Psi(v_0) - \Psi(u_0) < -3\varepsilon \|v_0 - u_0\| \, . \tag{3.4}$$

Proof. If (3.4) were not true, then we could find sequences $\{\varepsilon_n\} \subset \mathbb{R}$ and $\{u_n\} \subset X \setminus U$ such that $\varepsilon_n \downarrow 0$ and $I(u_n) \to c$ and

$$\Phi^0(u_n; v - u_n) + \Psi(v) - \Psi(u_n) \geq -3\varepsilon_n \|v - u_n\|, \ \forall \, v \in X.$$

Then properties $(\text{PS})_c$ and (i) of Propositon 3.1 imply that up to a subsequence $u_n \to u \in K_c(I)$. Thus $u_n \in U$ if n is sufficiently large, which is a contradiction.□

Lemma 3.2 Under the assumptions of Lemma 3.1 let an $\varepsilon \in (0, \bar{\varepsilon})$ be fixed as there stated. Then for each $u_0 \in I_{c+\varepsilon} \setminus U$, there exist $v_0 \in X$ and a neighborhood U_0 of u_0 in X such that Φ is Lipschitz continuous on U_0 of Lipschitz constant $K > 0$ and the relations below hold:

$$\Phi^0(u; v_0 - u) + \Psi(v_0) - \Psi(w) \leq K(\|u - v_0\| + \|w - v_0\|), \ \forall \, u, w \in U_0, \tag{3.5}$$

and

$$\begin{gathered}\Phi^0(u; v_0 - u) + \Psi(v_0) - \Psi(w) \leq -3\varepsilon \|v_0 - w\|, \\ \forall \, u, w \in U_0 \ \text{ with } \ I(w) \geq c - \varepsilon.\end{gathered} \tag{3.6}$$

If $u_o \in K(I)$ one can take $v_0 = u_0$. Otherwise the data v_0, U_0 and a number $\delta_0 > 0$ can be chosen so that $v_0 \notin \overline{U}_0$ and

$$\Phi^0(u; v_0 - u) + \Psi(v_0) - \Psi(w) \leq -\delta_0 \|v_0 - w\|, \quad \forall \, u, w \in U_0. \tag{3.7}$$

Proof. Case n° 1: $u_0 \in K(I)$. By (3.1) if follows that

$$\Phi^0(u; u_0 - u) + \Psi(u_0) - \Psi(w) \leq \Phi^0(u; u_0 - u) + \Phi^0(u_0; w - u_0), \quad \forall \, u, w \in X.$$

Thus for all u close enough to u_0,

$$\Phi^0(u; u_0 - u) + \Psi(u_0) - \Psi(w) \leq K(\|u - u_0\| + \|w - u_0\|).$$

Therefore (3.5) holds with $v_0 = u_0$.

We claim that if $u_0 \in K(I)$ with $u_0 \in I_{c+\varepsilon} \setminus U$, then one has

$$I(u_0) < c - \varepsilon. \tag{3.8}$$

Suppose that this is not true. Then we would have $u_0 \in I^{-1}([c-\varepsilon, c+\varepsilon])\backslash U$. Then Lemma 3.1 yields a point $v_0 \in X$ satisfying (3.4). This contradicts the assumption that $u_0 \in K(I)$ and the claim (3.8) is proved. In order to justify (3.6) with $v_0 = u_0$, when $u_0 \in K(I)$, we suppose that every neighborhood of u_0 contains points u with $I(u) \geq c - \varepsilon$. Let us show that there exist a constant $d > 0$ and a neighborhood U_0 of u_0 such that

$$\Psi(u) - \Psi(u_0) \geq d, \quad \forall\, u \in U_0 \ \text{ with } \ I(u) \geq c - \varepsilon. \tag{3.9}$$

Assuming the contrary there would exist a sequence $u_n \to u_0$ in X that fulfills $\Psi(u_n) \to \Psi(u_0)$ and $I(u_n) \geq c - \varepsilon$. Then, due to the continuity of Φ, we obtain

$$c - \varepsilon \leq \Phi(u_n) + \Psi(u_n) < c - \varepsilon$$

for n large enough, where the last inequality above follows from (3.8). The contradiction ensures that (3.9) is true.

Taking into account (3.9), we obtain

$$\Phi^0(u; u_0 - u) + \Psi(u_0) - \Psi(w) \leq \Phi^0(u; u_0 - u) - d,$$

$\forall\, u, w \in U_0,\ I(w) \geq c - \varepsilon$.

Moreover, for a possibly smaller neighborhood U_0 of u_o we have also

$$\Phi^0(u; u_0 - u) - d \leq -3\varepsilon \|u_0 - w\|, \ \forall\, u, w \in U_0,\ I(w) \geq c - \varepsilon.$$

Indeed, if we suppose the contrary then we can find sequences $u_n \to u_0$ and $w_n \to u_0$ such that $I(w_n) \geq c - \varepsilon$ and

$$\Phi^0(u_n; u_0 - u_n) > -3\varepsilon \|u_0 - w_n\| + d.$$

Taking the lim sup above and using the upper semicontinuity of $\Phi^0(\cdot;\cdot)$ we obtain $d \leq 0$, which is a contradiction. Thus (3.6) is checked .

<u>Case n°2</u>: $u_0 \notin K(I)$. Let us treat firstly the case $I(u_0) < c - \varepsilon$. According to (3.1) the assumption $u_0 \notin K(I)$ implies the existence of $v_0 \in X$ such that

$$\Phi^0(u_0; v_0 - u_0) + \Psi(v_0) - \Psi(u_0) < 0. \tag{3.10}$$

Notice that the point $v_0 \in X$ in (3.10) can be chosen arbitrarily close to u_0. To see this we use the convexity of Ψ and the positive homogeneity of $\Phi^0(u_0;\cdot)$ (cf. Proposition 1.3 in Chapter 1), together with (3.10). Thus we can write

$$\Phi^0(u_0; tv_0 + (1-t)u_0 - u_0) + \Psi(tv_0 + (1-t)u_0) - \Psi(u_0) \leq$$

$$\leq t(\Phi^0(u_0; v_0 - u_0) + \Psi(v_0) - \Psi(u_0)) < 0, \quad \forall\, t \in (0,1).$$

This shows that v_0 in (3.10) can be taken arbitrarily close to u_0. We choose it so that $\|v_0 - u_0\| \leq \min\{\frac{d}{3K}, \frac{d}{18\varepsilon}\}$. From (3.10), we get

$$\Psi(v_0) - \Psi(u_0) < -\Phi^0(u_0; v_0 - u_0) \le K\|v_0 - u_0\| \le \frac{d}{3}, \tag{3.11}$$

whenever $\|v_0 - u_0\|$ is small enough, where d stands for the positive constant in (3.9). Then (3.9) (which holds true since we have supposed that $I(u_0) < c - \varepsilon$) and (3.11) allow to write

$$\begin{aligned}
&\Phi^0(u; v_0 - u) + \Psi(v_0) - \Psi(w) \\
&\quad\le K\|v_0 - u\| + \Psi(v_0) - \Psi(u_0) + \Psi(u_0) - \Psi(w) \\
&\quad\le K\|v_0 - u\| + \frac{d}{3} - d,
\end{aligned}$$

for all $u, w \in U_0$ with $I(w) \ge c - \varepsilon$ and $\|v_0 - u_0\|$ sufficiently small. Thus with U_0 possibly smaller, we can say that

$$K\|v_0 - u\| - \frac{2d}{3} \le -3\varepsilon\|v_0 - w\|.$$

If not we obtain sequences $u_n \to u_0$ and $w_n \to u_0$ such that

$$K\|v_0 - u_n\| > \frac{2d}{3} - 3\varepsilon\|v_0 - w_n\|.$$

and thus

$$\frac{d}{3} \ge K\|v_0 - u_0\| \ge \frac{2d}{3} - 3\varepsilon\|v_0 - u_0\| \ge \frac{2d}{3} - 3\varepsilon\left(\frac{d}{18\varepsilon}\right) = \frac{d}{2},$$

which is a contradiction. Thus

$$\Phi^0(u; v_0 - u) + \Psi(v_0) - \Psi(w) \le -3\varepsilon\|v_0 - w\|,$$

for all $u, w \in U_0$ with $I(w) \ge c - \varepsilon$. This is just (3.6). We pass to justify (3.7) in the case where $u_0 \notin K(I)$ and $I(u_0) < c - \varepsilon$. In this respect, from (3.10) it follows that we can admit $v_0 \notin \overline{U}_0$, and a constant $\delta_0 > 0$ can be found to have

$$\Phi^0(u_0; v_0 - u_0) + \Psi(v_0) - \Psi(u_0) < -\delta_0\|v_0 - u_0\|. \tag{3.12}$$

We establish (3.7) with this $\delta_0 > 0$ arguing by contradiction. To this end let $u_n \to u_0$ and $w_n \to u_0$ fulfill

$$\Phi^0(u_n; v_0 - u_n) + \Psi(v_0) - \Psi(w_n) > -\delta_0\|v_0 - w_n\|.$$

Then the lower semicontinuity of Ψ and the upper semicontinuity of $\Phi^0(\cdot;\cdot)$ (see Proposition 1.3) enable us to conclude

$$\Phi^0(u_0; v_0 - u_0) + \Psi(v_0) - \Psi(u_0) \ge -\delta_0\|v_0 - u_0\|.$$

Since we arrived at a contradiction with (3.12), the proof of (3.7) is complete in the case specified above. We proceed now to the situation where $u_0 \notin K(I)$ and $I(u_0) \ge c - \varepsilon$. Notice that we are in position to apply Lemma 3.1 since $u_0 \in I_{c+\varepsilon}$

by assumption. Then arguing as above on the basis of (3.4) in place of (3.12) one obtains (3.6) and (3.7) with $\delta_0 = -3\varepsilon$. This completes the proof of Lemma 3.2.□

Theorem 3.1 Let the functional $I : X \to (-\infty, +\infty]$ and the number $c \in \mathbb{R}$ be such that conditions (H) and $(\mathrm{PS})_c$ are satisfied. Let U be a neighborhood of $K_c(I)$ and let $\bar{\varepsilon} > 0$ be a fixed number. Then there exists $\varepsilon \in (0, \bar{\varepsilon})$ such that for every compact subset $A \subset X \backslash U$ with

$$c \leq \sup_A I \leq c + \varepsilon \tag{3.13}$$

there exist a closed neighborhood W of A in X and a deformation $\alpha : W \times [0, \bar{s}] \to X$, $\bar{s} > 0$, i.e. a continuous map with $\alpha(\cdot, 0) = id_W$, satisfying

$$\|u - \alpha(u, s)\| \leq s, \quad \forall\, u \in W, \quad \forall\, s \in [0, \bar{s}]; \tag{3.14}$$

$$I(\alpha(u, s)) - I(u) \leq Ms, \quad \forall\, u \in W, \quad \forall\, s \in [0, \bar{s}], \tag{3.15}$$

with a constant $M > 0$ independent of u and s;

$$I(\alpha(u, s)) - I(u) \leq -2\varepsilon s, \tag{3.16}$$

for all $u \in W$ with

$$I(u) \geq c - \varepsilon, \quad \forall\, s \in [0, \bar{s}];$$

$$\sup_{u \in A} I(\alpha(u, s)) - \sup_{u \in A} I(u) \leq -2\varepsilon s, \quad \forall\, s \in [0, \bar{s}]. \tag{3.17}$$

Furthermore, if W_0 is a closed subset of X with $W_0 \cap K(I) = \emptyset$, W and α can be chosen so that

$$I(\alpha(u, s)) - I(u) \leq 0, \quad \forall\, u \in W \cap W_0, \quad \forall\, s \in [0, \bar{s}]. \tag{3.18}$$

Proof. Consider an $\varepsilon \in (0, \bar{\varepsilon})$ determined by Lemma 3.2. By Lemma 3.2 and relation (3.13), for each point $u_0 \in A$ there exists an open neighborhood U_0 of u_0 in X and some $v_0 \in X$ verifying the properties stated in Lemma 3.2. In addition, we can suppose

$$u_0 \in K(I) \Longrightarrow U_0 \cap W_0 = \emptyset. \tag{3.19}$$

Take also another neighborhood U_0' of u_0 with $\overline{U}_0' \subset U_0$.

The compactness of A insures the existence of finite open coverings $\{U_i\}_{1 \leq i \leq m}$ and $\{U_i'\}_{1 \leq i \leq m}$ of A in X, where every pair of sets U_i, U_i' corresponds to a point $u_i \in A$ in the same way as does U_0, U_0' for u_0 above and $\overline{U}_i' \subset U_i$. Let v_i denote the point corresponding to u_i as v_0 was obtained by means of u_0. Taking, if necessary, a finer covering of A in X we may assume

$$u_i \in K(I) \Longrightarrow \operatorname{dist}\{u_i, U_j\} > 0, \quad \forall\, j \neq i\,. \tag{3.20}$$

We introduce

$$V = \bigcup_{i=1}^{m} U_i \tag{3.21}$$

and, for each $i = 1, ..., m$, $\sigma_i : V \to \mathbf{R}$ by

$$\sigma_i = \rho_i / \sum_{j=1}^{m} \rho_j \ , \tag{3.22}$$

where $\rho_i : X \to \mathbf{R}$ is given by

$$\rho_i(x) = \text{dist}\ \{x, X \backslash U_i\}, \quad \forall\, x \in X.$$

Now, we define $\alpha : V \times [0, +\infty) \to X$. If $u \in U_{i_0} \backslash \cup_{j \neq i_0} U_j$ with $u_{i_0} \in A \cap K(I)$ we put

$$\alpha(u, s) := \begin{cases} u + s(u_{i_0} - u)/\|u_{i_0} - u\|, & 0 \leq \ s < \|u_{i_0} - u\| \\ u_{i_0}, & s \geq \|u_{i_0} - u\| \end{cases} \tag{3.23}$$

and for all other $u \in V$

$$\alpha(u, s) = u + s \sum_i \sigma_i(u)(v_i - u)/\|v_i - u\| \ . \tag{3.24}$$

In the sum of (3.24) we count only the terms with $\sigma_i(u) \neq 0$, i.e. $u \in U_i$. If $u_i \notin K(I)$ then from Lemma 3.2, $v_i \notin \overline{U}_i$. If $u_i \in K(I)$ then $v_i = u_i$ and from (3.20), we obtain $v_i \notin U_j$, $j \neq i$. Thus (3.24) is well-defined. Moreover, we note that for

$$u \in U_{i_0} \backslash \bigcup_{j \neq i_o} U_j$$

and

$$u \neq u_{i_0} \in A \cap K(I) \ ,$$

by (3.21) and (3.22), the expression of $\alpha(u, s)$ in (3.24) makes sense and, if $s < \|u_{i_0} - u\|$, coincides with (3.23). Simple computations show the existence of $\bar{s} > 0$ such that $\alpha : V \times [0, \bar{s}] \to X$ is continuous. Clearly, the relations $\alpha(\cdot, 0) = id_V$ and (3.14) are satisfied. We check now (3.15), (3.16) and (3.18) in the case where $\alpha(u, s)$ is expressed by the formula (3.24). Let us write (3.24) in the form

$$\alpha(u, s) = u + s\overline{w}, \tag{3.25}$$

where

$$\overline{w} = \sum_i \sigma_i(u)(v_i - u)/\|v_i - u\|.$$

The Lebourg's mean value theorem (see Theorem 1.1), Proposition 1.4 (ii) and relation (3.25) imply

$$\begin{aligned} I(\alpha(u, s)) \ &= \ \Phi(u + s\overline{w}) + \Psi(u + s\overline{w}) \\ &\leq \ \Phi(u) + s\Phi^0(u + r\overline{w}; \overline{w}) + \Psi(u + s\overline{w}), \end{aligned} \tag{3.26}$$

for some $r = r(s) \in (0, s)$. From the subadditivity and the positive homogeneity of $\Phi^0(u + r\overline{w}; \cdot)$ we get

$$\begin{aligned}\Phi^0(u+r\overline{w};\overline{w}) &\le \sum_i \frac{\sigma_i(u)}{\|v_i-u\|}\Phi^0(u+r\overline{w};v_i-u)\\ &\le \sum_i \frac{\sigma_i(u)}{\|v_i-u\|}(\Phi^0(u+r\overline{w};v_i-u-r\overline{w})+r\Phi^0(u+r\overline{w};\overline{w})).\end{aligned}$$

Since Φ is Lipschitz continuous on the compact set A one finds a closed neighborhood W of A with $W \subset \cup_{i=1}^m U_i'$ and a constant $C > 0$ such that

$$\Phi^0(u+r\overline{w};\overline{w}) \le \sum_i \frac{\sigma_i(u)}{\|v_i-u\|}\Phi^0(u+r\overline{w};v_i-u-r\overline{w})+Cs, \tag{3.27}$$

$$\forall\, u \in W,\quad \forall\, 0 \le s \le \bar{s},$$

provided $\bar{s} > 0$ is small enough. The convexity of Ψ allows us to write

$$\begin{aligned}\Psi(u+s\overline{w}) &= \Psi((1-s\sum_i \frac{\sigma_i(u)}{\|v_i-u\|})u + s\sum_i \frac{\sigma_i(u)}{\|v_i-u\|}v_i)\\ &\le \Psi(u)+s\sum_i \frac{\sigma_i(u)}{\|v_i-u\|}(\Psi(v_i)-\Psi(u)),\end{aligned} \tag{3.28}$$

where $0 \le s \le \bar{s}$, with $\bar{s}$ sufficiently small to have actually in (3.28) a convex combination. Then (3.26)-(3.28) lead to

$$\begin{aligned}I(\alpha(u,s)) \le I(u) \;&+\; s\sum_{i=1}^m \frac{\sigma_i(u)}{\|v_i-u\|}(\Phi^0(u+r\overline{w};v_i-u-r\overline{w})\\ &+\; \Psi(v_i)-\Psi(u))+Cs^2\,.\end{aligned} \tag{3.29}$$

We can choose W and $\bar{s} > 0$ so small that if $x \in U_i' \cap W$, $0 \le s \le \bar{s}$ and $\|y\| \le 1$, then $x + sy \in U_i$ for all $i = 1, ..., m$. Thus using (3.29) together with (3.5), we obtain:

$$I(\alpha(u,s))-I(u) \le s\sum_{i=1}^m \frac{\sigma_i(u)}{\|v_i-u\|}K(\|u+r\overline{w}-v_i\|+\|u-v_i\|)+Cs^2\,.$$

Thus (3.15) is satisfied. Analogously, from (3.6) and (3.29) we infer that (3.16) holds. It is clear that combining (3.7), (3.19) and (3.29) one obtains (3.18).

Let us check now (3.15), (3.16), (3.18) in the case where $\alpha(u,s)$ is given by (3.23). In view of (3.19) it is clear that (3.18) is vacuous in this situation. Notice that if $s < \|u_{i_0} - u\|$ in (3.23), $\alpha(u,s)$ is expressed by (3.25) with

$$\overline{w} = \|u_{i_0}-u\|^{-1}(u_{i_0}-u)\,.$$

A careful examination of arguments centered around (3.26) - (3.29) shows readily that using the same procedure we deduce (3.15), (3.16). We suppose that $s \ge \|u_{i_0} - u\|$ holds in (3.23). Then we get

$$\begin{aligned}I(\alpha(u,s))-I(u) &= I(u_{i_0})-I(u)\\ &= \Phi(u_{i_0})-\Phi(u)+\Psi(u_{i_0})-\Psi(u)\\ &\le \Phi^0(u_{i_0}+\lambda(u-u_{i_0});u_{i_0}-u)+\Psi(u_{i_0})-\Psi(u),\end{aligned}$$

where $\lambda \in (0,1)$ is given by the Lebourg's mean value theorem. Thus

$$\begin{aligned} I(\alpha(u,s)) - I(u) &\leq \Phi^0(u_{i_0} + \lambda(u - u_{i_0}); u_{i_0} - (u_{i_0} + \lambda(u - u_{i_0})) \\ &\quad + (1-\lambda)(u_{i_0} - u)) + \Psi(u_{i_0}) - \Psi(u) \\ &\leq K\|u - u_{i_0}\| + \Phi^0(u_{i_0} + \lambda(u - u_{i_0}); u_{i_0} \\ &\quad - (u_{i_0} + \lambda(u - u_{i_0}))) + \Psi(u_{i_0}) - \Psi(u), \end{aligned}$$

where $K > 0$ denotes the Lipschitz constant of Φ on U_{i_0}. Then, by using relation (3.5) appropriately, we obtain:

$$I(\alpha(u,s)) - I(u) \leq K(\lambda\|u - u_{i_0}\| + \|u - u_{i_0}\|)$$

which shows that (3.15) holds true because $\|u - u_{i_0}\| \leq s$. Since $u_{i_0} \in K(I)$, by (3.8) it is known that

$$I(\alpha(u,s)) = I(u_{i_0}) < c - \varepsilon,$$

so for a sufficiently small $s > 0$,

$$I(u_{i_0}) + 2\varepsilon s < c - \varepsilon.$$

Now, with the additional assumption that $I(u) \geq c - \varepsilon$ are arrive at (3.16).

It remains to prove (3.17). The argument is the same as in Szulkin [32]. We supply it for the sake of completeness. If we have

$$\sup_{u \in A} I(\alpha(u,s)) \leq c - \frac{\varepsilon}{2},$$

then taking $s \leq \frac{1}{4}$, by the first inequality of (3.13), it results that

$$\sup_{u \in A} I(\alpha(u,s)) - \sup_{u \in A} I(u) \leq c - \frac{\varepsilon}{2} - c \leq -\frac{\varepsilon}{2}4s = -2\varepsilon s.$$

If the inequality below occurs

$$\sup_{u \in A} I(\alpha(u,s)) > c - \frac{\varepsilon}{2},$$

then by virtue of (3.15), (3.16) and with s small enough it is seen that

$$\begin{aligned} c - \frac{1}{2}\varepsilon &< \sup_{u \in A} I(\alpha(u,s)) = \sup_{\substack{u \in A \\ I(u) \geq c - \varepsilon}} I(\alpha(u,s)) \\ &\leq \sup_{\substack{u \in A \\ I(u) \geq c - \varepsilon}} I(u) - 2\varepsilon s \leq \sup_{u \in A} I(u) - 2\varepsilon s\ . \end{aligned}$$

Consequently, (3.17) follows. This completes the proof of Thorem 3.3. □

Corollary 3.1 Assume in additon to the requirements of Theorem 3.1 that the functionals Φ and Ψ are even, i.e., $\Phi(u) = \Phi(-u)$ and $\Psi(u) = \Psi(-u)$ for all

$u \in X$. If the set A is symmetric with respect to the origin, i.e., $A = -A$ then W can be chosen to be symmetric with respect to the origin and the deformation $\alpha : W \times [0, \bar{s}] \to X$ can be constructed so that $\alpha(\cdot, s) : W \to X$ is odd for each $s \in [0, \bar{s}]$, i.e., $\alpha(-u, s) = -\alpha(u, s)$, $\forall\, u \in W$.

Proof. Since I is even, its critical points arise in symmetric pairs $(u, -u)$. Thus, in view of the symmetry of A, the elements U_i and U_i' of the open coverings considered in the proof of Theorem 3.3 can be taken pairwise symmetric with respect to the origin. Then V in (3.21) satisfies $V = -V$, hence W can be chosen with the same property.

Now, starting with a defomation α as given by Theorem 3.1, we define $\beta : W \times [0, \bar{s}] \to X$ as

$$\beta(u, s) = \frac{1}{2}(\alpha(u, s) - \alpha(-u, s)), \quad \forall\, (u, s) \in W \times [0, \bar{s}] \,. \tag{3.30}$$

Then one can follow an analogous reasoning to the one developed in the proof of Theorem 3.1 based in fact on Lebourg's mean value theorem applied to Φ and the convexity of Ψ (that can be used since β above appears as a convex combination). We omit the details of the effective verification of assertions (3.14)-(3.18) with $\alpha(u, s)$ replaced by $\beta(u, s)$. □

Remark 3.2 Following the pattern of Corollary 3.1, under appropriate conditions, we can construct deformations $\alpha(u, s)$ with properties (3.14)-(3.18) such that each mapping $\alpha(\cdot, s)$ is equivariant with respect to the action of a topological group G on X, i.e.

$$\alpha(gu, s) = g\alpha(u, s), \qquad \forall\, u \in X, \quad \forall\, g \in G \,.$$

The situation treated in Corollary 3.1 corresponds to the group $G = \mathbf{Z}_2 = \{id_X, -id_X\}$. Related results for the pure locally Lipschitz case, i.e. $\Psi = 0$, are given in Chapter 2 as well as in Mironescu and Radulescu [14].

3.4 Minimax Principles for Functionals of Type (H)

Based on the deformation result in Theorem 3.1 we are now able to provide a general minimax principle for the nonsmooth functionals verifying hypothesis (H). We recall the definition of linking (cf. Du [4]). A slightly more general definition of linking was given in Chapter 2.

Definition 3.3 Let S be a nonempty subset of the Banach space X and let Q be a compact topological submanifold of X with nonempty boundary ∂Q (in the sense of manifolds with boundary). We say that S and Q link if the next properties hold

$$S \cap \partial Q = \emptyset \tag{3.31}$$

and

$$f(Q) \cap S \neq \emptyset \tag{3.32}$$

whenever $f \in \Gamma$, where

$$\Gamma := \{f \in C(Q,X) : \ f|_{\partial Q} = id_{\partial Q}\} . \tag{3.33}$$

The theorem below is our main result of this section. It extends and unifies various nonsmooth (and a fortiori smooth) minimax principles.

Theorem 3.2 Let the functional $I : X \to (-\infty, +\infty]$ on the Banach space X satisfy assumptions (H) and (PS). Let S and Q link in the sense of Definition 3.3. Assume further that

$$\sup_Q I \in \mathbb{R}, \quad b := \inf_S I \in \mathbb{R}, \quad a := \sup_{\partial Q} I < b . \tag{3.34}$$

Then the number

$$c := \inf_{f \in \Gamma} \sup_{x \in Q} I(f(x)), \tag{3.35}$$

where

$$\Gamma = \{f \in C(Q,X) : \ f|_{\partial Q} = id_{\partial Q}\} , \tag{3.36}$$

is a critical value of I with

$$c \geq b . \tag{3.37}$$

This means in particular that $K_c(I)$ is nonempty.

Proof. According to the linking property of Definition 3.3 we know that $f(Q) \cap S \neq \emptyset$ for all $f \in \Gamma$. By (3.34) and (3.35) this ensures that (3.37) is true. In addition, the first relation in (3.34) yields $c < \infty$ by putting $f = id_Q$ in (3.35), (3.36).

We have to show that c is a critical value of I. Arguing by contradiction we assume that $K_c(I) = \emptyset$. Then we can take $U = \emptyset$ in Theorem 3.1. Applying Theorem 3.1 with

$$\bar{\varepsilon} = c - a, \tag{3.38}$$

which is a positive number by (3.34), (3.37), we obtain an $\varepsilon \in (0, \bar{\varepsilon})$. Notice that

$$\partial Q \subset I_{c-\frac{\varepsilon}{2}} . \tag{3.39}$$

Indeed, (3.39) follows easily from (3.34), (3.37) and (3.38).

The class Γ in (3.36) does not generally satisfy the property $\alpha(f(\cdot), s) \in \Gamma$ whenever $f \in \Gamma$, where α is the deformation in Theorem 3.1. Therefore it is necessary to enlarge Γ to fulfill the foregoing requirement. Let us set

$$\Gamma_1 := \{f \in C(Q,X) : \ f|_{\partial Q} \text{ and } id_{\partial Q} \text{ are homotopic as maps from } \partial Q \text{ into } I_{c-\frac{\varepsilon}{4}} \text{ and } f(\partial Q) \subset I_{c-\frac{\varepsilon}{2}}\} . \tag{3.40}$$

Taking into account (3.39), the family (3.40) is well defined and nonempty containing at least id_Q. Corresponding to the family Γ_1 in (3.40) we introduce the minimax value

$$c_1 = \inf_{f \in \Gamma_1} \sup_{x \in Q} I(f(x)) \,. \tag{3.41}$$

We claim that

$$c = c_1 \,. \tag{3.42}$$

Since $\Gamma \subset \Gamma_1$, from (3.35), (3.41) it is clear that $c_1 \le c$. Assuming by contradiction that $c_1 < c$, there would exist $f \in \Gamma_1$ provided

$$\sup_{x \in Q} I(f(x)) < c \,.$$

Recalling that $f|_{\partial Q}$ and $id_{\partial Q}$ are homotopic in $I_{c-\frac{\varepsilon}{4}}$, the (absolute) Homotopy Extension Property (see Proposition I.9.2 of Hu [12] or Spanier [31]) guarantees that $id_{\partial Q}$ can be extended to some $g \in C(Q, X)$, so $g \in \Gamma$, such that

$$\sup_{x \in Q} I(g(x)) < c \,.$$

This happens because $(Q, \partial Q)$ is a (finitely) triangulable pair in view of the fact that Q is a compact topological manifold with boundary ∂Q (see Munkres [23]). This fact is in contradiction with the definition of c in (3.35). Hence the claim (3.42) is verified.

The next objective is to show that Γ_1 in (3.40) is a closed subset of the Banach space $C(Q, X)$ with respect to the usual uniform norm (recall that Q is compact in X), i.e.

$$\|f\| = \sup_{x \in Q} \|f(x)\|, \quad \forall\, f \in C(Q, X).$$

To verify this assertion let us take a sequence $\{f_n\} \subset \Gamma_1$ which converges in $C(Q, X)$, say $f_n \to f$. By (3.40) one has

$$I(f_n(x)) \le c - \frac{\varepsilon}{2}, \quad \forall\, x \in \partial Q, \quad \forall\, n \ge 1.$$

Since I is l.s.c., we derive

$$I(f(x)) \le \liminf_{n \to \infty} I(f_n(x)) \le c - \frac{\varepsilon}{2}, \quad \forall\, x \in \partial Q. \tag{3.43}$$

Using Lebourg's mean value theorem (see Theorem 1.1) for Φ and the convexity of Ψ it is seen that

$$\begin{aligned}
&I(tf_n(x) + (1-t)f(x)) = \\
&\quad = \Phi(tf_n(x) + (1-t)f(x)) + \Psi(tf_n(x) + (1-t)f(x)) \\
&\quad \le \Phi(f(x)) + \bar{t}\Phi^0(f(x) + t(f_n(x) - f(x)); f_n(x) - f(x)) + \\
&\qquad + t\Psi(f_n(x)) + (1-t)\Psi(f(x)), \\
&\qquad\qquad\qquad \forall\, x \in \partial Q, \quad \forall\, t \in [0,1],
\end{aligned}$$

for some $\bar{t} \in (0,1)$. Since $f(\partial Q)$ is compact, the sequence $\{f_n\}$ converges uniformly to f and Φ being locally Lipschitz, Φ is Lipschitz continuous near $f(\partial Q)$ of Lipschitz constant $K > 0$, it turns out that

$$I(tf_n(x)+(1-t)f(x)) \leq \Phi(f(x))+K\|f_n(x)-f(x)\| +t\Psi(f_n(x))+(1-t)\Psi(f(x)), \quad \forall\, x\in\partial Q,\ \ \forall\, t\in[0,1].$$

Taking into account the Lipschitz continuity of Φ around $f(\partial Q)$ we obtain that

$$\Phi(f(x))\leq\Phi(f_n(x))+K\|f_n(x)-f(x)\|,\ \ \forall\, x\in\partial Q.$$

It results that

$$\begin{aligned} I(tf_n(x)+(1-t)f(x)) &\leq t\Phi(f_n(x))+(1-t)\Phi(f(x)) \\ &\quad +2K\|f_n(x)-f(x)\|+t\Psi(f_n(x))+(1-t)\Psi(f(x)) \\ &= tI(f_n(x))+(1-t)I(f(x))+2K\|f_n(x)-f(x)\| \end{aligned}$$

for all $x\in\partial Q$ and $t\in[0,1]$. Then, using that $f_n\in\Gamma_1$ and (3.43), we obtain

$$I(tf_n(x)+(1-t)f(x))\leq c-\frac{\varepsilon}{2}+2K\|f_n(x)-f(x)\|,\ \ \forall\, x\in\partial Q\ .$$

Since $\{f_n\}$ converges uniformly to f it follows

$$I(tf_n(x)+(1-t)f(x))\leq c-\frac{\varepsilon}{4},\ \ \forall\, x\in\partial Q\ , \tag{3.44}$$

provided that n is sufficiently large. From (3.44) we deduce that $f_n|_{\partial Q}$ and $f|_{\partial Q}$ are homotopic in $I_{c-\frac{\varepsilon}{4}}$. Hence, by (3.40) and $f_n\in\Gamma_1$, we conclude that $f|_{\partial Q}$ and $id_{\partial Q}$ are homotopic in $I_{c-\frac{\varepsilon}{4}}$. This property combined with (3.43) entails that $f\in\Gamma_1$. Consequently, we proved that Γ_1 is a closed subset of $C(Q,X)$, so a complete metric space.

Let us introduce the functional $\Pi : C(Q,X)\to(-\infty,+\infty]$ by

$$\Pi(f)=\sup_{x\in Q} I(f(x)),\ \ \forall\, f\in C(Q,X)\ . \tag{3.45}$$

It is straightforward to see that Π is l.s.c. This is the consequence of the lower semicontinuity of I. Therefore we know that $\Pi : \Gamma_1\to(-\infty,+\infty]$ is l.s.c.. As established previously Γ_1 is a complete metric space being a closed subset of $C(Q,X)$. It is allowed to apply to $\Pi : \Gamma_1\to(-\infty,+\infty]$ Ekeland's variational principle (see Ekeland [5],[6]). Notice that Π on Γ_1 is bounded from below by $c\in\mathbb{R}$ because (3.41), (3.42) hold. Ekeland's variational principle gives rise to some $f\in\Gamma_1$ satisfying

$$c\leq\Pi(f)\leq c+\varepsilon \tag{3.46}$$

and

$$\Pi(g)-\Pi(f)\geq-\varepsilon\|g-f\|,\ \ \forall\, g\in\Gamma_1\ . \tag{3.47}$$

Now we invoke Theorem 3.1 that provides the deformation $\alpha : W\times[0,\bar{s}]\to X$ corresponding to the compact set $A=f(Q)$. We point out that the specified set A satisfies the required condition (3.13) due to (3.46). Let us show that for $\bar{s}>0$ small enough we have

$$\alpha(f(\cdot), s) \in \Gamma_1, \quad \forall\, s \in [0, \bar{s}] . \tag{3.48}$$

Looking at (3.40), we see that in order to prove (3.48) it suffices to establish that $\alpha(f(\cdot), s)|_{\partial Q}$ and $f|_{\partial Q}$ are homotopic in $I_{c-\frac{\varepsilon}{2}}$. Clearly, a homotopy in X between the involved mappings is $(x, t) \in \partial Q \times [0, 1] \to \alpha(f(x), ts) \in X$. Therefore, it is sufficient to show that

$$I(\alpha(f(x), s)) \leq c - \frac{\varepsilon}{2}, \quad \forall\, x \in \partial Q, \ \ \forall\, 0 \leq s \leq \bar{s} . \tag{3.49}$$

If one has for $x \in \partial Q$ that

$$I(f(x)) \in [c - \varepsilon, c - \frac{\varepsilon}{2}] ,$$

then by (3.16) we get

$$I(\alpha(f(x), s)) \leq I(f(x)) - 2\varepsilon s \leq c - \frac{\varepsilon}{2} - 2\varepsilon s < c - \frac{\varepsilon}{2} .$$

If for $x \in \partial Q$ holds

$$I(f(x)) < c - \varepsilon,$$

then (3.15) implies

$$I(\alpha(f(x), s)) \leq I(f(x)) + Ms < c - \varepsilon + Ms \leq c - \frac{\varepsilon}{2}$$

provided that $\bar{s} > 0$ is sufficiently small. We thus justified that (3.49) is valid, so (3.48) holds true.

From (3.14), (3.17), (3.45), (3.47) and (3.48) we then deduce

$$\begin{aligned} -2\varepsilon s &\geq \Pi(\alpha(f(\cdot), s)) - \Pi(f) \\ &\geq -\varepsilon \|\alpha(f(\cdot), s) - f\| \geq -\varepsilon s, \quad \forall\, 0 \leq s \leq \bar{s}. \end{aligned}$$

We arrived at a contradiction which proves that our initial assumption that c in (3.35) is not a critical value of I is false. Thus the proof of Theorem 3.2 is completed. □

Corresponding to different situations of linking in Definition 3.3, Theorem 3.2 implies useful minimax results for functionals satisfying hypothesis (H).

Corollary 3.2 (Mountain Pass Theorem). Assume that the functional $I : X \to (-\infty, +\infty]$ on the Banach space X satisfies (H), (PS), $I(0) = 0$,

(i) there exist constants $\alpha > 0$ and $\rho > 0$ such that $I(u) \geq \alpha$ for all $\|u\| = \rho$;

(ii) there exists $e \in X$ with $\|e\| > \rho$ and $I(e) \leq 0$.

Then the number

$$c = \inf_{f \in \Gamma} \sup_{t \in [0,1]} I(f(t)),$$

where

$$\Gamma = \{f \in C([0,1], X) : f(0) = 0, f(1) = e\},$$

is a critical value of I with $c \geq \alpha$.

Proof. It is sufficient to apply Theorem 3.2 for

$$S = \{u \in X : \ \|u\| = \rho\}$$

and

$$Q = [0; e] = \{te : \ t \in [0,1]\}.$$

The first condition in (3.34) holds by the convexity of Ψ. □

Corollary 3.3 (Saddle-Point Theorem). Let a Banach space X with $X = X_1 \oplus X_2$, $\dim X_1 < \infty$, and let a functional $I : X \to (-\infty, +\infty]$ satisfy (H), (PS),

(i) there exist constants $\rho > 0$ and $\alpha \in \mathbb{R}$ such that $I(u) \leq \alpha$, $\forall\, u \in X$, with $\|u\| = \rho$;

(ii) there exists a constant $\beta \in \mathbb{R}$ such that $\beta > \alpha$ and $I(u) \geq \beta$, $\forall\, u \in X_2$.

Then the number

$$c = \inf_{f \in \Gamma} \sup_{x \in Q} I(f(x)), \tag{3.50}$$

where

$$\Gamma = \{f \in C(Q, X) : \ f|_{\partial Q} = id_{\partial Q}\} \tag{3.51}$$

with

$$Q = \{x \in X_1 : \ \|x\| \leq 1\} \ \text{ and } \partial Q = \{x \in X_1 : \ \|x\| = 1\},$$

is a critical value of I satisfying $c \geq \beta$.

Proof. One applies Theorem 3.2 by taking $S = X_2$ and Q as specified in the statement. The linking property in Definition 3.3 follows by applying the topological degree argument as in Rabinowitz [30]. □

Corollary 3.4 (Generalized Mountain Pass Theorem). Assume that $X = X_1 \oplus X_2$ with $\dim X_1 < \infty$ and the functional $I : X \to (-\infty, +\infty]$ satisfies (H), (PS),

(i) there exist constants $\rho > 0$ and $\alpha > 0$ such that $I(u) \geq \alpha$ for all $u \in X_2$ with $\|u\| = \rho$;

(ii) there exist a constant $r > \rho$ and a point $e \in X_2$, $\|e\| = 1$ such that $I(u) \leq 0$ for all $u \in \partial Q$, where $Q = \{u \in X_1 : \ \|u\| \ \leq r\} \oplus \{te : \ t \in [0, r]\}$.

Then the number

$$c = \inf_{f \in \Gamma} \sup_{x \in Q} I(f(x)),$$

where

$$\Gamma = \{f \in C(Q, X) : \ f|_{\partial Q} = id_{\partial Q}\},$$

is a critical value of I with $c \geq \alpha$.

Proof. Theorem 3.2 is applied for

$$S = \{u \in X_2 : \ \|u\| = \rho\}$$

and Q as described in the statement. The fact that S and Q link in the sense of Definition 3.3 can be proved by means of an argument relying on the topological degree as in Rabinowitz [30]. □

In the same spirit one can derive from Corollary 3.1 minimax principles with additional multiplicity information for even functionals of type (H).

Corollary 3.5 Assume that $I : X \to (-\infty, +\infty]$ satisfies (H), (PS), $I(0) = 0$ and the functions Φ and Ψ entering (H) are even. Assume also that

$$c_j := \inf_{A \in \Gamma_j} \sup_{u \in A} I(u) \in (-\infty, 0), \quad \forall\, j = 1, ..., k, \tag{3.52}$$

where Γ_j denotes the closure with respect to the Hausdorff metric of the collection of all nonempty compact symmetric subsets A of X for which $0 \notin A$ and the (Krasnoselski) genus $\gamma(A)$ is $\geq j$. Then I possesses at least k distinct pairs of symmetric nontrivial critical points (whose corresponding critical values are the numbers c_j).

Proof. Using the odd deformation $\alpha(\cdot, s)$ in Corollary 3.1 we can follow the same reasoning as in Theorem 4.3 of Szulkin [32]. □

Corollary 3.6 (Symmetric version of Mountain Pass Theorem). Assume that $I : X \to (-\infty, +\infty]$ satisfies (H) with Φ and Ψ even, (PS), $I(0) = 0$ and the conditions

(i) there exists a subspace X_1 of X of finite codimension and numbers $\alpha > 0$, $\rho > 0$ such that

$$I(u) \geq \alpha \quad \text{whenever} \quad u \in X_1, \ \|u\| = \rho; \tag{3.53}$$

(ii) there exists a finite dimensional subspace X_2 of X, $\mathrm{dim} X_2 > \mathrm{codim} X_1$, such that $I(u) \to -\infty$ as $\|u\| \to \infty$, $u \in X_2$.

Then I has at least $\mathrm{dim} X_2 - \mathrm{codim} X_1$ distinct pairs of symmetric nontrivial critical points. In particular, if (ii) is replaced by

(ii$'$) there exists a k-dimensional subspace X_2 as in (ii) for each positive integer k, then I admits infinitely many distinct pairs $(u, -u)$ of nontrivial critical points.

Proof. Since we have at our disposal the deformation result of Theorem 2.3 for the general functionals in (H), we may follow the same lines as in Szulkin [32] to obtain the desired conclusion. □

Remark 3.3 The results in this Section extend the minimax principles in Szulkin [32] to the case where the functional Φ of (H) is assumed to be locally

Lipschitz instead continuously differentiable. We stress that we did not impose additional assumptions in comparison with those in Szulkin [32]. Another feature of our approach is that Theorem 3.2, on the basis of linking property of Definition 3.3, unifies various specific minimax results, and in particular the ones of Szulkin [32]. If we take $\Psi = 0$, then we find the minimax results of Chang [2] for locally Lipschitz functionals (see Theorems 3.3, 3.4 in Chang [2]), again presented in a unifying manner by means of the linking concepts in Definition 3.3. Therefore our minimax principle in Theorem 3.2 contains as special cases the two principal nonsmooth critical points theories, namely the ones of Chang [2] and Szulkin [32]. Consequently, this is true for the classical minimax principle of the smooth critical point theory (see Ambrosetti and Rabinowitz [1] and Rabinowitz [29], [30]).

The problem of extending Szulkin's critical point results [32] to include the locally Lipschitz functionals was considered firstly by Lefter [13], but his statement is constructed on an argument that is not satisfactorily relying on a deformation lemma which is too restrictive to be applied in the case of l.s.c. functionals. Towards the general formulation in Theorem 3.2 regarding this problem, a partial answer has been offered in Goeleven and Motreanu [11], where it is demanded in addition the continuity of Ψ on its effective domain. Inspired by recent minimax principles for smooth functionals (see Du [4] and Motreanu [16]) it will be the objective of a future work to drop the strict inequality sign in the last relation of (3.34) by allowing the possibility $a = b$. An interesting related question is the location of the obtained critical points by minimax principles. Some results of this nature for the smooth case can be found in Ghoussoub [7] and Motreanu [16]. A general class of locally Lipschitz functionals encountered in applications is discussed in Section 2.4 (see also [15]).

3.5 A Variational - Hemivariational Inequality

Let Ω be a bounded domain in $\mathbb{R}^N$ with a C^1 boundary $\partial\Omega$. Consider the Sobolev space $H_0^1(\Omega)$ which is a Hilbert space with respect to the scalar product

$$(u, v)_{H_0^1(\Omega)} = \int_\Omega \nabla u \cdot \nabla v dx, \quad \forall\, u, v \in H_0^1(\Omega).$$

The cone of nonnegative functions in $H_0^1(\Omega)$, i.e.,

$$K = \{u \in H_0^1(\Omega) : \ u(x) \geq 0 \ \text{ for a.e. } \ x \in \Omega\} \tag{3.54}$$

forms a convex and closed subset of $H_0^1(\Omega)$. Denoting by $\Psi : \ H_0^1(\Omega) \to (-\infty, +\infty]$ its indicator function, that is

$$\Psi(u) = \begin{cases} 0, & u \in K \\ +\infty, & u \notin K, \end{cases} \tag{3.55}$$

it follows that Ψ is proper, convex and l.s.c., as required in assumption (H).

There are given $g \in L^2(\Omega)$ satisfying

$$g < 0 \quad \text{a.e. in } \Omega \tag{3.56}$$

and a (Carathéodory) function $j: \Omega \times \mathbb{R} \to \mathbb{R}$ such that
(a) $j(\cdot, y): \Omega \to \mathbb{R}$ is measurable, $\forall\, y \in \mathbb{R}$;
(b) $j(x, \cdot): \mathbb{R} \to \mathbb{R}$ is locally Lipschitz.

In view of (b) it makes sense to consider the generalized directional derivative of $j(x, y)$ with respect to the second variable $y \in \mathbb{R}$, i.e.,

$$j^0(x, y; z) = \limsup_{\substack{\bar{y} \to y \\ t \downarrow 0}} \frac{1}{t}(j(x, \bar{y} + tz) - j(x, \bar{y}))$$

and the corresponding generalized gradient $\partial j(x, y)$ of j with respect to $y \in \mathbb{R}$, namely

$$\partial j(x, y) = \{\xi \in \mathbb{R} : \ \xi z \leq j^0(x, y; z), \ \forall\, z \in \mathbb{R}\}$$

(see Chapter 1).

The function $j: \Omega \times \mathbb{R} \to \mathbb{R}$ is required to fulfill the conditions below

(j_1) $|\xi| \leq c(1 + |y|^{p-1})$, $\forall\, x \in \Omega$, $y \in \mathbb{R}$, $\xi \in \partial j(x, y)$, with constants $c > 0$ and $1 \leq p < \frac{2N}{N-2}$ if $N \geq 3$ and an arbitrary $p \geq 1$ if $N = 1$ or $N = 2$;

(j_2) $\liminf_{y \to 0} y^{-2} j(x, y) \geq 0$ uniformly with respect to $x \in \Omega$, $j(x, 0) = 0$;

(j_3) $\mu^{-1} j^0(x, y; y) \leq j(x, y)$ for all $x \in \Omega$ and $y \in \mathbb{R}$, $y \geq 0$ where μ is a constant with $\mu > 2$;

(j_4) there is $e \in H_0^1(\Omega) \setminus \{0\}$ such that

$$\int_\Omega j(x, e(x))dx < 0.$$

Denote by λ_1 the first eigenvalue of Δ on $H_0^1(\Omega)$.

Theorem 3.3 Under the assumptions (3.56) for $g \in L^2(\Omega)$ and (j_1)-(j_4) for $j: \Omega \times \mathbb{R} \to \mathbb{R}$, for every $\lambda < \lambda_1$ the variational-hemivariational inequality: find $u \in K$ in (3.54) such that

$$\int_\Omega \nabla u \cdot (\nabla v - \nabla u)dx + \int_\Omega j^0(x, u(x); v(x) - u(x))dx \geq$$

$$\geq \lambda \int_\Omega u(v - u)dx + \int_\Omega g(v - u)dx, \quad \forall\, v \in K, \tag{3.57}$$

has a nontrivial solution.

Proof. The idea is to apply Corollary 3.2 for $I = \Phi + \Psi: \ H_0^1(\Omega) \to (-\infty, +\infty]$, with Ψ determined by (3.55) and $\Phi: \ H_0^1(\Omega) \to \mathbb{R}$ expressed as follows

$$\Phi(u)=\frac{1}{2}\|u\|^2_{H^1_0}+\int_\Omega j(x,u(x))dx-\frac{\lambda}{2}\|u\|^2_{L^2}-\int_\Omega gudx,\quad \forall\, u\in H^1_0(\Omega)\ . \tag{3.58}$$

It is clear from (j_1) that Φ in (3.58) is locally Lipschitz (see Chang [2] and Clarke [3] as well as Chapter 1), so hypothesis (H) is fulfilled. Observe also that by (j_2) one has $I(0)=0$. For a later use we introduce the locally Lipschitz functional $J:\ H^1_0(\Omega)\to\mathbb{R}$ by

$$J(u)=\int_\Omega j(x,u(x))dx,\quad \forall\, u\in H^1_0(\Omega)\ . \tag{3.59}$$

The first task is to show that our functional I satisfies condition (PS) in the sense of Definition 3.2. To this end let a sequence $\{u_n\}\subset K$ satisfy $I(u_n)\to c\in\mathbb{R}$ and (3.2) for some $\varepsilon_n\to 0$ in $\mathbb{R}$. In our case of (3.55), (3.58), inequality (3.2) reads

$$\begin{aligned}\Phi^0(u_n;v-u_n) &= \int_\Omega \nabla u_n\cdot\nabla(v-u_n)dx+J^0(u_n;v-u_n)\\ &\quad -\lambda\int_\Omega u_n(v-u_n)dx-\int_\Omega g(v-u_n)dx\\ &\geq -\varepsilon_n\|v-u_n\|_{H^1_0},\quad \forall\, v\in K.\end{aligned} \tag{3.60}$$

In particular, by setting $v=2u_n$ in (3.60), we derive

$$\Phi^0(u_n;u_n)\geq -\varepsilon_n\|u_n\|_{H^1_0}\ .$$

Then, for any sufficiently large n, we can write in view of (3.58), (3.60), (j_3) and Rayleigh - Ritz variational characterization of λ_1 that

$$\begin{aligned}c+1+\frac{1}{\mu}\|u_n\|_{H^1_0} &\geq \Phi(u_n)+\frac{1}{\mu}\varepsilon_n\|u_n\|_{H^1_0}\\ &\geq \Phi(u_n)-\frac{1}{\mu}\Phi^0(u_n;u_n)\geq(\frac{1}{2}-\frac{1}{\mu})\|u_n\|^2_{H^1_0}\\ &\quad +\lambda(\frac{1}{\mu}-\frac{1}{2})\|u_n\|^2_{L^2}+(\frac{1}{\mu}-1)\int_\Omega gu_ndx\\ &\quad +\int_\Omega(j(x,u_n)-\frac{1}{\mu}j^0(x,u_n;u_n))dx\\ &\geq (\frac{1}{2}-\frac{1}{\mu})(1-\lambda_1^{-1}\lambda)\|u_n\|^2_{H^1_0}+(\frac{1}{\mu}-1)\,\lambda_1^{-1}\,\|g\|_{L^2}\,\|u_n\|_{H^1_0}\ .\end{aligned}$$

Since we suppose that $\mu>2$ and $\lambda<\lambda_1$ it follows the boundedness of $\{u_n\}$ in $H^1_0(\Omega)$. Then there exists $\bar{u}\in K$ such that $u_n\to\bar{u}$ weakly in $H^1_0(\Omega)$ and $u_n\to\bar{u}$ strongly in $L^2(\Omega)$. Therefore from (3.60) with $v=\bar{u}$ we get

$$\limsup_{n\to\infty}\|u_n\|^2_{H^1_0}\leq\|\bar{u}\|^2_{H^1_0}+\limsup_{n\to\infty}J^0(u_n;\bar{u}-u_n)\ . \tag{3.61}$$

A basic property of generalized gradient in the sense of Clarke (see Proposition 1.4 (ii)) ensures that there exists $z_n\in\partial J(u_n)\subset H^{-1}(\Omega)$ satisfying

$$\langle z_n,\bar{u}-u_n\rangle=J^0(u_n;\bar{u}-u_n)\ . \tag{3.62}$$

The boundedness of $\{u_n\}$ in $H_0^1(\Omega)$, the compactness of the embedding $H_0^1(\Omega) \subset L^p(\Omega)$, assumption (j$_1$) and the fact that J in (3.59) is locally Lipschitz on $L^p(\Omega)$ imply that, up to a subsequence, $\{z_n\}$ converges strongly in $H^{-1}(\Omega)$. From (3.61) and (3.62) we obtain that

$$\limsup_{n\to\infty} \|u_n\|_{H_0^1}^2 \le \|\bar{u}\|_{H_0^1}^2 .$$

This yields that $u_n \to \bar{u}$ strongly in $H_0^1(\Omega)$ along a subsequence which enables us to conclude that $I = \Phi + \Psi$ satisfies condition (PS).

Let us check that condition (i) in Corollary 3.2 holds. We carry out the proof as follows. Fix an $\varepsilon > 0$. Assumption (j$_2$) assures that some $\delta > 0$ exists to have

$$|y|^{-2} j(x,y) \ge -\varepsilon, \quad \forall\, x \in \Omega,\ y \in \mathbb{R},\ |y| \le \delta . \tag{3.63}$$

Lebourg's mean value theorem and the growth condition (j$_1$) imply

$$|j(x,y)| = |j(x,y) - j(x,0)| \le c(1 + |y|^{p-1})|y|,\ \forall\ (x,y) \in \Omega \times \mathbb{R} .$$

In particular, we see that

$$|j(x,y)| \le c(\delta^{-p+1} + 1)|y|^p,\ \forall\ x \in \Omega,\ y \in \mathbb{R},\ |y| \ge \delta .$$

Combining it with (3.63) we find

$$-j(x,y) \le \varepsilon |y|^2 + c(\delta^{-p+1} + 1)|y|^p,\ \forall\ (x,y) \in \Omega \times \mathbb{R} .$$

Thus we obtain the next estimate for the functional J in (3.59)

$$-J(u) \le \varepsilon \|u\|_{L^2}^2 + c(\delta^{-p+1} + 1)\|u\|_{L^p}^p \le \varepsilon B \|u\|_{H_0^1}^2 \tag{3.64}$$

provided $\|u\|_{H_0^1}$ is sufficiently small, where $B > 0$ denotes a constant independent of ε. The second inequality in (3.64) is due essentially to the fact that without loss of generality we may assume $p > 2$ (see (j$_1$)). Then (3.56),(3.58) and (3.64) allow to write for $u \in K$ that

$$I(u) = \Phi(u) \ge \frac{1}{2}(1 - \lambda \lambda_1^{-1} - \varepsilon B)\|u\|_{H_0^1}^2$$

if $\|u\|_{H_0^1}$ is small enough. Since $\varepsilon > 0$ can be chosen arbitrary small, property (i) in Corollary 3.2 is verified.

We pass to the proof of condition (ii) in Corollary 3.2. For this we need the formula below involving the generalized gradient ∂_t with respect to $t \in \mathbb{R}$

$$\partial_t(t^{-\mu} j(x,ty)) = \mu t^{-1-\mu}(\mu^{-1} ty \partial j(x,ty) - j(x,ty))$$

for all $x \in \Omega$, $y \in \mathbb{R}$ and $t > 0$. By Lebourg's mean value theorem and the previous relation we see that

$$t^{-\mu} j(x,ty) - j(x,y) \le \mu \tau^{-1-\mu}(\mu^{-1} j^0(x,\tau y; \tau y) - j(x,\tau y))(t-1)$$

for all $x \in \Omega,\ y \in \mathbb{R},\ t > 1$, with some $\tau \in (1, t)$. Then assumption (j_3) implies

$$j(x, ty) \leq t^{\mu} j(x, y) \quad \text{for all } x \in \Omega,\ y \in \mathbb{R},\ t > 0\ . \tag{3.65}$$

Now from (3.65) and assumption (j_4) we deduce

$$I(te) = \Phi(te) \leq \frac{t^2}{2}(\|e\|^2_{H^1_0} - \lambda\, \|e\|^2_{L^2}) + t^{\mu} \int_{\Omega} j(x, e(x))dx -$$

$$-t \int_{\Omega} g(x)e(x)dx \longrightarrow -\infty \quad \text{as } t \to +\infty$$

because $\mu > 2$. It follows that (ii) in Corollary 3.2 is satisfied.

Applying Corollary 3.2 we obtain a nontrivial critical point $u \in K$ of $I = \Phi + \Psi$. It remains only to justify that the given function $u \in K$ solves (3.57). By Definition 3.1 we know that $u \in K$ fulfills

$$\int_{\Omega} \nabla u \cdot (\nabla v - \nabla u)dx + J^0(u; v-u) \geq \lambda \int_{\Omega} u(v-u)dx + \int_{\Omega} g(v-u)dx, \quad \forall\, v \in K, \tag{3.66}$$

where $\lambda < \lambda_1$. On the other hand assumption (j_1) and a basic result of Clarke (see Theorem 1.3) insures

$$J^0(u; w) \leq \int_{\Omega} j^0(x, u(x); w(x))dx, \quad \forall\, u, w \in L^p(\Omega)\ . \tag{3.67}$$

Combining (3.66) and (3.67) we arrive at (3.57). This completes the proof. □

Corollary 3.7 Under the same assumptions as in Theorem 3.3 and with the modification that (j_3) is replaced by the stronger

(j_3') $\mu^{-1} j^0(x, y; y) \leq j(x, y) - a|y|^q, \quad \forall\, x \in \Omega,\ y > 0$, with constants $\mu > 2,\ a > 0$ and $q > 2$,

the conclusion of Theorem 3.3 holds for every $\lambda \in \mathbb{R}$.

Proof. The first difference from the proof of Theorem 3.3 lies in the fact that (j_3') allows to deduce the boundedness of sequence $\{u_n\} \subset K$ in (3.60) for every $\lambda \in \mathbb{R}$, so condition (PS) holds whenever $\lambda \in \mathbb{R}$. The second difference from the proof of Theorem 3.2 consists in the argument to establish condition (i) of Corollary 3.3. Here we carry out the proof arguing by contradiction. Assume there exists a sequence $\{u_n\}$ in K such that $u_n \to 0$ in $H^1_0(\Omega)$ and

$$I(u_n) = \Phi(u_n) \leq \|u_n\|^2_{H^1_0}/n\ .$$

Without loss of generality we may suppose $p > 2$ (see (j_1)). Denoting $w_n = u_n/\|u_n\|_{H^1_0}$ one has

$$\Phi(u_n)\|u_n\|^{-2}_{H^1_0} = \frac{1}{2} - \frac{1}{2}\lambda\, \|w_n\|^2_{L^2} +$$

$$+\|u_n\|_{H_0^1}^{-2}\int_\Omega j(x,u_n(x))dx - \|u_n\|_{H_0^1}^{-1}\int_\Omega gw_n dx \le \frac{1}{n} . \tag{3.68}$$

Admit for moment that

$$\liminf_{n\to\infty} \|u_n\|_{H_0^1}^{-2}\int_\Omega j(x,u_n(x))dx \ge 0 . \tag{3.69}$$

Passing to a subsequence if necessary, it is allowed to assume that $w_n \to w$ weakly in $H_0^1(\Omega)$ and strongly in $L^2(\Omega)$ for some $w \in K$. Notice that $w \neq 0$ because otherwise (3.68), (3.69) and (3.56) would imply

$$\liminf_{n\to\infty} \Phi(u_n)\|u_n\|_{H_0^1}^{-2} \ge \frac{1}{2}$$

which contradicts the choice of the sequence $\{u_n\}$. Then, applying again (3.56), we find

$$\int_\Omega gwdx < 0 .$$

Passing to limit in (3.68) for $n \to \infty$ and taking into account (3.69) we are led to the conclusion

$$\lim_{n\to\infty} \Phi(u_n)\|u_n\|_{H_0^1}^{-2} = +\infty .$$

This contradicts the choice of the sequence $\{u_n\}$. To complete the proof of (i) in Corollary 3.2 we have to prove (3.69). At this point we apply (3.64), which was established on the basis of (j_4), to obtain

$$\liminf_{\substack{u\to 0\\ u\in H_0^1(\Omega)}} J(u)\|u_n\|_{H_0^1}^{-2} \ge -B\varepsilon .$$

Since $\varepsilon > 0$ is arbitrary, inequality (3.69) follows.

Condition (ii) of Corollary 3.2 can be checked exactly as in the proof of Theorem 3.3. Then Corollary 3.2 implies directly the desired conclusion. □

Remark 3.4 Theorem 3.3 and Corollary 3.7 are extensions of Theorem 5.1 of Szulkin [32]. A relevant example of locally Lipschitz function $j : \mathbb{R} \to \mathbb{R}$ satisfying conditions (j_1)-(j_4) is the following

$$j(t) = \max\{-\frac{1}{p}|t|^p, -\frac{1}{r}|t|^r\}, \quad \forall\, t \in \mathbb{R}, \tag{3.70}$$

with constants $p, r \in (2, \frac{2N}{N-2})$ if $N \ge 3$. The theory of variational - hemivariational inequalities has been initiated by P.D.Panagiotopoulos ([25]-[27]). Various applications of nonsmooth critical point theory in this field (pure hemivariational inequalites) have been done by Goeleven, Motreanu, Naniewicz and Panagiotopoulos [9]-[11], [17] - [22], [24] - [27].

Remark 3.5 Two adhesively connected v. Kármán plates subjected to elastoplastic boundary conditions, or to unilateral contact boundary conditions, or to friction boundary conditions lead to variational - hemivariational inequalities [28]. The corresponding buckling problem leads to an eigenvalue problem of the type studied here.

3.6 Semicoercive Unilateral Problems and Periodic Solutions

In this Section we apply the multiplicity results in Corollary 3.6 to a problem related to the one considered in Section 1.5 in Chapter 1.

On the reflexive Banach space X we consider a proper, convex, l.s.c. functional $\Psi : X \to (-\infty, +\infty]$, a symmetric bounded linear operator $T : X \to X^*$ and a locally Lipschitz functional $C : X \to \mathbb{R}$. We assume that Ψ and C are even, Ψ is α - positively homogeneous and C is β - positively homogeneous, where the numbers α and β satisfy

$$\beta \geq 2^\alpha - 1, \quad \beta > 2, \quad \alpha > 1 . \tag{3.71}$$

Concerning the operator $T : X \to X^*$ we suppose that

$$\dim \mathrm{Ker}T < \infty \tag{3.72}$$

and T is semicoercive, that is,

$$\langle Tu, u \rangle \geq c\|Pu\|^2, \quad \forall\, u \in X , \tag{3.73}$$

with a constant $c > 0$ and a projection $P : X \to Y$ relative to a direct sum decomposition $X = \mathrm{Ker}T \oplus Y$.

The locally Lipschitz function $C : X \to \mathbb{R}$ is assumed to verify the following properties: there exist a finite-dimensional linear subspace E of X with

$$\dim E > \dim \mathrm{Ker}T \tag{3.74}$$

such that for constants $k_1 > 0, \ c_1 \in \mathbb{R}$,

$$\begin{cases} C(u) \leq -k_1 & \forall\, u \in E; \ \ \|u\| = 1, \\ \\ C(u) \geq c_1, & \forall\, u \in Y, \ \ \|u\| = 1, \end{cases} \tag{3.75}$$

and $\limsup\limits_{n\to\infty} C^0(u_n; u - u_n) \leq 0$ whenever $u_n \to u$ weakly in X.

For the function Ψ we assume that

$$\begin{cases} \Psi(u) > 0, & \forall\, u \in \mathrm{Ker}T, \ \ u \neq 0 \\ \\ \Psi(u) \geq 0, & \forall\, u \in X, \ \ \Psi \text{ bounded above on the unit sphere of } E. \end{cases} \tag{3.76}$$

Theorem 3.4 Assume that the foregoing hypotheses hold. Then the semicoercive variational – hemivariational inequality

$$u \in X : \ \ \langle Tu, v - u \rangle + C^0(u; v - u) + \Psi(v) - \Psi(u) \geq 0, \ \ \forall\, v \in X, \tag{3.77}$$

admits at least $\dim E - \dim \mathrm{Ker}T$ distinct pairs $(-u, u)$ of nontrivial solutions.

Proof. Define $\Phi : X \to \mathbb{R}$ by

$$\Phi(u) = \frac{1}{2}\langle Tu, u\rangle + C(u), \quad \forall\, u \in X .$$

Then we are in the setting of Corollary 3.6 for $I = \Phi + \Psi$.

Let us check that I verifies condition (PS). To this end let $\{u_n\} \subset X$ be a sequence such that $I(u_n) \to c \in \mathbb{R}$ and

$$\langle Tu_n, v-u_n\rangle + C^0(u_n; v-u_n) + \Psi(v) - \Psi(u_n) \geq -\varepsilon_n\|v-u_n\|, \quad \forall\, v \in X, \tag{3.78}$$

with $\varepsilon_n \to 0^+$.

Firstly, we prove that (u_n) is bounded. If not, we put $v = 2u_n$ in (3.78) that yields

$$\langle Tu_n, u_n\rangle + C^0(u_n; u_n) + (2^\alpha - 1)\Psi(u_n) \geq -\varepsilon_n\|u_n\| . \tag{3.79}$$

On the other hand differentiating with respect to $t > 0$ the equality

$$C(tu) = t^\beta C(u), \quad \forall\, u \in X,\ \forall\, t > 0,$$

we get for $t = 1$ that

$$C^0(u; u) = \beta C(u), \quad \forall\, u \in X . \tag{3.80}$$

Introducing (3.80) in (3.79) one obtains

$$\langle Tu_n, u_n\rangle + \beta C(u_n) + (2^\alpha - 1)\Psi(u_n) \geq -\varepsilon_n\|u_n\| .$$

It follows that for a sufficiently large n one has

$$\beta I(u_n) - \langle Tu_n, u_n\rangle - \beta C(u_n) - (2^\alpha - 1)\Psi(u_n) \leq \beta(c+1) + \|u_n\|,$$

or

$$(\beta + 1 - 2^\alpha)\Psi(u_n) + (\frac{\beta}{2} - 1)\langle Tu_n, u_n\rangle \leq \beta(c+1) + \|u_n\| . \tag{3.81}$$

From (3.71),(3.76) and (3.81) we can write

$$(\frac{\beta}{2} - 1)\langle Tv_n, v_n\rangle \leq \beta(c+1)\|u_n\|^{-2} + \|u_n\|^{-1} , \tag{3.82}$$

where we have put

$$v_n := u_n/\|u_n\| .$$

Passing to a subsequence we can assume that $v_n \to v \in X$ weakly in X. Then (3.71), (3.73) and (3.82) show

$$\langle Tv, v\rangle = 0 .$$

By (3.73) we have $v \in \mathrm{Ker}T$. Since (3.73) and (3.82) ensure that

$$\liminf_{n\to\infty} \|Pv_n\| = 0,$$

along a subsequence we can assume $Pv_n \to 0$ as $n \to \infty$. In view of (3.72) we conclude that $v_n \to v$ strongly in X up to a subsequence, thus $\|v\| = 1$. From (3.73) and (3.81) we deduce

$$(\beta + 1 - 2^\alpha)\Psi(v_n) \leq \beta(c+1)\|u_n\|^{-\alpha} + \|u_n\|^{1-\alpha} .$$

Letting $n \to \infty$ it is seen from (3.71) that $\Psi(v) \leq 0$. This contradicts (3.76). So $\{u_n\}$ is a bounded sequence in X.

Taking eventually a subsequence we may assume that $u_n \to u \in X$ weakly in X. For $v = u$ in (3.78) we know that

$$\langle Tu_n, u - u_n\rangle + C^0(u_n; u - u_n) + \Psi(u) - \Psi(u_n) \geq -\varepsilon_n\|u - u_n\| .$$

By virtue of the imposed hypotheses we derive

$$\liminf_{n\to\infty}\langle Tu_n, u_n - u\rangle \leq 0 .$$

Therefore one has

$$\liminf_{n\to\infty}\langle T(u_n - u), u_n - u\rangle = \liminf_{n\to\infty}\langle Tu_n, u_n - u\rangle \leq 0 .$$

By (3.73) it results that $Pu_n \to Pu$ strongly in X and then, using (3.72), we find that a subsequence of $\{u_n\}$ converges to u in X. Condition (PS) is thus verified.

Condition (i) in Corollary 3.6 holds for $X_1 = Y$ (cf. the direct sum $X = \mathrm{Ker}T \oplus Y$). It follows from (3.75), (3.76) that

$$\begin{aligned} I(u) &= \frac{1}{2}\langle Tu, u\rangle + C(u) + \Psi(u) \\ &\geq \frac{c}{2}\|u\|^2 + \|u\|^\beta C\left(\frac{u}{\|u\|}\right) \geq \frac{c}{2}\|u\|^2 + c_1\|u\|^\beta, \quad \forall\, u \in X_1\setminus\{0\}, \end{aligned}$$

with a constant c_1. Since $\beta > 2$, condition (i) of Corollary 3.6 is valid.

To check condition (ii) of Corollary 3.6 we make use of assumptions (3.74)-(3.76). Hence, if we set $X_2 = E$, we infer that

$$\begin{aligned} I(u) &\leq \|T\|\|u\|^2 + \|u\|^\beta C\left(\frac{u}{\|u\|}\right) + \|u\|^\alpha \Psi\left(\frac{u}{\|u\|}\right) \\ &\leq \|T\|\|u\|^2 - k_1\|u\|^\beta + k_2\|u\|^\alpha, \quad \forall\, u \in E\setminus\{0\}, \end{aligned}$$

with constants $k_1,\ k_2 \in \mathbb{R},\ k_1 > 0$. By (3.71) we see that (ii) of Corollary 3.6 is satisfied.

Applying Corollary 3.6 we finish the proof of Theorem 3.4. □

A suitable framework for the application of Theorem 3.4 is formed by the periodic unilateral problems.

Example 3.1 For a given $T > 0$, let $X = H^1(\mathbb{R}/T\mathbb{Z}, \mathbb{R})$ the usual Sobolev space of T - periodic functions endowed with the Hilbertian norm

$$\|u\|^2 = \int_0^T (u^2(t) + u'(t)^2)dt, \quad \forall\, u \in X\,.$$

Fix a constant $\alpha > 1$ and a function $g \in L^\infty([0,T])$ nonnegative and $\not\equiv 0$.

Let $V : \mathbb{R} \times \mathbb{R} \to \mathbb{R}$ be a Carathéodory function such that for every $t \in \mathbb{R}$, $V(t,\cdot)$ is locally Lipschitz whose generalized directional derivative and generalized gradient are denoted $V_u^0(t,\cdot;\cdot)$ and $\partial_u V(\cdot)$, respectively.

Employing the notations above we state the next periodic unilateral problem

$$u \in X : \int_0^T u'(v' - u')dt + \int_0^T V_u^0(t, u(t); v(t) - u(t))dt +$$

$$+ \int_0^T g(t)(|v(t)|^\alpha - |u(t)|^\alpha)dt \geq 0 \quad \forall\, v \in X\,. \tag{3.83}$$

The function $V : \mathbb{R} \times \mathbb{R} \to \mathbb{R}$ satisfies in addition the growth condition

$$|V(t,x)| \leq a|x|^p + b(t), \quad \forall\, x \in \mathbb{R}, \text{ a.e. } t \in \mathbb{R}, \tag{3.84}$$

for $a \in \mathbb{R}$, $p \geq 1$ and $b \in L^{p/(p-1)}([0,T])$, the sign condition

$$\exists\, y \in \mathbb{R} \text{ such that } V(\cdot, y) < 0 \text{ on a nonzero measure subset,} \tag{3.85}$$

the homogeneity condition,

$$V \text{ is } \beta - \text{positively homogeneous with respect to the second variable for some } \beta \text{ satisfying (3.71),} \tag{3.86}$$

and the symmetry condition

$$V(t,\cdot) \text{ is even}, \quad \forall\, t \in \mathbb{R}\,. \tag{3.87}$$

If the assumptions above hold, then the variational inequality (3.83) possesses infinitely many distinct pairs $(u, -u)$ of nonconstant periodic solutions.

To arrive at this conclusion let us define $T : X \to X^*$ by

$$\langle Tu, v\rangle = \int_0^T \nabla u \cdot \nabla v dt, \quad \forall\, u, v \in X\,,$$

$C : X \to \mathbb{R}$ by

$$C(u) = \int_0^T V(t, u(t))dt, \quad \forall\, u \in V\,,$$

and $\Psi : X \to \mathbb{R}$ by

$$\Psi(u) = \int_0^T g(t)|u(t)|^\alpha dt, \quad \forall\, u \in V\,.$$

Condition (3.85), ensures that (3.74), (3.75) are satisfied for linear subspaces E of X of arbitrary dimension. Assertion $\limsup_{n\to\infty} C^0(u_n; u - u_n) \leq 0$ if $u_n \to u$

weakly in X holds by the compactness of the inclusion $X \subset C([0,T],\mathbb{R})$ and the upper semicontinuity of $C^0(\cdot,\cdot)$. The compactness of the embedding $X \subset L^p([0,T])$ and the continuity of C on $L^p([0,T])$ assure that C is bounded on the unit sphere of X. The requirements in (3.76) for Ψ are easily verified taking into account the properties of g and that Ker$T = \mathbb{R}$. Assumptions (3.86), (3.87) give rise to the same properties for the functional C. Thus Theorem 3.4 provides the desired result.

Remark 3.6 Theorem 3.4 and Example 3.1 are extensions for the corresponding results in Goeleven, Nguyen and Willem [8].

References

[1]A. Ambrosetti and P.H. Rabinowitz, Dual Variational Methods in Critical Point Theory and Applications, *J. Func. Anal.* **14** (1973), 349-381.

[2]K.C. Chang, Variational Methods for Nondifferentiable Functionals and Applications to Partial Differential Equations, *J. Math. Anal. Appl.* **80** (1981), 102-129.

[3]F.H.Clarke, *Optimization and Nonsmooth Analysis*, John Wiley and Sons, New York, 1983.

[4]Y. Du, A deformation Lemma and some Critical Point Theorems, *Bull. Austral. Math.* **43** (1991), 161-168.

[5]I. Ekeland, On the Variational Principle, *J. Math. Anal. Appl.* **47** (1974), 324-353.

[6]I. Ekeland, Nonconvex Minimization Problems, *Bull. (New Series) Amer. Math.* **1** (1979), 443-474.

[7]N. Ghoussoub, Location, Multiplicity and Morse Indices of Min-Max Critical Points, *J. reine angew. Math.* **417** (1991), 27-76.

[8]D. Goeleven, V.H. Nguyen and M. Willem, Existence and Multiplicity Results for Semicoercive Unilateral Problems, *Bull. Austral. Math. Soc.* **49** (1994), 489-497.

[9]D. Goeleven, D. Motreanu and P.D.Panagiotopoulos, Multiple Solutions for a Class of Eigenvalue Problems in Hemivariational Inequalities, *Nonlinear Anal. TMA* **20** (1997), 9-16.

[10]D. Motreanu and P.D.Panagiotopoulos, General Minimax Methods for Variational-Hemivariational Inequalities, submitted.

[11]D. Goeleven and D. Motreanu, Eigenvalue and Dynamic Problems for Variational and Hemivariational Inequalities, *Comm. Appl. Nonl. Analysis* **3** (1996), 1-21.

[12]L.T. Hu, *Homotopy Theory*, Academic Press, New York, 1959.

[13]G. Lefter, Critical Point Theorms for Lower Semicontinuous Functions, University of Iasi, unpublished manuscript.

[14]P. Mironescu and V. Radulescu, A Multiplicity Theorem for Locally Lipschitz Periodic functionals, *J. Math. Anal. Appl.* **195** (1995), 621-637.

[15]D. Motreanu, Existence of Critical Points in a General Setting, *Set-Valued Anal.* **3** (1995), 295-305.

[16]D. Motreanu, A Multiple Linking Minimax Principle, *Bull. Austral. Math. Soc.* **53** (1996), 39-49.

[17]D. Motreanu and Z. Naniewicz, Discontinuous Semilinear Problems in Vector-Valued Function Spaces, *Differ. Int. Equations* **9** (1996), 581-598.

[18]D. Motreanu and P.D.Panagiotopoulos, Nonconvex Energy Functions, Related Eigenvalue Hemivariational Inequalities on the Sphere and Applications, *J. Global Optimization* **6** (1995), 163-177.

[19]D. Motreanu and P.D.Panagiotopoulos, An Eigenvalue Problem for a Hemivariational Inequality Involving a Nonlinear Compact Operator, *Set-Valued Anal.* **3** (1995), 157-166.

[20]D. Motreanu and P.D. Panagiotopoulos, A Minimax Approach to the Eigenvalue Problem of Hemivariational Inequalities and Applications, *Applicable Anal.* **58** (1995), 157-166.

[21]D. Motreanu and P.D.Panagiotopoulos, On the Eigenvalue Problem for Hemivariational Inequalities: Existence and Multiplicity of Solutions, *J. Math. Anal. Appl.* **197** (1996), 75-89.

[22]D. Motreanu and P.D.Panagiotopoulos, Double Eigenvalue Problems for Hemivariational Inequalities, *Arch. Rational Mech. Anal.* **140** (1997), 225-251.

[23]J.R. Munkres, Elementary Differential Topology, Princeton University Press, Princeton, 1966.

[24]Z. Naniewicz, Hemivariational Inequalities with Functions Fulfilling Directional Growth Condition, *Applicable Anal.* **55** (1994), 259-285.

[25]Z. Naniewicz and P.D.Panagiotopoulos, *Mathematical Theory of Hemnivariational Inequalities and Applications*, Marcel Dekker, Inc., New York, 1995.

[26]P.D.Panagiotopoulos, *Inequality Problems in Mechanics and Applications. Convex and Nonconvex Energy Functions*, Birkhäuser Verlag, Basel, Boston, 1985 (Russian Translation MIR Publ. Moscow, 1989).

[27]P.D.Panagiotopoulos, *Hemivariational Inequalites. Applications in Mechanics and Engineering*, Springer-Verlag, Berlin, 1993.

[28]P.D.Panagiotopoulos, G. Stavroulakis: A Variational - Hemivariational Inequality Approach to the Laminated Plate Theory under Subdifferential Boundary Conditions, *Quart. Appl. Math.* **46** (1988), 409-430.

[29]P.H. Rabinowitz, Some Minimax Theorems and Applications to Nonlinear Partial Differential Equations, in: Nonlinear Analysis: A collection of papers in honor of Erich Röthe, Academic Press, New York, 1978.

[30]P.H. Rabinowitz, Minimax methods in critical point theory with applications to differential equations. CBMS Reg. Conf. Ser. Math. 65, *Amer. Math. Soc.*, Providence, R.I., 1986.

[31]E.H. Spanier, *Algebraic topology*, McGraw-Hill, New York, 1966.

[32]A. Szulkin, Minimax Principles for Lower Semicontinuous Functions and Applications to Nonlinear Boundary Value Problems, Ann. Inst. Henri Poincaré. *Analyse non linéaire* **3** (1986), 77-109.

4. Eigenvalue Problems for Hemivariational Inequalities

The aim of the present chapter is to formulate and study two types of eigenvalue problems for hemivariational inequalities. The first type is the most classical one, whereas the second is an eigenvalue problem for a hemivariational inequality involving a nonlinear compact operator. After giving certain existence results, we illustrate the theory with applications.

4.1 Formulation of the First Problem and Prerequisites

Here we consider the following eigenvalue problem for a simple hemivariational inequality (problem (P)):
Find $u \in V \subset L^2(\Omega)$ and $\lambda \in \mathbf{R}$ satisfying the inequality

$$a(u,v) + \int_{\Omega'} j^0(u;v)dx \geq \lambda(u,v)_V \quad \forall v \in V. \tag{4.1}$$

We denote by Ω an open bounded subset of $\mathbf{R}^n$, Ω' a subdomain of Ω, V a real Hilbert space with the scalar product $(.,.)_V$, which is continuously imbedded into $L^2(\Omega)$, $a : V \times V \to \mathbf{R}$ a continuous symmetric bilinear form on V and by $j : \mathbf{R} \to \mathbf{R}$ a function defined by means of (cf. Section 1.3)

$$j(t) = \int_0^t \beta(s)ds, \quad t \in \mathbf{R}, \tag{4.2}$$

where $\beta \in L^\infty_{\text{loc}}(\mathbf{R})$. A possible method to treat problem (P) is to regularize the nonsmooth integral term and then to pass to limit (cf. e.g. [7][8]). Here, the use of the critical point theory permits us to obtain additionally certain qualitative information for the solution.

Thus we write problem (P) in the form (problem (P'))

$$0 \in \partial I_\lambda(u), \quad u \in V, \tag{4.3}$$

which is called after Rockafellar [26] a substationarity form, where $I_\lambda : V \to \mathbf{R}$ denotes a suitably defined locally Lipschitz functional depending on λ, which is associated to problem (P), and $\partial I_\lambda(u)$ represents Clarke's generalized gradient of I_λ at $u \in V$. Problem (P') expresses that $u \in V$ is a critical point of the locally Lipschitz functional I_λ in the sense of Chang [2] or a substationarity point according to R.T.Rockafellar [26]. As we know problem (P) may have a larger number of solutions that problem (P') [12][14].

Since we are looking here for at least one solution of (P) it is sufficient to prove that (P') has at least one solution. Thus the study of problem (P) is reduced to finding a (the) critical point(s) of the corresponding functional I_λ on V. To prove their existence we apply two minimax results due to Chang [2]. The existence of nontrivial solutions of problem (P) is proved by showing the existence of a global minimum or a minimax point of an appropriately defined functional I_λ. Moreover we prove that there exist solutions of problem (P) in spaces $L^{2+d}(\Omega)$ with $d \geq 0$ in the case where the function β satisfies an appropriate growth condition. We base our mathematical study to some special cases of the theorems proved in the two previous chapters. These theorems, due to Chang [2] and Rabinowitz [24] were actually the starting points of our research.

Let X be a Banach space with norm $\|\cdot\|$ and let X^* be its dual space. The locally Lipschitz functional $f : X \to \mathbf{R}$ is said to satisfy the Palais-Smale condition if every sequence $\{x_n\}$ in X for which $\{f(x_n)\}$ is bounded in $\mathbf{R}$ and

$$\min_{w \in \partial f(x_n)} \|w\|_{X^*} \to 0 \text{ as } n \to \infty,$$

contains a convergent subsequence in X.

We recall the following nonsmooth version of the Palais-Smale Minimization Theorem [2].

Proposition 4.1 Suppose that the Banach space X is reflexive and the locally Lipschitz functional $f : X \to \mathbf{R}$ is bounded from below and satisfies the Palais-Smale condition. Then $\inf_X f$ is a critical value of f, i.e., there exists a point $x \in X$ such that

$$\inf_X f = f(x) \text{ and thus } 0 \in \partial f(x).$$

The next result is a slightly modified version of a result of Chang [4] and it has a similarity with the classical Mountain Pass Theorem (see [24], p.7). More general minimax principles for nonsmooth functionals can be found in Chapters 2 and 3.

Proposition 4.2 Let the locally Lipschitz functional $f : X \to \mathbf{R}$ on the reflexive Banach space X satisfy the Palais-Smale condition together with the following assumptions

(i) $f(0) = 0$ and there exist positive constants ρ, d such that

$$f(x) \geq d \text{ for each } x \in X \text{ with } \|x\| = \rho; \tag{4.4}$$

(ii) there is a point $e \in X$ such that

$$\|e\| > \rho \quad \text{and} \quad f(e) \leq 0. \tag{4.5}$$

Then there exists a critical point $x_0 \in X$ of f satisfying

$$f(x_0) = \inf_{\gamma \in \Gamma} \max_{t \in [0,1]} f(\gamma(t)) \tag{4.6}$$

where

$$\Gamma = \{\gamma \in C([0,1]), X) | \ \gamma(0) = 0, \quad \gamma(1) = e\}. \tag{4.7}$$

Proof. The proof follows the same steps as in the Mountain Pass Theorem of Ambrosetti and Rabinowitz (cf. [24]). The sphere $\|x\| = \rho$ separates in X the points 0 and e, and thus

$$c = \inf_{\gamma \in \Gamma} \max_{t \in [0,1]} f(\gamma(t)) \geq d. \tag{4.8}$$

We will show that the number c in (4.8) is a critical value of f, i.e. that there is a critical point x_0 of f with $f(x_0) = c$. Let us assume first that c is not a critical value of f. Then the nonsmooth Deformation Lemma of Chang ([2], Theorem 3.1) yields the existence of a real number $\varepsilon \in (0, d/2)$ and of a homeomorphism $h : X \to X$ such that

$$h(x) = x \text{ for every } x \notin \{y \in X | \, |f(x) - c| < d/2\} \tag{4.9}$$

and

$$f(h(x)) \leq c - \varepsilon \text{ for every } x \in X \text{ with } f(x) \leq c + \varepsilon. \tag{4.10}$$

From (4.8) one can find a $\gamma \in \Gamma$ such that

$$f(\gamma(t)) \leq c + \varepsilon \quad \forall t \in [0,1]. \tag{4.11}$$

Let

$$g(t) = h(\gamma(t)), \quad t \in [0,1]. \tag{4.12}$$

From (4.9) and the assumptions (i) and (ii) it follows that $g \in \Gamma$, while from inequalities (4.10), (4.11) we obtain that

$$f(g(t)) \leq c - \varepsilon, \quad t \in [0,1]. \tag{4.13}$$

Thus we arrive at the contradiction

$$c \leq \max_{t \in [0,1]} f(g(t)) \leq c - \varepsilon. \tag{4.14}$$

This completes the proof of Prop. 4.2. □

Concerning problem (P) we deal with functionals $J : L^{\sigma+1}(\Omega) \to \mathbb{R}$ of the type

$$J(u) = \int_{\Omega'} j(u(x))dx, \quad u \in L^{\sigma+1}(\Omega), \qquad \sigma \geq 1 \tag{4.15}$$

where Ω' is a subdomain of Ω and $j : \mathbb{R} \to \mathbb{R}$ is the function fulfilling (4.2).

Now we must impose upon β the following growth condition

$$|\beta(t)| \le c_1 + c_2 |t|^\sigma, \quad t \in \mathbb{R}, \tag{4.16}$$

for some positive constants c_1, c_2 independent of $t \in \mathbb{R}$, in order to define J on $L^{\sigma+1}(\Omega)$. We can easily verify that under assumption (4.16) the functional J is locally Lipschitz. Then using the same argument as in Proposition 1.7, we obtain the following characterization of the generalized gradient $\partial J(u)$.

Proposition 4.3 If $w \in \partial J(u)$, then $w \in (L^{\sigma+1}(\Omega))^*$ is such that $w(x) \in [\underline{\beta}(u(x)), \overline{\beta}(u(x))]$, for a.e. $x \in \Omega'$. The functions $\overline{\beta}$ and $\underline{\beta}$ are defined as follows for $t \in \mathbb{R}$

$$\begin{aligned} \underline{\beta}(t) &= \lim_{\delta \to 0} \operatorname*{essinf}_{|s-t|<\delta} \beta(s), \\ \overline{\beta}(t) &= \lim_{\delta \to 0} \operatorname*{esssup}_{|s-t|<\delta} \beta(s). \end{aligned} \tag{4.17}$$

If β admits the one-sided limits $\beta(t \pm 0)$ at $t \in \mathbb{R}$, we have

$$\begin{aligned} \underline{\beta}(t) &= \min(\beta(t-0), \beta(t+0)) \\ \overline{\beta}(t) &= \max(\beta(t-0), \beta(t+0)). \end{aligned} \tag{4.18}$$

4.2 The Existence of Solutions of the Eigenvalue Problem in $L^2(\Omega)$.

Let us make now the following assumptions for problem (P):

(H_1) the function $\beta \in L^\infty_{\text{loc}}(\mathbb{R})$ satisfies the growth condition

$$|\beta(t)| \le c_1 + c_2|t| \quad \forall t \in R, \tag{4.19}$$

where c_1 and c_2 are positive constants independent of t.

(H_2) Ω is an open bounded subset of some Euclidean space $\mathbb{R}^n$, V is a real Hilbert space with the scalar product $(\cdot,\cdot)_V$ and the induced norm $\| \|_V$ such that V is continuously and densely imbedded as a linear subspace of $L^2(\Omega)$, and $a : V \times V \to \mathbb{R}$ is a continuous symmetric bilinear form on V. Let us denote by $A : V \to V^*$ the linear operator which corresponds to the bilinear form $a(\cdot,\cdot)$ (i.e. $a(u,v) = \langle Au, v\rangle_{V,V^*}$ $\forall\, u, v \in V$, where $\langle .,. \rangle$ denotes the duality pairing).

(H_3) Ω' is a subdomain of Ω such that $W = \{u|_{\Omega'} \;|u \in V\}$ is a reflexive Banach space with the property that the imbedding $W \subset L^2(\Omega')$ is compact and the restriction mapping $r : V \to W$ given by $r(u) = u|_{\Omega'}$ $\forall u \in V$ is continuous.

Now let us consider for each number $\lambda \in \mathbb{R}$ the functional $I_\lambda : V \to \mathbb{R}$

$$I_\lambda(u) = \frac{1}{2}(a(u,u) - \lambda \|u\|_V^2) + J(u), \quad u \in V, \tag{4.20}$$

where $J : L^2(\Omega) \to \mathbb{R}$ is defined in (4.15) by means of the function $j : \mathbb{R} \to \mathbb{R}$ with β satisfying (H_1). Assumptions (H_1) and (H_2), imply that the functional I_λ is locally Lipschitz on V.

Proposition 4.4 Assume that the hypotheses $(H_1), (H_2), (H_3)$ hold. If $u_0 \in V$ is a critical point of the functional $I_\lambda : V \to \mathbb{R}$, for some $\lambda \in \mathbb{R}$, then the pair (u_0, λ) solves problem (P).

Proof. I_λ is locally Lipschitz. It follows from (4.20) and (H_2) that the generalized gradient $\partial I_\lambda(u)$ is given by

$$\partial I_\lambda(u) = Au - \lambda \Lambda u + \partial(J|_V)(u), \tag{4.21}$$

where $\Lambda : V \to V^*$ is the duality mapping

$$(\Lambda u)(v) = (u,v)_V \quad \forall v \in V. \tag{4.22}$$

Hypothesis (H_1) and relations (4.17) yield the estimate

$$|w| \le \max(|\underline{\beta}(t)|, |\overline{\beta}(t)|) \le c_1 + c_2|t| \tag{4.23}$$

$$\forall t \in \mathbb{R} \text{ and } w \in \partial j(t),$$

which by means of Theorem 1.3 (cf. also [4] p.83) implies that

$$\partial(J|_{L^2(\Omega')})(u) \subset \int_{\Omega'} \partial j(u(x))dx \;\; \forall u \in L^2(\Omega'). \tag{4.24}$$

The restriction map $r : V \to W$ is continuously differentiable and onto. Thus

$$\partial(J|_V)(u) = \partial(J|_W)(u|_{\Omega'}) \circ r \quad \forall u \in V. \tag{4.25}$$

Due to (H_2) W is dense in $L^2(\Omega')$. This property and hypothesis (H_3) allow us to apply a theorem of Chang (cf. Corollary 1.2), which implies that

$$\partial(J|_W)(u) \subset \partial(J|_{L^2(\Omega')})(u) \;\; \forall u \in W. \tag{4.26}$$

Let us now consider the critical point $u_0 \in V$ of I_λ for some $\lambda \in \mathbb{R}$; thus u_0 is a solution of the problem (P'). Relations (4.21), (4.24), (4.25) and (4.26) lead to the relation

$$\begin{aligned} \lambda(u_0,v)_V - a(u_0,v) &\le \int_{\Omega'} (\max_{z \in \partial j(u_0(x))} (z, v(x)))dx \\ &= \int_{\Omega'} j^0(u_0(x), v(x))dx \qquad \forall v \in V \end{aligned} \tag{4.27}$$

due to Proposition 1.4(ii). Relation (4.27) shows that $u_0 \in V$ is a solution of problem (P), which corresponds to the real number λ appearing in (P'). □

Now we show the following technical result which is necessary in order to verify the Palais-Smale condition for the functional I_λ.

Proposition 4.5 Assume that conditions (H_1), (H_2), (H_3) hold and let us choose $\lambda \in \mathbb{R}$ such that

$$\lambda < \inf\{a(u,u)/\|u\|_V^2 \;\big|\; u \in V,\ u \neq 0\}. \tag{4.28}$$

If a sequence $\{u_n\}$ is bounded in V and a sequence $w_n \in \partial(J|_V)(u_n)$ exists such that

$$Au_n - \lambda \Lambda u_n + w_n \to 0 \text{ in } V' \text{ as } n \to \infty, \tag{4.29}$$

then $\{u_n\}$ contains a subsequence which is convergent in V.

Proof. Since $\{u_n\}$ is bounded in V and the imbedding $W \subset L^2(\Omega')$ is compact, there exists a convergent subsequence in $L^2(\Omega')$ that we denote again by $\{u_n\}$. From (4.25) and (4.26) we obtain that

$$w_n|_{\Omega'} \in \partial(J|_{L^2(\Omega')})(u_n|_{\Omega'}). \tag{4.30}$$

Since $J|_{L^2(\Omega')}$ is locally Lipschitz and $\{u_n\}$ converges in $L^2(\Omega')$, it results from (4.30) that $\{w_n\}$ is a bounded sequence in $L^2(\Omega')$. By hypothesis (H_3) it turns out that the imbedding $L^2(\Omega') \subset W^*$ is compact. Thus $\{w_n\}$ contains a convergent subsequence in W^*, which is denoted again by $\{w_n\}$. The continuity of the linear mapping $r : V \to W$ and relations (4.30), (4.25) imply the convergence of the subsequence $\{w_n\}$ in V^*. The duality mapping $\Lambda : V \to V^*$ is a topological linear isomorphism. Thus we obtain from (4.29) that

$$\Lambda^{-1}Au_n - \lambda u_n \text{ converges in } V, \tag{4.31}$$

where $\{u_n\}$ denotes the subsequence of $\{u_n\}$ corresponding, due to (4.30), to the subsequence $\{w_n\}$. We fix now some $\lambda \in \mathbb{R}$ satisfying (4.31). Then, using Schwarz inequality we can write the relation

$$\begin{aligned}
&\Big(\inf_{\substack{u\in V\\ u\neq 0}} \frac{a(u,u)}{\|u\|_V^2} - \lambda\Big)\|u_n - u_m\|_V^2 \\
&\quad\leq a(u_n - u_m, u_n - u_m) - \lambda\|u_n - u_m\|_V^2 \\
&\quad\leq \|\Lambda^{-1}A(u_n - u_m) - \lambda(u_n - u_m)\|_V \, \|u_n - u_m\|_V,
\end{aligned} \tag{4.32}$$

holding for all n and m integers. From (4.28),(4.31) and (4.32) we obtain that $\{u_n\}$ converges in V. □

Now we can prove the first main result of this section.

Theorem 4.1 Suppose that hypotheses (H_1), (H_2), (H_3) are verified. Let $C(\Omega)$ be a positive constant such that

$$\|u\|_{L^2(\Omega)} \leq C(\Omega)\|u\|_V \quad \forall u \in V. \tag{4.33}$$

Then, for every $\lambda \in \mathbb{R}$ satisfying

$$\lambda < \inf_{\substack{u\in V\\ u\neq 0}} \frac{a(u,u)}{\|u\|_V^2} - c_2(C(\Omega))^2, \quad c_2 > 0, \tag{4.34}$$

where c_2 is the same as in (H_1), there exists a point $u_0 \in V$ such that

$$\inf_{u\in V} I_\lambda(u) = I_\lambda(u_0). \tag{4.35}$$

Thus the point $u_0 \in V$ is a critical point of the functional $I_\lambda : V \to \mathbb{R}$, and the pair (u_0, λ) is a solution of problem (P).

Proof. The constant $C(\Omega)$ exists because of assumption (H_2). First we prove that $I_\lambda : V \to \mathbb{R}$ is bounded from below. Let λ be a real number verifying (4.34). Then

$$\gamma = \inf_{\substack{u\in V\\ u\neq 0}} \frac{a(u,u)}{\|u\|_V^2} - c_2(C(\Omega))^2 - \lambda > 0. \tag{4.36}$$

Relations (4.20), (4.33), (4.36), assumption (H_1) and the Schwarz inequality imply that

$$\begin{aligned} I_\lambda(u) &\geq \frac{1}{2}(\gamma + c_2(C(\Omega))^2)\|u\|_V^2 - \int_\Omega \Big|\int_0^{u(x)} \beta(t)dt\Big|dx \\ &\geq \frac{1}{2}\gamma\|u\|_V^2 - c_1(\text{mes}\,(\Omega))^{1/2}C(\Omega)\|u\|_V \qquad \forall u \in V. \end{aligned} \tag{4.37}$$

From (4.37) the boundedness from below of the functional $I_\lambda : V \to \mathbb{R}$ results. Let us now prove that if λ satisfies (4.34), the corresponding functional $I_\lambda : V \to \mathbb{R}$ verifies the Palais Smale condition. To prove this let $\{u_n\}$ be a sequence in V such that $\{I_\lambda(u_n)\}$ is bounded and

$$\min_{w\in\partial I_\lambda(u_n)} \|w\|_{V^*} \to 0 \quad \text{as } n \to \infty. \tag{4.38}$$

We may show that the boundedness of $\{I_\lambda(u_n)\}$ in $\mathbb{R}$ implies the boundedness of $\{u_n\}$ in V: indeed, let $M > 0$ denote a constant such that

$$I_\lambda(u_n) \leq M \quad \forall n. \tag{4.39}$$

Then from (4.37) and (4.39) we obtain the existence of a positive constant b such that the following inequality holds

$$M \geq \frac{1}{2}\gamma\|u_n\|_V^2 - b\|u_n\|_V \quad \forall n. \tag{4.40}$$

Because of (4.36) inequality (4.40) implies that the sequence $\{u_n\}$ is bounded in V. From (4.38) we obtain that (4.29) holds. Thus Proposition 4.5 implies the existence of a subsequence of $\{u_n\}$ converging in V. Accordingly $I_\lambda : V \to \mathbb{R}$ satisfies the Palais-Smale condition.

We proved that, if $\lambda \in \mathbb{R}$ verifies (4.34), $I_\lambda : V \to \mathbb{R}$ satisfies all the hypotheses of Proposition 4.1. This Proposition implies the existence of a point $u_0 \in V$ satisfying (4.35); thus u_0 is a critical point of I_λ. Proposition 4.4 then implies that the pair (u_0, λ) is a solution of problem P. □

4.3 The Existence of Solutions of the Eigenvalue Problem in $L^{2+d}(\Omega)$

With a modification in the assumptions (H_1), (H_2) and (H_3) we can prove the existence of minimax solutions of the eigenvalue problem in $L^{2+d}(\Omega)$ with $d > 0$. We assume that:

(H_1') the function $\beta \in L^\infty_{\text{loc}}(\mathbb{R})$ satisfies the growth condition

$$|\beta(t)| \leq c_1 + c_2|t|^{1+d} \quad \forall t \in \mathbb{R}, \tag{4.41}$$

where $d > 0$ and c_1, $c_2 \geq 0$ are constants independent of $t \in \mathbb{R}$. (Note that (4.19) implies the validity of (4.41) with the appropriate changes in constants c_1, c_2).

(H_2') Ω is an open bounded subset of the space $\mathbb{R}^n$, V is a real Hilbert space which is continuously imbedded as a dense linear subspace of $L^{2+d}(\Omega)$, and $a : V \times V \to \mathbb{R}$ is a continuous symmetric bilinear form on V.

(H_3') Ω' is a subdomain of Ω such that $W = \{u|_{\Omega'} \;|u \in V\}$ is a reflexive Banach space with the property that the imbedding $W \subset L^{2+d}(\Omega')$ is compact and the restriction map $r : V \to W$, $r(u) = u|_{\Omega'}$ $\forall u \in V$ is continuous.

Hypothesis (H_1') implies that the functional J introduced in (4.15) is defined in $L^{2+d}(\Omega)$. Furthermore, taking $\sigma = 1 + d$ in Proposition 4.3 we find that, if $w \in \partial J(u)$ with $u \in L^{2+d}(\Omega)$ then

$$w(x) \in [\underline{\beta}(u(x)), \overline{\beta}(u(x))] \text{ a.e. for } x \in \Omega' \tag{4.42}$$

where $\underline{\beta}$ and $\overline{\beta}$ are defined by (4.17). Here we have to study the problem (P) in the space V on the assumptions (H_2'), (H_3'). Since we work in the more general setting of hypothesis (H_1'), the approach of the previous section does not hold. Therefore we proceed by using the minimax Proposition 4.2 which extends a smooth version which can be found in Rabinowitz ([24], Chap. 2).

(H_3') implies that $I_\lambda : V \to \mathbb{R}, \lambda \in \mathbb{R}$, defined again as in (4.20) is locally Lipschitz. The result below shows that Proposition 4.4 remains true.

Proposition 4.6 Suppose that (H_1'), (H_2'), (H_3') hold. If $u_0 \in V$ is a critical point of $I_\lambda : V \to \mathbb{R}$ for some $\lambda \in \mathbb{R}$, then the pair $(u_0, \lambda) \in V \times R$ is a solution of problem (P).

Proof. Hypothesis (H_1') guarantees that in place of (4.23) the following inequality holds

$$|w| \leq \max(|\underline{\beta}(t)|, |\overline{\beta}(t)|) \leq c_1 + c_2|t|^{1+d} \quad \forall t \in \mathbb{R} \text{ and } w \in \partial j(t), \tag{4.43}$$

where $j : \mathbb{R} \to \mathbb{R}$ is the function defined in (4.2). Applying Theorem 1.3 (cf. also [4] p.83, Theorem 2.7.5) we may write that

$$\partial(J|_{L^{2+d}(\Omega')})(u) \subset \int_{\Omega'} \partial j(u(x))dx \;\; \forall u \in L^{2+d}(\Omega'). \tag{4.44}$$

By means of (H_2'), (H_3'), and with the same arguments as in Proposition 4.4 we can verify (4.25) and the inclusion

$$\partial(J|_W)(u) \subset \partial(J|_{L^{2+d}(\Omega')})(u) \;\; \forall u \in V. \tag{4.45}$$

Since $u_0 \in V$ is a critical point of I_λ, we obtain from the above relations that the inequality (4.27) holds for each $v \in V$. □

Further, a proposition analogous to Proposition 4.5 holds.

Proposition 4.7 Assume that (H_1'), (H_2'), (H_3') hold, and let λ be a real number satisfying (4.28). If $\{u_n\}$ is a bounded sequence in V such that there exists a sequence $w_n \in \partial(J|_V)(u_n)$ in V^* satisfying (4.29), then $\{u_n\}$ has a convergent subsequence in V.

Proof. The proof is analogous to the proof of Proposition 4.5 with the difference that the imbedding $V \subset L^2(\Omega)$ is replaced by $V \subset L^{2+d}(\Omega)$ and (H_1'), (H_2'), (H_3') are applied instead of (H_1), (H_2), (H_3). □

Now we can prove the following theorem .

Theorem 4.2 In addition to the assumptions (H_1'), (H_2'), (H_3') we suppose that

(i) $\lim_{t \to 0} \beta(t)t^{-1} = 0;$ (4.46)

(ii) there exist real numbers $c_1 > 0$ and c_2 such that

$$\int_0^t \beta(s)ds \leq -c_1|t|^{2+d} + c_2 \;\; \text{for } t < 0, \tag{4.47}$$

where $d > 0$ is the same constant as in (H_1');

(iii) there exist real numbers $\mu > 2$ and $n_0 > 0$ such that

$$\int_0^t \beta(s)ds \geq \mu^{-1} \max(\overline{\beta}(t)t, \underline{\beta}(t)t) \quad \forall |t| > n_0. \tag{4.48}$$

Then for every $\lambda \in \mathbb{R}$ satisfying (4.28) and every constant c such that

$$0 < c < 2^{-1}(\inf_{\substack{u \in V \\ u \neq 0}} (a(u,u)/\|u\|_V^2) - \lambda), \tag{4.49}$$

we obtain that the locally Lipschitz functional $I_\lambda : V \to \mathbb{R}$ has a critical point $u_\lambda \in V$ such that $I_\lambda(u_\lambda) \geq c$. In particular problem (P) has a nontrivial solution u_λ in V for every $\lambda \in \mathbb{R}$ verifying (4.28).

Proof. Fix some $\varepsilon > 0$. Due to (i), there is $\delta = \delta(\varepsilon) > 0$ such that $|\beta(t)| \leq \varepsilon|t|$ for each $|t| \leq \delta$. Then the function $j : \mathbb{R} \to \mathbb{R}$ defined in (4.2) satisfies the relation

$$|j(t)| \leq \frac{1}{2}\varepsilon|t|^2 \quad \text{for } |t| \leq \delta. \tag{4.50}$$

From (H_1') we can determine a constant $A(\delta) > 0$ such that

$$|j(t)| \leq A(\delta)|t|^{2+d} \quad \text{for } |t| > \delta. \tag{4.51}$$

From (4.50) and (4.51) it results that

$$|j(t)| \leq (\frac{1}{2}\varepsilon + A(\delta)|t|^d)t^2 \quad \forall t \in \mathbb{R}. \tag{4.52}$$

Inequality (4.52) yields the following estimate

$$\begin{aligned} |J(u)| &\leq \frac{\varepsilon}{2}\|u\|^2_{L^2(\Omega)} + A(\delta)\|u\|^{2+d}_{L^{2+d}(\Omega)} \\ &\leq C(\frac{\varepsilon}{2} + A(\delta)\|u\|_V^d)\|u\|_V^2 \qquad \forall u \in V, \end{aligned} \tag{4.53}$$

where $C > 0$ denotes a constant depending only on the continuous imbedding $V \subset L^{2+d}(\Omega) \subset L^2(\Omega)$. If we take in (4.53)

$$\|u\|_V < (\frac{\varepsilon}{2A(\delta)})^{1/d},$$

we obtain

$$|J(u)| \leq \varepsilon C\|u\|_V^2,$$

and thus

$$\lim_{u \to 0}(J(u)/\|u\|_V^2) = 0. \tag{4.54}$$

One can write that(cf. eq. (4.20))

$$I_\lambda(u) \geq \frac{1}{2}(\inf_{\substack{u\in V\\ u\neq 0}} \frac{a(u,u)}{\|u\|_V^2} - \lambda)\|u\|_V^2 + (J|_V)(u). \tag{4.55}$$

Relation (4.54) implies that

$$I_\lambda(u) \geq c \quad \text{for } \|u\|_V \quad \text{sufficiently small.} \tag{4.56}$$

Hence condition (i) in Proposition 4.2 is satisfied in the case of the functional $I_\lambda : V \to \mathbb{R}$. In order to check the condition (ii) of the same proposition let us fix some function $u \in V$, such that $u(x) > 0$ a.e. in Ω. Hypothesis (H_2') assures the existence of such function u. Then, by assumption (4.47) we obtain

$$\begin{aligned} I_\lambda(tu) \;&\leq\; \frac{1}{2}(a(u,u) - \lambda\|u\|_V^2)t^2 \\ &-c_1 \int_\Omega u(x)^{2+d}dx|t|^{2+d} + c_2(\text{meas}(\Omega), \quad \forall t < 0. \end{aligned} \tag{4.57}$$

Because $c_1 > 0$, it turns out from (4.57) that $I(tu) \to -\infty$ as $t \to -\infty$. This property shows that condition (ii) of Proposition 4.2 is verified for the functional $f = I_\lambda : V \to \mathbb{R}$.

We verify now that for every $\lambda \in \mathbb{R}$ as in (4.28) the corresponding functional $I_\lambda : V \to \mathbb{R}$ satisfies the Palais-Smale condition: let $\{u_n\}$ be a sequence in V provided there is a constant $M > 0$ such that

$$|I_\lambda(u_n)| \leq M \quad \text{for every } n \tag{4.58}$$

and a sequence $w_n \in \partial(J|_V)(u_n)$ satisfying (4.29). From (4.58) and (4.29) one can verify that for each integer n sufficiently large the following inequality holds

$$I_\lambda(u_n) - \mu^{-1}(a(u_n,u_n) - \lambda\|u_n\|_V^2 + \langle w_n, u_n\rangle) \leq M + \mu^{-1}\|u_n\|_V. \tag{4.59}$$

where $\langle\cdot;\cdot\rangle$ denotes the duality pairing We replace in the left-hand side of (4.59) $I_\lambda(u_n)$ by its expression in (4.20) and we take into account (4.42). It results that there exists a constant K such that

$$\begin{aligned} &(\frac{1}{2} - \frac{1}{\mu})(a(u_n,u_n) - \lambda\|u_n\|_V^2) + \\ &+ \int_{|u_n(x)|\geq n_0} \left(\int_0^{u_n(x)} \beta(t)dt - \frac{1}{\mu}\max(\underline{\beta}(u_n(x))u_n(x), \overline{\beta}(u_n(x))u_n(x)) \right) dx \\ &\leq M + \frac{1}{\mu}\|u_n\|_V + K \qquad \text{for all positive integers } n. \end{aligned} \tag{4.60}$$

From (4.60), (4.48), and (4.28) we obtain that $\{u_n\}$ is bounded in V. Then Proposition 4.7 implies that one can extract from $\{u_n\}$ a convergent subsequence in V. The Palais-Smale condition holds for $I_\lambda : V \to \mathbb{R}$, where λ satisfies (4.28). Consequently, $I_\lambda : V \to \mathbb{R}$ satisfies the hypotheses of Proposition 4.2. It results that there exists a critical point $u_\lambda \in V$ of $I_\lambda : V \to \mathbb{R}$ such that

$$I_\lambda(u_\lambda) = \inf_{\gamma\in\Gamma} \max_{t\in[0,1]} I_\lambda(\gamma(t)) \tag{4.61}$$

where

$$\Gamma = \{\gamma \in C([0,1],V) \,\big|\, \gamma(0) = 0 \text{ and } \gamma(1) = e\} \tag{4.62}$$

for some fixed point $e \in V$ far enough from the origin ($e = tv$ for $t > 0$ sufficiently large). This minimax characterization of $I_\lambda(u_\lambda)$ and relation (4.56) show that $I_\lambda(u_\lambda) \geq c$. Finally, Proposition 4.6 implies that $u_\lambda \in V$ is a nontrivial solution of problem (P). □

Note that Theorem 4.1 holds also for the following case.

Let $u \in V \subset [L^2(\Omega)]^n$ and $L : V \to L^2(\Omega')$, $Lu = \hat{u}$ ($\hat{u}(x) \in \mathbb{R}$) is a linear continuous mapping. Then we consider instead of (4.1) the problem (Problem P_1): Find $u \in V$, $\lambda \in \mathbb{R}$ such that (cf. the abstract problem of Section 2.4)

$$a(u,v) + \int_{\Omega'} j^0(\hat{u},\hat{v})d\Omega \geq \lambda(u,v)_V \qquad v \in V. \tag{4.63}$$

For problem (P_1) we can prove a theorem analogous to Theorem 4.1. Then (H_3) must be replaced by $(\tilde{H}_3)$, whereas (H_1) and (H_2) remain as in the case of Theorem 4.1.

$(\tilde{H}_3)$ Ω' is a subdomain of Ω such that $W = \{Lu|_{\Omega'} : u \in V\}$ is a closed subspace in $L^2(\Omega')$ with the property that the imbedding $W \subset L^2(\Omega')$ is compact.

Then Theorem 4.1 holds, with $C(\Omega)$ replaced by $\tilde{C}(\Omega)$, where

$$\|\hat{u}\|_{L^2(\Omega')} \leq \tilde{C}(\Omega)\|u\|_V \qquad u \in V. \tag{4.64}$$

Analogously we proceed in the case of the eigenvalue problem (P_2) resulting from (4.63) by replacing $\int_{\Omega'} j^0(\hat{u},\hat{v})d\Omega$ by $\int_\Gamma j^0(\hat{u},\hat{v})d\Gamma$, where Γ is the boundary of Ω (or a part of it) and $L : u \to \hat{u} : V \to L^2(\Gamma)$ ($\hat{u}(x) \in \mathbb{R}$) is assumed to be linear and compact. Then (H_3) must be replaced by $(\tilde{H}_3')$:

$(\tilde{H}_3')$ $W = \{Lu = \hat{u} \in L^2(\Gamma),\ u \in V\}$ is a closed subspace of $L^2(\Gamma)$ Theorem 4.1 holds also with $C(\Omega)$ replaced by $\tilde{\tilde{C}}(\Omega)$ where

$$\|\hat{u}\|_{L^2(\Gamma)} \leq \tilde{\tilde{C}}(\Omega)\|u\|_V \qquad u \in V. \tag{4.64a}$$

Note that the foregoing results do not need for their proofs the coercivity of the bilinear form $a : V \times V \to \mathbb{R}$ but they need the growth assumption (4.19) or (4.41).

The regularization approach to the eigenvalue problem of hemivariational inequalities [7],[8] allows to treat cases not covered by the growth assumptions used here, but leads only to the existence of at least one solution and not to minimax properties.

4.4 Eigenvalue Problem for a Hemivariational Inequality involving a Nonlinear Compact Operator

Let V be a real Hilbert space with the scalar product $(\cdot,\cdot)$ and the associated norm $||\cdot||$ which is densely and compactly imbedded into $L^p(\Omega;\mathbb{R})$ for some $p \geq 2$. Here Ω is a bounded domain in $\mathbb{R}^n$. Let us consider a continuous symmetric bilinear form $\alpha : V \times V \to \mathbb{R}$ not necessarily coercive, a nonlinear compact operator $C : V \to V$ and a function $\beta : \mathbb{R} \to \mathbb{R}$ belonging to $L^\infty_{\text{loc}}(\mathbb{R})$. We introduce further the locally Lipschitz function j by (4.2).

We formulate the following eigenvalue problem for a hemivariational inequality: find $u \in V$ and $\lambda \in \mathbb{R}$ such that (problem (P))

$$\alpha(u,v) + (C(u),v) + \int_\Omega j^0(u(x);v(x))dx \geq \lambda(u,v) \quad \text{for all } v \in V. \tag{4.65}$$

The motivation to study problem (P) comes from Mechanics, precisely from the buckling theory of adhesively connected von Kármán plates allowing for delamination effects. The equilibrium problem for such plates has been studied in [15][16] and the corresponding dynamic problems in [14].

The aim of the present section is to solve the eigenvalue problem (P) by critical point arguments. This is not straightforward because of the nonlinear operator $C : V \to V$ which makes the inequality (P) to be of a non potential type. To deal with this Ekeland's variational principle [5] is combined with a technique suggested by the nonsmooth version of Palais-Smale minimization theorem as developed by Chang [2]. We mention that the case $C = 0$ in (P) has been treated in the previous Section.

The advantages of this approach are the following: (i) our hypotheses are weaker and much easier to be verified than the usual ones even of the classical convex case. Indeed we do not demand any pseudomonotonicity, tangency, homogeneity and coerciveness assumptions (see e.g. hypotheses (i)-(vii) in Panagiotopoulos [17], p. 229 for comparison where the eigenvalue problem of variational inequalities arising in the von Kármán plate theory is studied); (ii) we prove the existence of eigenvalues on the whole real-line or on a whole half-line, not of just only one as it was previously obtained even for the convex case; in addition our result ensures more regularity for the solutions since we search them in $L^p(\Omega;\mathbb{R})$. (iii) the method exploits in a complete way the inequality form of problem (P).

$C_p(\Omega)$ denotes in this section a positive constant given by the continuous injection $V \subset L^p(\Omega)$

$$||u||_{L^p} \leq C_p(\Omega)||u|| \quad \text{for all } u \in V. \tag{4.66}$$

Let $\alpha : V \times V \to \mathbb{R}$ be a continuous symmetric bilinear form without any coerciveness assumption. Let $C : V \to V$ be a (nonlinear) compact operator satisfying:

(H_1) there exist a continuously differentiable function $g : V \to \mathbb{R}$ and a lower semicontinuous convex function $f : V \to \mathbb{R}$ such that

$$(C(u), v-u) \geq (g'(u), v-u) + f(v) - f(u) \quad \text{for all } u, v \in V.$$

Let the function $\beta \in L^\infty_{\text{loc}}(\mathbb{R})$ satisfy:

(H_2) there exist constants $\alpha_1, \alpha_2 \in \mathbb{R}$ with $\alpha_2 > 0$ such that the following growth condition holds

$$|\beta(t)| \leq \alpha_1 + \alpha_2 |t|^{p-1} \quad \text{for all } t \in \mathbb{R}.$$

Concerning the functions f and g we assume in addition that

(H_3) there exist constants $\beta_1, \beta_2 \in \mathbb{R}$ with $\beta_2 > (1/p)\alpha_2(C_p(\Omega))^p$ such that

$$f(u) + g(u) \geq \beta_1 + \beta_2 ||u||^p \quad \text{for all } u \in V.$$

We state the following existence result for solutions of the eigenvalue problem (P).

Theorem 4.3 Assume that the hypotheses $(H_1), (H_2), (H_3)$ hold. Let $p > 2$. Then for every $\lambda \in \mathbb{R}$ which is in the resolvent set of the bilinear form a, there exists $u \in V$ satisfying problem (P). Let $p = 2$. Then for every $\lambda \in \mathbb{R}$ with $\lambda < 2\beta_2 - \alpha_2(C_2(\Omega))^2 - ||a||$ which belongs to the resolvent set of a, there exists $u \in V$ solving (P).

Proof. Let us fix some $\lambda \in \mathbb{R}$ as in the statement of the theorem. Then the linear mapping $u \in V \to a(u, \cdot) - \lambda(u, \cdot) \in V^\star$ is a topological isomorphism. Corresponding to this λ let $h : V \to \mathbb{R}$ be the functional defined by

$$h(u) = \frac{1}{2}a(u,u) + \int_\Omega j(u(x))dx - \frac{\lambda}{2}||u||^2 + g(u), \quad u \in V, \tag{4.67}$$

and let the functional $I : V \to \mathbb{R}$ given by

$$I(u) = h(u) + f(u), \quad u \in V. \tag{4.68}$$

We introduce also the functional $J : L^p(\Omega) \to \mathbb{R}$

$$J(u) = \int_\Omega j(u(x))dx, \quad u \in L^p(\Omega). \tag{4.69}$$

Because of (H_2) J is well-defined and is locally Lipschitz on $L^p(\Omega)$, or on V (cf. [2]).

To prove the theorem we proceed in several steps.

First we show that the functional I is bounded from below on V, thus there exists

$$c = \inf_V I > -\infty. \tag{4.70}$$

Indeed, from hypotheses $(H_2), (H_3)$ and (4.66) one obtains the estimate

$$\begin{aligned} I(u) &\geq -\frac{1}{2}||\alpha||||u||^2 - \alpha_1(\text{meas}(\Omega))^{1/2}||u||_{L^2} \\ &\quad -\frac{1}{p}\alpha_2||u||^p_{L^p} - \frac{\lambda}{2}||u||^2 + g(u) + f(u) \\ &\geq (\beta_2 - \frac{1}{p}\alpha_2(C_p(\Omega)^p)||u||^p - \frac{1}{2}(||\alpha|| + \lambda)||u||^2 \\ &\quad -\alpha_1(\text{meas}(\Omega_1))^{1/2}C_2(\Omega)||u|| + \beta_1 \qquad \text{for all } u \in V. \end{aligned} \tag{4.71}$$

The relations between the constants involved in (4.71) imply (4.70). Further we show that there exists a sequence $\{u_n\}$ in V with the properties

$$I(u_n) \leq c + \frac{1}{n} \quad \text{for all } n. \tag{4.72}$$

and

$$a(u_n, v) - \lambda(u_n, v) + (C(u_n), v) + J^0(u_n; v) \geq -\frac{1}{n}||v|| \quad \text{for all } v \in V. \tag{4.73}$$

In order to prove it we apply to $I : V \to \mathbb{R}$ Ekeland's variational principle [5]. This is possible because I is lower semicontinuous and bounded from below on V due to (4.70). According to Ekeland's result we obtain a sequence $\{u_n\}$ in V such that (4.72) is valid and the following inequality holds

$$I(w) - I(u_n) \geq -\frac{1}{n}||w - u_n|| \quad \text{for all } w \in V. \tag{4.74}$$

Taking $w = (1-t)u_n + tv$ for every $0 < t < 1$ and $v \in V$, (4.74) yields

$$h(u_n + t(v - u_n)) - h(u_n) + f((1-t)u_n + tv) - f(u_n) \geq -\frac{t}{n}||v - u_n||$$

$$\text{for all } v \in V, \quad 0 < t < 1, \text{ and } n \text{ positive integer.}$$

Then the convexity of function $f : V \to \mathbb{R}$ implies

$$\frac{1}{t}(h(u_n + t(v - u_n)) - h(u_n)) + f(v) - f(u_n) \geq -\frac{1}{n}||v - u_n|| \tag{4.75}$$

$$\text{for all } v \in V, \ 0 < t < 1 \text{ and } n.$$

Passing to limit for $t \to 0$ in (4.75) and using the continuous differentiability of $g : V \to \mathbb{R}$ we obtain

$$\begin{aligned} &a(u_n, v - u_n) - \lambda(u_n, v - u_n) + J^0(u_n; v - u_n) + \\ &+(g'(u_n), v - u_n) + f(v) - f(u_n) \geq -\frac{1}{n}||v - u_n|| \end{aligned} \tag{4.76}$$

$$\text{for all } v \in V \text{ and } n.$$

Hypothesis (H_1) then implies that relation (4.73) is true.

We show now that corresponding to the sequence $\{u_n\} \subset V$ constructed in the previous step, i.e. in (4.72),(4.73), there exist sequences $\{q_n\}$ and $\{w_n\}$ in $V^\star$ such that $q_n \to 0$ in $V^\star$, $w_n \in \partial(J|_V)(u_n)$ and

$$-\alpha(u_n, \cdot) + \lambda \Lambda u_n - \Lambda C(u_n) + q_n = w_n \quad \text{for all } n, \tag{4.77}$$

where ∂ denotes Clarke's generalized gradient and $\Lambda : V \to V^\star$ the duality mapping $\Lambda u(v) = (u, v)$ for all $u, v \in V$.

Since the left-hand side of inequality (4.73) is with respect to v a lower semicontinuous convex function on V vanishing at 0, Lemma 1.3 in Szulkin [27] shows that one can find $z_n \in V^\star$ with $||z_n|| \leq 1$ so that

$$\alpha(u_n, v) - \lambda(u_n, v) + (C(u_n), v) + J^0(u_n; v) \geq \frac{1}{n}(z_n, v) \quad \text{for all } v \in V. \tag{4.78}$$

By choosing $q_n = n^{-1} z_n$ we have that $q_n \to 0$ in $V^\star$ and

$$-\alpha(u_n, \cdot) + \lambda \Lambda u_n - \Lambda C(u_n) + q_n \in \partial(J|_V)(u_n) \quad \text{for all } n.$$

Thus we have shown that there exists $w_n \in \partial(J|_V)(u_n)$ satisfying (4.77).

Further we prove that the sequence $\{u_n\} \subset V$ of the two previous steps possesses a convergent subsequence in V.

First we prove that $\{u_n\}$ is bounded in V. This results easily from relations (4.71), (4.72) and the imposed conditions upon the constants α_2, β_2. To derive the convergence of a subsequence of $\{u_n\}$ we consider (4.77). Here the boundedness of $\{u_n\}$ in V and the compactness of mapping $C : V \to V$ implies the covergence of $C(u_n)$ along a subsequence in V denoted again by $C(u_n)$. The compactness of the imbedding $V \subset L^p(\Omega; \mathbb{R})$ assures that a subsequence of $\{u_n\}$ also denoted by $\{u_n\}$ converges in $L^p(\Omega; \mathbb{R})$. On the other hand the density of V in $L^p(\Omega; \mathbb{R})$ implies that

$$w_n \in \partial(J|_V)(u_n) \subset \partial J(u_n) \quad \text{for all } n \tag{4.79}$$

(cf. Corollary 1.2). The fact that J is locally Lipschitz on $L^p(\Omega; \mathbb{R})$ and (4.79) ensure that $\{w_n\}$ is bounded in $L^p(\Omega; \mathbb{R})$. The compactness of the imbedding $(L^p(\Omega; \mathbb{R}))^\star \subset V^\star$ implies that $\{w_n\}$ converges along a subsequence in $V^\star$. For $n \to \infty$ in (4.77) we obtain the convergence along a subsequence of $\{-\alpha(u_n, \cdot) + \lambda \Lambda u_n\}$ in $V^\star$. Due to the fact that λ does not belong to the spectrum of the symmetric bilinear form $\alpha : V \times V \to \mathbb{R}$ we can conclude that a subsequence of $\{u_n\}$ converges in V.

We show now that $u = \lim_{n \to \infty} u_n \in V$, which is limit of the convergent subsequence $\{u_n\}$, satisfies the inequality

$$\alpha(u, v) - \lambda(u, v) + (C(u), v) + J^0(u; v) \geq 0 \quad \text{for all } v \in V. \tag{4.80}$$

The continuity of $h : V \to \mathbb{R}$ and relation (4.72) imply the existence of $\lim_{n \to \infty} f(u_n)$. We have also that

$$\lim_{n\to\infty} f(u_n) = \lim_{n\to\infty} (I(u_n) - h(u_n)) = c - h(u).$$

Then the lower semicontinuity of $f : V \to \mathbb{R}$ allows us to write

$$f(u) \le \liminf_{n\to\infty} f(u_n) = \lim_{n\to\infty} f(u_n) = c - h(u).$$

It results that

$$I(u) = h(u) + f(u) = c = \inf_V I. \tag{4.81}$$

To verify (4.80) we see first that the following inequality holds

$$\begin{aligned} \alpha(u, v-u) \quad &- \quad \lambda(u, v-u) + J^0(u; v-u) + \\ &+ \quad (g'(u), v-u) + f(v) - f(u) \ge 0 \qquad \text{for all } v \in V. \end{aligned} \tag{4.82}$$

Due to (4.81) we may write

$$I(u + t(v-u)) \ge I(u) \quad \text{for all } v \in V \text{ and } 0 < t < 1. \tag{4.83}$$

The convexity of $f : V \to \mathbb{R}$ implies that

$$\frac{1}{t}(h(u + t(v-u)) - h(u) + f(v) - f(u) \ge 0 \tag{4.84}$$

$$\text{for all } v \in V \text{ and } 0 < t < 1.$$

For $t \to 0$ the inequality (4.82) is obtained, which together with hypothesis (H_1) implies that inequality (4.80) holds.

We prove now that the element $u = \lim_{n\to\infty} u_n$ of V is a solution of problem (P) corresponding to the fixed number $\lambda \in \mathbb{R}$.

For the generalized gradient of $J|_V$ the following relation holds

$$\partial(J|_V)(y) \subset \int_\Omega \partial j(w(x))dx \quad \text{for all } y \in V. \tag{4.85}$$

Indeed, if $w \in \partial J(u)$, then $w \in (L^p(\Omega))^* = L^{p/(p-1)}(\Omega)$ and

$$w(x) \in [\underline{\beta}(u(x)), \bar{\beta}(u(x))], \quad \text{for a.e. } x \in \Omega,$$

with $\underline{\beta}$ and $\bar{\beta}$ given by (4.17) (cf. Proposition 1.7). Then (H_2) implies that

$$|z| \le \max(\underline{\beta}(t), \bar{\beta}(t)) \le \alpha_1 + \alpha_2 |t|^{p-1} \quad \text{for all } t \in \mathbb{R} \text{ and } z \in \partial j(t). \tag{4.86}$$

This estimate shows that (1.14) is verified. Hence Theorem 1.3 and inclusion (4.79) with u_n replaced by y imply (4.85).

From (4.80) and (4.85), recalling the definition of the generalized gradient and Proposition 1.4 (ii), we conclude that $u = \lim_{n\to\infty} u_n$ satisfies the inequality

$$\begin{aligned} -\alpha(u,v) + \lambda(u,v) - (C(u), v) \;&\le\; \int_\Omega \max_{z\in\partial j(u(x))} z(v(x))dx \\ &=\; \int_\Omega j^0(u(x); v(x))dx \quad \text{for all } v \in V. \end{aligned} \tag{4.87}$$

Consequently, $u = \lim_{n\to\infty} u_n$ is a solution of the hemivariational eigenvalue problem (P) corresponding to the given real number λ. This completes the proof of the theorem. □

The solution $u \in V$ of problem (P) is obtained by a limiting process in V. Thus the proof we have given may serve as an archetype for the convergence proof of certain types of numerical approximations. Additionally, the solution is characterized as a minimum point of an appropriate functional I on V (cf. the relation (4.81)).

Note that the given proof is not exactly a critical point method but it can be characterized in fact as a "subcritical" one. The technique has been inspired from the work of Szulkin [27]. The minimax methods (cf. e.g. Rabinowitz [25]) cannot be applied here.

A specific feature of our approach is that we made use of the inequality form of problem (P) obtaining in fact the solution for an entire class of other eigenvalue problems. Following essentially the same lines we can deal with the eigenvalue problems (P), where the integral is taken on a subdomain $\Omega' \subset \Omega$. We consider here the case $\Omega' = \Omega$ for the sake of simplicity.

The case $C = 0$ has been studied in [11] (cf. also Sect. 4.2, 4.3). Note that the method of proof may be applied for the study of hemivariational inequalities with more general nonlinear operators (cf. also [10]).

It is worth-noting that in the proof of Theorem 4.3 we have obtained an existence result for the variational-hemivariational eigenvalue problem: find $u \in V$ and $\lambda \in \mathbb{R}$ such that

$$a(u, v-u) + \int_\Omega j^0(u(x); v(x))dx + f(v) - f(u) \geq \lambda(u, v-u), \quad \forall u \in V, \tag{4.88}$$

(see [14],[12],[18] for such problems).

Note that Theorem 4.3 holds also for the following case. We assume that $p = 2$ and let $u \in V \subset [L^2(\Omega)]^n$. Let $L : V \to L^2(\Omega')$, $Lu = \hat{u}(\hat{u}(x) \in \mathbb{R})$ be a linear compact mapping. Then we consider instead of (4.65) the problem (P_1): Find $u \in V$, $\lambda \in \mathbb{R}$ such that

$$a(u,v) + (C(u), v) + \int_{\Omega'} j^0(\hat{u}, \hat{v})d\Omega \geq \lambda(u,v) \qquad \forall\, v \in V. \tag{4.89}$$

One can easily verify that Theorem 4.3 holds for $p = 2$, where $C_2(\Omega)$ is replaced by the constant $\tilde{C}(\Omega)$ of the inequality (4.64).

Analogously we proceed in the case of the eigenvalue problem (P_2) resulting from (4.89) by replacing $\int_{\Omega'} j^0(\hat{u}, \hat{v})d\Omega$ by $\int_\Gamma j^0(\hat{u}, \hat{v})d\Gamma$. Here Γ is the boundary of Ω, or a part of the boundary, and the linear operator $L : u \to \hat{u} : V \to L^2(\Gamma)(\hat{u}(x) \in \mathbb{R})$ is assumed to be compact. Again one can easily verify Theorem 4.3 for $p = 2$ where $C_2(\Omega)$ is replaced by the constant $\tilde{\tilde{C}}$ of (4.64a).

We leave the corresponding problems (P_1) and (P_2) for $p > 2$ as an exercise for the reader.

4.5 Applications

4.5.1 An Eigenvalue Inclusion Problem

The existence result of Section 4.4 leads also to the existence of solutions for the following nonlinear eigenvalue problem.

$$(- (\tilde{A} + C)u + \lambda u)(x) \in [\underline{\beta}(u(x)), \bar{\beta}(u(x))] \quad \text{for a.e. } x \in \Omega, \tag{4.90}$$

where $\tilde{A} : V \to V$ is the linear operator given by $(\tilde{A}u, v) = \alpha(u, v), u, v \in V$ (cf. Lefter and Motreanu [9] for this type of problems).

4.5.2 On the Buckling of Adhesively Connected von Kármán Plates

The delamination effect of adhesively connected plates is one of the main causes of their failure. The mechanical behaviour of the binding adhesive material is described by a nonmonotone possibly multivalued law between the bonding forces and the corresponding displacements. Such problems have been already studied by Panagiotopoulos in [15],[16],[19],[20]. Here we study the corresponding buckling problem and therefore a large deflection plate theory, i.e. the von Kármán plate theory, must be taken into account. In this section we consider only one plate adhesively bonded with a support.

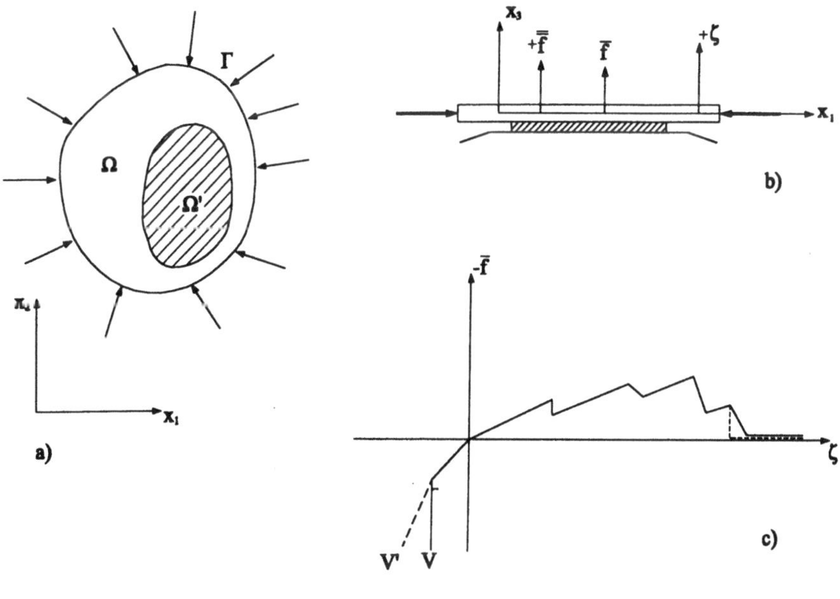

Fig. 4.1 Notation and geometry of the plate.

Consider a plate Ω and the binding material on $\Omega' \subset \Omega$ (Fig. 4.1a,b). In the undeformed state the middle surface of the plate occupies an open, bounded subset Ω of $\mathbb{R}^2$. It is referred to a fixed right-handed Cartesian coordinate system $Ox_1x_2x_3$. We denote by Γ the boundary of the plate which is assumed to be appropriately regular (in general, a Lipschitz boundary $C^{0,1}$ is sufficient). Let also the thin sheet of the binding material occupy a domain Ω' such that $\Omega' \subset \Omega$ and $\bar{\Omega}' \cap \Gamma = \emptyset$. We denote by $\zeta(x)$ the vertical deflection of the point $x \in \Omega$ of the plate, and by $f = (0, 0, f_3(x))$ the distributed vertical loading.

Further, let $u = \{u_1, u_2\}$ be the in-plane displacement of the plate. We assume that the plate has constant thickness h. Moreover, we assume that the plate obeys the von Kármán plate theory, i.e. it is a thin plate having large deflections. Then the following system of differential equations holds (the comma denotes the partial derivation):

$$K\Delta\Delta\zeta - h(\sigma_{\alpha\beta}\zeta_{,\beta})_{,\alpha} = f \quad \text{in } \Omega, \tag{4.91}$$

$$\sigma_{\alpha\beta,\beta} = 0 \quad \text{in } \Omega, \tag{4.92}$$

and

$$\sigma_{\alpha\beta} = C_{\alpha\beta\gamma\delta}(\varepsilon_{\gamma\delta}(u) + \frac{1}{2}\zeta_{,\gamma}\zeta_{,\delta}), \quad \varepsilon_{\alpha\beta} = \frac{1}{2}(u_{\alpha,\beta} + u_{\beta,\alpha}) \quad \text{in } \Omega. \tag{4.93}$$

Here the subscripts $\alpha, \beta, \gamma, \delta = 1, 2$ correspond to the coordinate directions; $\{\sigma_{\alpha\beta}\}$, $\{\varepsilon_{\alpha\beta}\}$, and $\{C_{\alpha\beta\gamma\delta}\}$ denote the stress, strain and elasticity tensors in the plane of the plate. The components of C are elements of $L^\infty(\Omega)$ and have the usual symmetry and ellipticity properties. Moreover, $K = Eh^3/12(1-\nu^2)$ is the bending rigidity of the plate with E the modulus of elasticity and ν the Poisson ratio. For the sake of simplicity, we consider here isotropic homogeneous plates of constant thickness.

In order to model the delamination effect f is split into $\bar{f}$, which describes the action of the adhesive and $\bar{\bar{f}} \in L^2(\Omega)$, which represents the external loading applied on the plate:

$$f = \bar{f} + \bar{\bar{f}} \quad \text{in } \Omega. \tag{4.94}$$

Further a phenomenological law is formulated connecting $\bar{f}$ with the corresponding deflection of the plate (see Fig. 4.1c). This law describes the action of the adhesive material [21].

We assume that

$$-\bar{f} \in \tilde{\beta}(\zeta) \quad \text{in } \Omega' \tag{4.95}$$

where $\tilde{\beta} = [\underline{\beta}, \overline{\beta}]$ is a multivalued function defined by (4.17) (i.e. roughly speaking by filling-in the jumps in the graph of a function $\beta \in L^\infty_{\text{loc}}(\mathbb{R})$). We note here that cracking as well as crushing effects of either a brittle or semi-brittle nature can be accounted for by means of this law. The impenetrability restriction would imply a vertical branch AV (Fig. 4.1c) in the final compression state of the binding layer. However a slightly inclined branch AV$'$ takes here into account the compressibility of the plate in the Ox_3-direction. To complete in a natural way the definition of $\bar{f}$ we pose

$$\bar{f} = 0 \quad \text{in } \Omega - \Omega'. \tag{4.96}$$

In order to obtain a variational formulation of the problem, we express relation (4.95) in a superpotential form [22][23]. If $\beta(\xi_{\pm 0})$ exists for every $\xi \in \mathbb{R}$ then (cf. [2],[14]) a locally Lipschitz (nonconvex) function $j : \mathbb{R} \to \mathbb{R}$ can be determined up to an additive constant, such that

$$\tilde{\beta}(\xi) = \partial j(\xi) \tag{4.97}$$

where ∂ is the generalized gradient of Clarke.

For the in-plane action we assume the boundary conditions

$$\sigma_{\alpha\beta} n_\beta = \lambda g_\alpha \quad \text{on } \Gamma \quad \alpha = 1,2 \tag{4.98}$$

where g_α are selfequilibrating forces. For the moment we assume that $g_\alpha = 0$ $\alpha = 1,2$.

We can now derive the variational formulation of the problem. From (4.91), by assuming sufficiently regular functions, multiplying by $z - \zeta$, integrating and applying the Green-Gauss theorem, we obtain the expressions:

$$\begin{aligned} &\alpha(\zeta, z-\zeta) + \int_\Omega h\sigma_{\alpha\beta}\zeta_{,\alpha}(z-\zeta)_{,\beta}\, d\Omega \\ &= \int_\Gamma h\sigma_{\alpha\beta}\zeta_{,\beta} n_\alpha (z-\zeta)\, d\Gamma + \int_\Omega \bar{\bar{f}}\,(z-\zeta)\, d\Omega \\ &\quad + \int_\Gamma K_n(\zeta)(z-\zeta)\, d\Gamma - \int_\Gamma M_n(\zeta)\frac{\partial(z-\zeta)}{\partial n}\, d\Gamma, \qquad \alpha,\beta = 1,2. \end{aligned} \tag{4.99}$$

Here n denotes the outward normal unit vector to Γ, K_n denotes the Kirchhoff shearing force at the boundary Γ and M_n the boundary bending moment, and the following relations hold

$$\alpha(\zeta, z) = K \int_\Omega [(1-\nu)\zeta_{,\alpha\beta} z_{,\alpha\beta} + \nu \Delta\zeta \Delta z] d\Omega, \qquad \alpha,\beta = 1,2,\ 0 < \nu < \frac{1}{2}, \tag{4.100}$$

$$M_n(\zeta) = -K[\nu \Delta\zeta + (1-\nu)(2n_1 n_2 \zeta_{,12} + n_1^2 \zeta_{,11} + n_2^2 \zeta_{,22})], \tag{4.101}$$

$$K_n(\zeta) = -K\left[\frac{\partial \Delta\zeta}{\partial n} + (1-\nu)\frac{\partial}{\partial \tau}[n_1 n_2(\zeta_{,22} - \zeta_{,11}) + (n_1^2 - n_2^2)\zeta_{,12}]\right]. \tag{4.102}$$

Here ν is the Poisson ratio and τ is the unit vector tangential to Γ, such that n, τ and the Ox_3-axis form a right-handed system. Applying the same technique to (4.92), implies the relation

$$\int_\Omega \sigma_{\alpha\beta}\varepsilon_{\alpha\beta}(v-u)\, d\Omega = \int_\Gamma \sigma_{\alpha\beta} n_\beta (v_\alpha - u_\alpha)\, d\Gamma, \quad \alpha,\beta = 1,2. \tag{4.103}$$

Further, the following notations are introduced:

$$R(m,k) = \int_\Omega C_{\alpha\beta\gamma\delta} m_{\alpha\beta} k_{\gamma\delta}\, d\Omega, \quad \alpha,\beta,\gamma,\delta = 1,2 \tag{4.104}$$

and

$$P(\zeta, z) = \{\zeta_{,\alpha} z_{,\beta}\}, \quad P(\zeta, \zeta) = P(\zeta), \tag{4.105}$$

where $m = \{m_{\alpha\beta}\}$ and $k = \{k_{\alpha\beta}\}, \alpha, \beta = 1, 2$ are 2×2 tensors. We assume that the plate has a simple hinge boundary condition, i.e.

$$\zeta = 0 \text{ and } M_n(\zeta) = 0 \text{ on } \Gamma \tag{4.106}$$

Let us also introduce a functional framework for the B.V.P. We assume that $u, v \in [H^1(\Omega)]^2$ and that $\zeta, z \in Z$, where

$$Z = \left\{ z \middle| z \in H^2(\Omega), \quad z = 0 \text{ on } \Gamma \right\}. \tag{4.107}$$

Taking into account the variational equalities (4.99), (4.103), the boundary conditions (4.98) with $g_a = 0$ and (4.106), and the inequalities defining the multivalued operator ∂ we obtain the following problem :
Find $u \in [H^1(\Omega)]^2$ and $\zeta \in Z$, $j = 1, 2$, such as to satisfy the hemivariational inequality

$$\alpha(\zeta, z - \zeta) + hR(\varepsilon(u) + \frac{1}{2}P(\zeta), P(\zeta, z - \zeta)) +$$

$$+ \int_{\Omega'} j^0(\zeta, z - \zeta) d\Omega \geq \int_\Omega \overline{\overline{f}} (z - \zeta) d\Omega, \quad \forall z \in Z, \tag{4.108}$$

and the variational equality

$$R(\varepsilon(u) + \frac{1}{2}P(\zeta), \varepsilon(v - u)) = 0, \quad \forall v \in [H^1(\Omega)]^2. \tag{4.109}$$

Further we eliminate the in-plane displacements of the plate. To this end we note first that $R(\cdot, \cdot)$ as defined in (4.104) is a continuous symmetric, coercive bilinear form on $[L^2(\Omega)]^4$, and that $P : [H^2(\Omega)]^2 \to [L^2(\Omega)]^4$ of (4.105) is a compact operator (cf. e.g. [21] p.219). Thus (4.109) and the Lax-Milgram theorem imply that to every deflection $\zeta \in Z$, there corresponds a plane displacement $u(\zeta) \in [H^1(\Omega)]^2$. Indeed, due to Korn's inequality $R(\varepsilon(u), \varepsilon(v))$ is a bilinear coercive form on the quotient space $[H^1(\Omega)]^2/\bar{\mathcal{R}}$,where $\bar{\mathcal{R}}$ is the space of in-plane rigid-plate displacements defined by

$$\bar{\mathcal{R}} = \{\bar{r} \middle| \bar{r} \in [H^1(\Omega)]^2, \quad \bar{r}_1 = \alpha_1 + bx_2, \bar{r}_2 = \alpha_2 - bx_2, \alpha_1, \alpha_2, b \in \mathbb{R}\}. \tag{4.110}$$

From (4.109) it results that

$$\varepsilon(u(\zeta)) : Z \to [L^2(\Omega)]^4 \tag{4.111}$$

is uniquely determined and is a compact quadratic function of ζ, since $\varepsilon(u(\zeta))$ is a linear continuous function of $P(\zeta)$. We also introduce the compact quadratic function $G : Z \to [L^2(\Omega)]^4$ which is defined by

$$\zeta \to G(\zeta) = \varepsilon(u(\zeta)) + \frac{1}{2}P(\zeta) \tag{4.112}$$

and satisfy the equation

$$R(G(\zeta), \varepsilon(u(\zeta))) = 0. \tag{4.113}$$

We now define the operators $A : Z \to Z^*$ and $C : Z \to Z$ such that

$$\alpha(\zeta, z) = \langle A\zeta, z\rangle \tag{4.114}$$

and

$$hR(G(\zeta), P(\zeta, z)) = (C(\zeta), z), \tag{4.115}$$

A is a continuous symmetric linear operator, C is a compact operator, $\langle\cdot,\cdot\rangle$ denotes the duality pairing between Z and Z^* and $(\cdot,\cdot)$ denotes the scalar product in Z. Thus the following variational problem results:
Find $\zeta \in Z$, so as to satisfy the hemivariational inequality

$$\alpha(\zeta, z-\zeta)+(C(\zeta), z-\zeta)+\int_{\Omega'} j^0(\zeta, z-\zeta)d\Omega \geq \int_{\Omega} \overline{\overline{f}}\,(z-\zeta)d\Omega \quad \forall z \in Z. \tag{4.116}$$

The last variational - hemivariational inequality characterizes the position of equilibrium of the problem studied.

Now we are in the position to formulate the corresponding eigenvalue problem.

The buckling is caused by boundary loading in the plane of the plate, of the form (4.98), where $g = g(x), x \in \Gamma$, is a self-equilibrating compressive load distribution on Γ, i.e.,

$$\int_\Gamma g_a d\Gamma = 0 \;\text{ and }\; \int_\Gamma (g_1 x_2 - g_2 x_1) d\Gamma = 0, \tag{4.117}$$

and λ is a real number measuring the magnitude of the load distribution.

For $g_a \in L^2(\Gamma)$, the following variational formulation is obtained for the problem defined by (4.91-4.96), (4.98),(4.106),(4.117) with $\overline{\overline{f}} = 0$.
Find $\zeta \in H^2(\Omega)$, $u \in [H^1(\Omega)]^2$ and $\lambda \in \mathbb{R}$ which satisfy

$$\alpha(\zeta, z-\zeta) + \int_{\Omega'} j^0(\zeta, z-\zeta)d\Omega + \int_\Omega h\sigma_{a\beta}\zeta_{,a}(z-\zeta)_{,\beta}d\Omega \geq \lambda\int_\Gamma h g_a \zeta_{,a}(z-\zeta)d\Gamma,$$

$$\forall\, z \in H^2(\Omega) \tag{4.118}$$

and

$$\int_\Omega \sigma_{a\beta}\,\varepsilon_{a\beta}(v)d\Omega = \lambda\int_\Gamma g_a v_a d\Gamma, \quad \forall\, v \in [H^1(\Omega)]^2, \tag{4.119}$$

where $\sigma = \{\sigma_{a\beta}\}$ and $\varepsilon = \{\varepsilon_{a\beta}\}$ are given by (4.93).

The triplet (ζ, u_1, u_2) is called the variational solution corresponding to λ. Since $\zeta = 0$ must be a solution for every λ, as is obvious from the nature of the mechanical problem, we assume that $0 \in \partial j(0)$. In order to obtain a simpler form of the problem we introduce the auxiliary quantities $\sigma^0 = \{\sigma^0_{a\beta}\}$, $\varepsilon^0 = \{\varepsilon^0_{a\beta}\}$

and $u^0 = \{u_a^0\}$, satisfying (4.92),the second relation in (4.93) , the stress-strain law

$$\sigma_{a\beta}^0 = C_{a\beta\gamma\delta}\,\varepsilon_{\gamma\delta}^0(u^0) \tag{4.120}$$

and the boundary condition (4.98) for $\lambda = 1$. For the resulting plane elasticity problem there exists, due to (4.117) (cf., also [6]) a unique solution $u^0 \in [H^1(\Omega)]^2/\bar{\mathcal{R}}$ such that

$$\int_\Omega \sigma_{a\beta}^0\,\varepsilon_{a\beta}(v)d\Omega = \int_\Gamma g_a v_a d\Gamma, \quad \forall\, v \in [H^1(\Omega)]^2. \tag{4.121}$$

Every other solution has the form $u^0 + \bar{r}$, $\bar{r} \in \bar{\mathcal{R}}$. Moreover, $\sigma_{a\beta}^0(u^0) \in L^2(\Omega)$ is uniquely determined.

Further, we consider the stress and strain tensor $\bar{\sigma} = \{\bar{\sigma}_{a\beta}\}$ and $\bar{\varepsilon} = \{\bar{\varepsilon}_{a\beta}\}$ and the displacement vector $\bar{u} = \{\bar{u}_a\}$ satisfying (4.92)(4.93) and the boundary conditions (4.98) for $g_a = 0$. From (4.109) we get that $\bar{\varepsilon} = \bar{\varepsilon}(\zeta)$ is unique and is a compact function from $H^2(\Omega)$ into $[L^2(\Omega)]^4$, and similarly for $\bar{\sigma} = \bar{\sigma}(\zeta)$. Then the following variational problem can be formulated:
Find $\zeta \in H^2(\Omega)$ and $\bar{u} \in [H^1(\Omega)]^2/\bar{\mathcal{R}}$ such that

$$\alpha(\zeta, z-\zeta) + \int_{\Omega'} j^0(\zeta, z-\zeta)d\Omega + \int_\Omega h\bar{\sigma}_{a\beta}\zeta_{,a}(z-\zeta)_{,\beta}d\Omega \geq$$

$$\geq -\lambda\int_\Omega h\sigma_{a\beta}^0\zeta_{,a}(z-\zeta)_{,\beta}d\Omega + \lambda\int_\Gamma hg_a\zeta_{,a}(z-\zeta)d\Gamma, \quad \forall\, z \in H^2(\Omega), \tag{4.122}$$

$$\int_\Omega \bar{\sigma}_{a\beta}\,\varepsilon_{a\beta}(v)d\Omega = 0, \quad \forall\, v \in [H^1(\Omega)]^2/\bar{\mathcal{R}}, \tag{4.123}$$

where $\bar{\sigma}_{a\beta}$ and $\bar{\varepsilon}_{a\beta}(\bar{u})$ satisfy (4.93).

We call a solution $(\zeta, \bar{u}_1, \bar{u}_2)$ of (4.122)(4.123) a reduced variational solution. It is easily verified that for a given λ the triplet (ζ, u_1, u_2) is a solution to (4.118), (4.119), if and only if $(\zeta, \bar{u}_1, \bar{u}_2)$ is a reduced variational solution, where u and $\bar{u}$ are related by

$$u_a = \bar{u}_a + \lambda u_a^0, \quad a = 1, 2 \tag{4.124}$$

and

$$\sigma_{a\beta} = \bar{\sigma}_{a\beta} + \lambda\sigma_{a\beta}^0, \quad a, \beta = 1, 2. \tag{4.125}$$

Note that for every $\lambda, (0,0,0)$ is a solution of (4.122),(4.123). A nontrivial reduced variational solution of the form $(0, \bar{u}_1, \bar{u}_2)$ is not possible for every λ, since then (4.123) would imply that $\bar{u} = 0$ as well.

It results that any solution of the buckling problem which is not trivial, i.e., $\zeta \neq 0$, may be written as the sum of a nontrivial reduced variational solution $(\zeta, \bar{u}_1, \bar{u}_2)$ and a plane displacement $\lambda(0, u_1^0 + \bar{r}_1, u_2^0 + \bar{r}_2)$ (cf. (4.110)).

The relations (4.122),(4.123) constitute an eigenvalue problem for hemivariational inequalites. Let us define a linear continuous mapping $B : Z \to Z$ such that

$$(B\zeta, z) = -\int_\Omega h\sigma_{a\beta}^0\zeta_{,a}z_{,\beta}d\Omega, \quad \forall\, \zeta, z \in Z \tag{4.126}$$

and let us consider now the boundary condition $\zeta = 0$ on Γ, i.e. (4.107). The triplet $(\zeta, u_1, u_2) \in Z \times ([H^1(\Omega)]^2/\bar{\mathcal{R}})$ is a reduced variational solution, if $\bar{u} = \bar{u}(\zeta)$ and

$$\alpha(\zeta, z-\zeta)+(C(\zeta), z-\zeta)+\int_{\Omega'} j^0(\zeta, z-\zeta)d\Omega \geq \lambda(B\zeta, z-\zeta), \quad \forall\, z \in Z. \quad (4.127)$$

Note that due to the boundary condition in (4.107) $\alpha(.,.)$ is coercive on Z.

Here B is a linear, selfadjoint, compact operator on Z. Analogous eigenvalue problems formulated in the framework of variational inequalities are solved (cf. [17] Sect. 7.2) on the additional technical assumption introduced in [13] that the boundary Γ_1 of every subdomain $\Omega_1 \subset \Omega$ of the plate is subjected to compressive forces, i.e. that a.e. on Γ_1 the inequality

$$\sigma^0_{ij}\bar{n}_i\bar{n}_j \leq 0 \quad (4.128)$$

holds, where $\bar{n} = \{\bar{n}_i\}$ is the unit normal outward vector on Γ_1. Then (4.128) implies that

$$(B\zeta, \zeta) > 0 \quad \forall \zeta \in Z \;\; \zeta \neq 0. \quad (4.129)$$

Such an assumption is not necessary, as we see in the next section.

4.5.3 Mathematical Study of the Buckling Problem

In the present section we study the problem formulated in Sect. 4.5.2. To this end we adapt the notation to the notation of Sect. 4.4. Let V be a real Hilbert space with the scalar product $(\cdot,\cdot)$ and the associated norm $||\cdot||$. Space V is densely and compactly imbedded into $L^p(\Omega;\mathbb{R})$ for some $p \geq 2$. Here Ω is a bounded domain in $\mathbb{R}^n$. Let us consider a continuous symmetric bilinear form $\alpha : V \times V \to \mathbb{R}$ not necessarily coercive, a nonlinear compact operator $C : V \to V$ and a function $\beta : \mathbb{R} \to \mathbb{R}$ belonging to $L^\infty_{\mathrm{loc}}(\mathbb{R})$. We introduce further the locally Lipschitz function j by means of (4.2) and we consider the problem: Find $v \in V$ and $\lambda \in \mathbb{R}$ such that

$$\alpha(u, v) + (C(u), v) + \int_\Omega j^0(u(x); v(x))dx \geq \lambda(Bu, v) \qquad \forall v \in V. \quad (4.130)$$

Throughout this Section $C_p(\Omega)$ denotes a positive constant given by the continuous imbedding $V \subset L^p(\Omega)$

$$||u||_{L_p} \leq C_p(\Omega)||u|| \quad \forall u \in V. \quad (4.131)$$

Let $A : V \to V^*$ be the isomorphism, with $a(u,v) = \langle Au, v\rangle \; \forall\, u, v \in V$. Let $C : V \to V$ be a (nonlinear) compact operator satisfying the assumptions (H_1),(H_2), (H_3) of Section 4.4.

We state the following existence result for solutions of the eigenvalue problem (P).

Theorem 4.4 Assume that B is a linear compact selfadjoint operator and that the hypotheses $(H_1), (H_2), (H_3)$ hold. Then if $p > 2$, for every $\lambda \in \mathbb{R}$ there exists $u \in V$ satisfying problem (P). If $p = 2$, for every $\lambda \in \mathbb{R}$ with $||B||\lambda < 2\beta_2 - \alpha_2(C_2(\Omega))^2 - ||\alpha||$, there exists $u \in V$ solving (P).

Proof. The proof follows the same steps as the proof of Theorem 4.3 ($||B||$ appears together with $||u||$). Moreover, at the end of the fourth step we use the compactness of B in order to pass to the strong limit. Let us fix some $\lambda \in \mathbb{R}$ and let $h : V \to \mathbb{R}$ be the functional defined by

$$h(u) = \frac{1}{2}\alpha(u,u) + \int_\Omega j(u(x))dx - \frac{\lambda}{2}(Bu,u) + g(u), \quad u \in V. \tag{4.132}$$

We define $I : V \to \mathbb{R}$ given by

$$I(u) = h(u) + f(u), \quad u \in V, \tag{4.133}$$

and $J : L^p(\Omega) \to \mathbb{R}$ given by

$$J(u) = \int_\Omega j(u(x))dx, \quad u \in L^p(\Omega). \tag{4.134}$$

Due to (H_2) J is well-defined and locally Lipschitz on $L^p(\Omega)$, or on V.

To prove the theorem we proceed in several steps.

i) First we show that the functional I is bounded from below on V, thus there exists

$$c = \inf_V I > -\infty. \tag{4.135}$$

Indeed, from hypotheses, $(H_2), (H_3)$ and (4.133) we obtain the estimate

$$\begin{aligned} I(u) \;\geq\; & -\frac{1}{2}||\alpha||||u||^2 - \alpha_1(\text{meas}(\Omega))^{1/2}||u||_L^2 \\ & -\frac{1}{p}\alpha_2||u||_{L^p}^p - \frac{\lambda}{2}||B||||u||^2 + g(u) + f(u) \geq \\ \geq\; & (\beta_2 - \frac{1}{p}\alpha_2(C_p(\Omega))^p)||u||^p - \frac{1}{2}(||\alpha|| + \lambda||B||)||u||^2 - \\ & -\alpha_1(\text{meas}(\Omega_1)^{1/2}C_2(\Omega)||u|| + \beta_1 \qquad \text{for all } u \in V. \end{aligned}$$

from which (4.135) results.

ii) In this step we show that there exists a sequence $\{u_n\}$ in V with the properties

$$I(u_n) \leq c + \frac{1}{n} \quad \text{for all } n. \tag{4.136}$$

and

$$\alpha(u_n, v) - \lambda(Bu_n, v) + (C(u_n), v) + J^0(u_n; v) \geq -\frac{1}{n}||v|| \quad \text{for all } v \in V. \tag{4.137}$$

We apply to $I : V \to \mathbb{R}$ Ekeland's variational principle in order to prove step ii). This is possible because I is lower semicontinuous and bounded from below on V due to step i). According to Ekeland's result we obtain a sequence $\{u_n\}$ in V such that (4.136) holds, as well as the following inequality

$$I(w) - I(u_n) \geq -\frac{1}{n}||w - u_n|| \quad \text{for all } w \in V. \tag{4.138}$$

Setting $w = (1-t)u_n + tv$ for every $0 < t < 1$ and $v \in V$ (4.138) yields

$$h(u_n + t(v - u_n)) - h(u_n) + f((1-t)u_n + tv) - f(u_n) \geq -\frac{t}{n}||v - u_n||$$

for all $v \in V$, $0 < t < 1$, and n positive integer.

Due to the convexity of function $f : V \to \mathbb{R}$ we have

$$\frac{1}{t}(h(u_n + t(v - u_n)) - h(u_n)) + f(v) - f(u_n) \geq -\frac{1}{n}||v - u_n||$$

for all $v \in V, 0 < t < 1$, and n.

Passing to limit for $t \to 0$ and using the continuous differentiability of $g : V \to \mathbb{R}$ implies

$$\alpha(u_n, v - u_n) - \lambda(Bu_n, v - u_n) + J^0(u_n; v - u_n) +$$

$$+(g'(u_n), v - u_n) + f(v) - f(u_n) \geq -\frac{1}{n}||v - u_n|| \quad \text{for all } v \in V \text{ and } n.$$

From (H_1) we obtain that (4.137) is true.

iii) We prove now that corresponding to the sequence $\{u_n\} \subset V$ constructed in ii) there exist sequences $\{q_n\}$ and $\{u_n\}$ in V^* such that $q_n \to 0$ in V^*, $w_n \in \partial(J|_V)(u_n)$ and

$$-\alpha(u_n, \cdot) + \lambda \Lambda Bu_n - \Lambda C(u_n) + q_n = w_n \quad \text{for all } n, \tag{4.139}$$

where ∂ denotes Clarke's generalized gradient and $\Lambda : V \to V^*$ is the duality mapping $\langle \Lambda u, v \rangle = (u, v)$ for all $u, v \in V$.

The left-hand side of inequality (4.137) is with respect to v a lower semicontinuous convex function on V vanishing at 0. Thus the Lemma 1.3 in Szulkin [27] implies that there is $z_n \in V^*$ with $||z_n|| \leq 1$ such that

$$\alpha(u_n, v) - \lambda(Bu_n, v) + (C(u_n), v) + J^0(u_n; v) \geq \frac{1}{n}(z_n, v) \quad \forall v \in V.$$

By choosing $q_n = n^{-1}z_n$ we have that $q_n \to 0$ in V^* and

$$-\alpha(u_n, \cdot) - \lambda \Lambda Bu_n - \Lambda C(u_n) + q_n \in \partial(J|_V)(u_n) \quad \text{for all } n.$$

Thus we have proved that there exists $w_n \in \partial(J|_V)(u_n)$ satisfying (4.139).

iv) We will prove further that the sequence $\{u_n\} \subset V$ of the steps ii) and iii) possesses a convergent subsequence in V.

First we prove that $\{u_n\}$ is bounded in V: this is a consequence of relations (4.136), the estimate in i) for $I(u)$ and the imposed conditions upon the constants α_2, β_2. To derive the convergence of a subsequence of $\{u_n\}$ we consider (4.139). Here the boundedness of $\{u_n\}$ in V and the compactness of B and of mapping $C : V \to V$ imply the convergence of $C(u_n)$ along a subsequence in V denoted again by $C(u_n)$. The compactness of the imbedding $V \subset L^p(\Omega; \mathbb{R})$ assures that a subsequence of $\{u_n\}$, also denoted by $\{u_n\}$, converges in $L^p(\Omega; \mathbb{R})$. On the other hand the density of V in $L^p(\Omega; \mathbb{R})$ implies that

$$w_n \in \partial(J|_V)(u_n) \subset \partial J(u_n) \quad \text{for all } n \tag{4.140}$$

(cf. Corollary 1.2). The fact that J is locally Lipschitz on $L^p(\Omega; \mathbb{R})$ and relation (4.140) ensure that $\{w_n\}$ is bounded in $L^p(\Omega; \mathbb{R})$. By the compactness of the imbedding $(L^p(\Omega; \mathbb{R}))^\star \subset V^\star$ it turns out that $\{w_n\}$ converges along a subsequence in $V^\star$. For $n \to \infty$ in (4.139) we obtain the convergence along a subsequence of $-\alpha(u_n, \cdot) + \lambda \Lambda B u_n$ in $V^\star$. The assumptions that B is compact and the operator A of the symmetric bilinear form $\alpha : V \times V \to \mathbb{R}$ is an isomorphism enables us to conclude that a subsequence of $\{u_n\}$ converges in V.

v) We show now that the limit $u = \lim_{n\to\infty} u_n \in V$ of the convergent subsequence $\{u_n\}$ of the previous step iv) satisfies the inequality

$$\alpha(u, v) - \lambda(Bu, v) + (C(u), v) + J^0(u; v) \geq 0 \quad \forall v \in V. \tag{4.141}$$

The continuity of $h : V \to \mathbb{R}$ and (4.138) imply the existence of $\lim_{n\to\infty} f(u_n)$ and of the equality

$$\lim_{n\to\infty} f(u_n) = \lim_{n\to\infty} (I(u_n) - h(u_n)) = c - h(u).$$

Due to the lower semicontinuity of $f : V \to \mathbb{R}$

$$f(u) \leq \liminf_{n\to\infty} f(u_n) = \lim_{n\to\infty} f(u_n) = c - h(u).$$

It results that

$$I(u) = h(u) + f(u) = c = \inf_V I. \tag{4.142}$$

To verify (4.141) we prove first that the inequality

$$\alpha(u, v-u) - \lambda(Bu, v-u) + J^0(u; v-u) +$$

$$+ (g'(u), v-u) + f(v) - f(u) \geq 0 \quad \forall\, v \in V \tag{4.143}$$

holds. Due to (4.142) we may write

$$I(u + t(v-u)) \geq I(u) \quad \forall v \in V \text{ and } 0 < t < 1.$$

Making use of the convexity of the function $f : V \to \mathbb{R}$ one finds that

$$\frac{1}{t}(h(u + t(v-u)) - h(u) + f(v) - f(u) \geq 0 \quad \forall v \in V \text{ and } 0 < t < 1.$$

Passing to the limit for $t \to 0$ implies the inequality (4.143), which with (H_1) implies the inequality (4.141).

vi) It remains to prove that the element $u = \lim_{n\to\infty} u_n$ of V constructed in step v) is a solution of (4.130) corresponding to the fixed number $\lambda \in \mathbb{R}$.

For the generalized gradient of $J|_V$ we can write the relation

$$\partial(J|_V)(y) \subset \int_\Omega \partial j(y(x))dx \quad \forall y \in V. \tag{4.144}$$

This can be proved as in the case of (4.85).

From (4.144) and (4.141), recalling the definition of the generalized gradient and Proposition 1.4 (ii), we conclude that $u = \lim_{n\to\infty} u_n$ satisfies the inequality

$$\begin{aligned} -a(u,v) + \lambda(Bu,v) - (C(u),v) &\leq \int_\Omega \max_{z\in\partial j(u(x))} z(u(x))dx \\ &= \int_\Omega j^0(u(x);v(x))dx \quad \forall v \in V. \end{aligned} \tag{4.145}$$

Consequently, $u = \lim_{n\to\infty} u_n$ is a solution of the hemivariational eigenvalue problem (P) corresponding to the given real number λ. □

Suppose further that we deal with the genuine problem of von Kármán plates as formulated in Sect. 4.5.2. Here the verification of the assumptions (H_1) and (H_3) is not an easy task, whereas (H_2) may be assumed to hold due to the properties of the adhesive. Setting $g(u) = \frac{1}{4}(C(u),u)$ and $f(u) = \frac{1}{2}a(u,u) - \frac{1}{2}(Bu,u)$ (assumed to be nonnegative) and applying the coercivity result of Ciarlet and Rabier ([3] p.94) we may verify (H_3). It is obvious that then (H_1) holds for a restricted class of von Kármán plates satisfying the relation

$$\forall\, u,v \in V \qquad (Bv,v) - (Bu,u) \geq a(v,v) - a(u,u) - (C(u), v-u) \tag{4.146}$$

One can also modify the mechanical model by introducing a term

$$\tilde{C}(\zeta) = C(\zeta) + a\zeta \tag{4.147}$$

where $a\zeta$ takes into account the deformability of the thickness of the plate and a is a given elasticity constant. Then (H_1) is verified as an equality for $\tilde{C}$ with $f = 0$ and $g(u) = \frac{1}{2}[(C(u),u) + a\|u\|^2]$. Obviously (H_3) holds. The case $a \to 0_+$ is still an open problem.

Remark 4.1 It is instructive for the reader to notice the differences in the seemingly similar proofs of the Theorems 4.3 and 4.4. These differences are concentrated in the compactness of B and in the isomorphism property of A in Theorem 4.4.

Remark 4.2 In the plate problem studied in this section $\alpha(.,.)$ is coercive due to the boundary condition $\zeta = 0$ on Γ. However, the proof of Theorem 4.4 needs only the assumption that A is an isomorphism, i.e. that the corresponding bilinear form $\alpha(\zeta, \zeta)$ is nondegenerate. Recall that $\alpha(\zeta, \zeta)$ is degenerate, if there exists $\zeta \neq 0$ such that $\alpha(\zeta, \zeta) = 0$. If α is coercive, then A is an isomorphism, but also if $\alpha(\zeta, \zeta) > 0 \ \forall \ \zeta \neq 0$. The isomorphism assumption for A can be relaxed by imposing that $\lambda \in \mathbb{R}$ does not belong to the spectrum of the linear problem $a(u, v) - \lambda(Bu, v) = 0, \ \forall v \in V$ (see also Theorem 4.3).

4.6 The Timoshenko Plate Buckling. The Nonconvex Superpotential Problem

Here we investigate the buckling on a rectangular Timoshenko plate which is simply supported along its sides. The plate is a rectangular $a \times b$ which is simply supported along the edges $x = 0$ and $x = a$, and $y = 0$ and $y = b$ (Fig. 4.2)

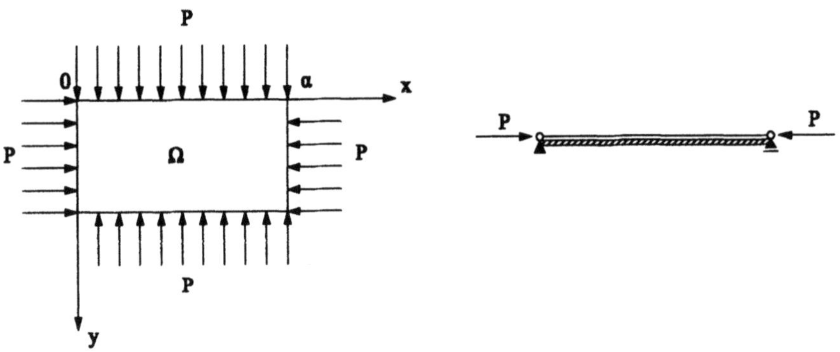

Fig. 4.2 Buckling of a Timoshenko plate.

According to [28] p.360 the problem obeys the following relations

$$K\Delta\Delta\zeta = -P\left(\frac{\partial^2\zeta}{\partial x^2} + \frac{\partial^2\zeta}{\partial y^2}\right) + q \tag{4.148}$$

where q denotes the adhesive force acting on the plate. It is derived by a nonconvex superpotential, i.e. it fulfills a multivalued law of the form (4.95),(4.97). Moreover, we introduce the admissible space Z for the deflections by setting

$$Z = \{z | z \in H^2(\Omega), \ z = 0 \text{ on } \Gamma\}. \tag{4.149}$$

From (4.148) we obtain by multiplying by $z - \zeta, \ z \in Z, \ \zeta \in Z$, intergrating over Ω, applying the Green-Gauss Theorem, taking into account the boundary condition $\zeta = 0$ and $M_n = 0$ on Γ as well as the inequality defining (4.95),(4.97), the following problem: Find $P \in \mathbb{R}$ and $\zeta \in Z$ such as to satisfy

$$a(\zeta, z-\zeta)-P\int_{\Omega}\text{grad}\zeta\cdot\text{grad}(z-\zeta)d\Omega+\int_{\Omega}j^{0}(\zeta, z-\zeta)d\Omega\geq 0 \quad \forall\, z\in Z. \tag{4.150}$$

Here $a(.,.)$ denotes the beding energy of the plate (see (4.100).

Note that $a(.,.)$ is on Z a norm equivalent to the H^2–norm and that we may write (here $(.,.)$ denotes the inner product in $H^2(\Omega)$)

$$\int \text{grad}\zeta\cdot\text{grad}z d\Omega=(L\zeta, z) \quad \forall\,\zeta, z\in Z, \tag{4.151}$$

where $L: Z\rightarrow Z$ is a linear compact operator. Thus the plate buckling problem takes the following form: Find $\lambda\in\mathbb{R}$ and $\zeta\in Z$ such that

$$a(\zeta, z-\zeta)+\int_{\Omega}j^{0}(\zeta, z-\zeta)d\Omega\geq\lambda(L\zeta, z-\zeta) \quad \forall\, z\in Z. \tag{4.152}$$

Here we can apply with minor modification the results of Section 4.2 and of Section 4.3 for the investigation of the existence of the solution of the hemivariational eigenvalue problem (4.152), because the compact operator L does not change the proofs of the Propositions of these Sections (cf. also the reasoning in the proof of Theorem 4.4 for problem (4.130)).

Note that analogously we can treat any other class of the plate buckling problem under the action of forces derived by nonconvex superpotentials (cf. [14] p.52) either acting in Ω or on Γ. Suppose, for instance, that the plate is subjected along the edges to the boundary conditions (cf. (4.101) for the definition of $\zeta\rightarrow M_n(\zeta)$)

$$\zeta=0 \quad\text{and}\quad M_n\in\partial j\left(\frac{\partial\zeta}{\partial n}\right), \tag{4.153}$$

where j is a locally Lipschitz functional of the boundary rotation $\frac{\partial\zeta}{\partial n}$ (n denotes the outward unit normal) and ∂ is the generalized gradient of Clarke. Then (4.148) holds for $q=0$ and the Green-Gauss theorem for plates (cf. (4.99) with $\sigma_{ab}=0$) leads to the eigenvalue problem: Find $P\in\mathbb{R}$ and $\zeta\in Z$ such as to satisfy

$$a(\zeta, z-\zeta)-P\int_{\Omega}\text{grad}\zeta\cdot\text{grad}(z-\zeta)d\Omega+\int_{\Gamma}j^{0}\left(\frac{\partial\zeta}{\partial n},\frac{\partial(z-\zeta)}{\partial n}\right)d\Gamma\geq 0\;\forall\, z\in Z. \tag{4.154}$$

Moreover, we assume that j results from $\beta\in L^{\infty}_{loc}(\mathbb{R})$ by filling in the jumps (cf. (4.2)). Then the eigenvalue problem reads: Find $\lambda\in\mathbb{R}$ and $\zeta\in Z$ such as to satisfy

$$a(\zeta, z-\zeta)+\int_{\Gamma}j^{0}\left(\frac{\partial\zeta}{\partial n},\frac{\partial(z-\zeta)}{\partial n}\right)d\Gamma\geq\lambda(L\zeta, z-\zeta) \quad \forall\, z\in Z. \tag{4.155}$$

The results of Section 4.2 hold in this case as well, with minor modifications. Indeed hypothesis (H_3) holds in the form (cf. also $(\tilde{H}_3)$ and $(\tilde{H}_3'))$

($\tilde{H}_3''$) $\{L_1\zeta = \frac{\partial\zeta}{\partial n} \in L^2(\Gamma),\ \zeta \in Z\}$ is a closed subspace of $L^2(\Gamma)$ and L_1 is a linear compact operator.

Then Theorem 4.1 holds also with $C(\Omega)$ replaced by $\tilde{\tilde{C}}'(\Omega)$, where

$$\left\| \frac{\partial\zeta}{\partial n} \right\|_{L^2(\Gamma)} \leq \tilde{\tilde{C}}'(\Omega)\|u\|_Z. \tag{4.156}$$

4.7 The Buckling of Cylindrical Shells Subjected to Adhesive Contact Conditions

Here we deal with the symmetrical buckling of cylindrical shells under the action of uniform axial compression (Fig. 4.3)

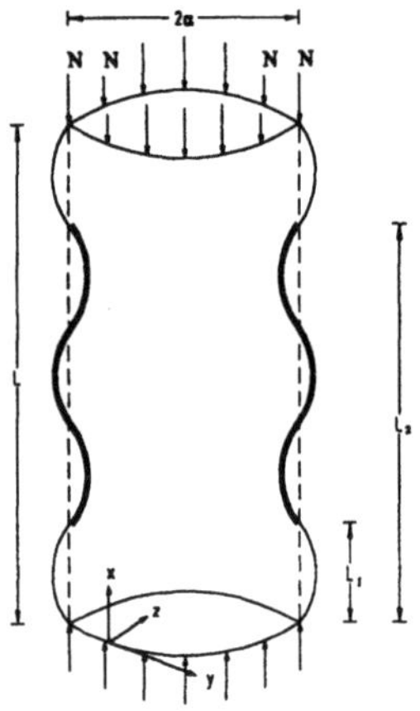

Fig. 4.3 Buckling of a Cylindrical Shell.

Due to the symmetry we look only for a function $x \to w(x)$ which fulfills the relation ([28] p. 457)

$$Dw'''' + N_x w'' + Eh\frac{w}{a^2} = q, \tag{4.157}$$

where $w' = \frac{dw}{dx}$, $D = \frac{Eh^3}{12(1-\nu^2)}$ is the flexural rigidity of the shell, E is the modulus of elasticity, ν is the Poisson's ratio, h is the thickness of the shell and a is the radius of the shell. Moreover we have that for $x = 0$ and $x = l$ the shell is simply supported, i.e.

$$w = 0 \quad \text{and} \quad \frac{\partial^2 w}{\partial x^2} = 0. \tag{4.158}$$

The second relation in (4.158) expresses the fact that the bending moment at $x = 0$ and $x = l$ is zero. In (4.157) q denotes the reaction of an adhesive support and a relation analogous to (4.95) holds. Note that N_x is positive if it is compressive and that the kinematically admissible space is

$$V = H^2(0,l) \cap \mathring{H}^1(0,l). \tag{4.159}$$

V is a Hilbert space with the inner product

$$a(w,v) = \int_0^l Dw''v''dx. \tag{4.160}$$

Moreover let the shell be subjected symmetrically to adhesive bonding conditions expressed by a relation of the form (cf. (4.95), (4.97))

$$-q \in \tilde{\beta}(w) \tag{4.161}$$

e.g. on the strip $l_1 \leq x \leq l_2$, $0 \leq l_1$, $l_2 < l$. From (4.157) we obtain the following eigenvalue problem by multiplying by $v - w$, integrating over $[0,l]$, applying the Green-Gauss theorem and using the inequality which holds by definition due to (4.95), (4.97), (4.161): Find $w \in V$, $\lambda \in \mathbf{R}$ such as to satisfy

$$a(w,v-w)+\frac{Eh}{a^2}\int_0^l w(v-w)dx+\int_{l_1}^{l_2} j^0(w,v-w)dx \geq \lambda\int_0^l w'(v'-w')dx \;\; \forall\, v \in V. \tag{4.162}$$

We define further a linear compact operator $L : V \to V$ such that (here $(.,.)$ denotes the inner product)

$$(Lv,w) = \int_0^l v'w'dx. \tag{4.163}$$

All the results of Sections 4.2, 4.3 hold for the above eigenvalue problem.

4.8 Eigenvalue Problems for Hemivariational Inequalities. The Mechanical Approach

i) The qualitative behaviour of the solution(s) of a hemivariational inequality may be studied by considering the corresponding eigenvalue problem. Let us consider instead of (4.130) the simplified hemivariational inequality: Find $u \in V$ such as to satisfy

$$a(u,v) + \int_{\Omega'} j^0(u,v)dx \geq (f,v) \quad v \in V. \tag{4.164}$$

Here all the quantities are defined as in Section 4.2 and $f \in V$ is the given applied loading. We may pose the following question: Which are those $\lambda \in \mathbf{R}$, for which we may have parallelism of the vectors f and u i.e. $f = \lambda u$, or more generally $f = \lambda Bu$? Here B is an appropriate linear operator $B : V \to V$.

Obviously the eigenvalue problem: find $(u,\lambda) \in V \times \mathbf{R}$ such as to satisfy

$$a(u,v) + \int_{\Omega'} j^0(u,v)dx \geq \lambda(u,v) \quad \forall\, v \in V, \tag{4.165}$$

gives the solution of this problem. Let us denote such a λ by λ_c. Then for (u,λ) not satisfying (4.165), we will not have the linearity relation $f = \lambda u$ with f, u

fulfilling (4.164), as it is obvious.

ii) Let us consider here a bifurcation problem similar to the loading-unloading problem studied in [14] p. 122. We consider a system Ω subjected to the loading $f = f(x)$. On the boundary part Γ_1 the system is fixed, i.e. $u_i = 0,\ i = 1,2,3$ and on the remaining part $\Gamma - \Gamma_1$ of the boundary we have a nonlinear reaction - displacement law (e.g. adhesive contact, or a fuzzy law, or a nonmonotone fricton law etc.) of the type (cf.[14])

$$-S \in \partial j(u) \quad S = \{S_i\},\ u = \{u_i\} \quad i = 1,2,3, \tag{4.166}$$

where j is a locally Lipschitz function. We assume that f is a function of $\lambda \in \mathbb{R}$ of the type $\lambda \to \lambda f(x)$. Due to the assumed nonlinearity of the system we will investigate the possibility that besides the solution u_0 varying with λ, i.e. $\lambda \to u_0(\lambda)$, another solution exists which intersects the path $\lambda \to u_0(\lambda)$ at $\lambda = \lambda_c$ and for $\lambda < \lambda_c$ coincides with it. The other solution can be written in the form $\lambda \to u_0(\lambda) + v(\lambda)$. Obviously

$$\lim_{\lambda \to \lambda_c} v(\lambda) = 0. \tag{4.167}$$

Let us write the potential energy of the structure in the form (cf. [1], p.41)

$$\Pi(u) = \frac{1}{2}a(u,u) - \langle S, u\rangle_{\Gamma-\Gamma_1} - \int_\Omega \lambda \Delta(u) d\Omega, \quad u \in V \tag{4.168}$$

where $u \to \Delta(u)$ denotes the "applied load-shortening" relation which is nonlinear due to the assumed nonlinearity of the behaviour of the system and $-\int_\Omega \lambda\Delta(u)d\Omega$ is the potential energy associated with the loading. Moreover $\langle S, u\rangle_{\Gamma-\Gamma_1}$ denotes the work of the nonlinear reaction-displacement law (4.166) on $\Gamma - \Gamma_1$ and $a(.,.)$ the strain energy of the system. The system can be a deformable body, or a discretized structure etc. Here we assume that $a(.,.)$ is a continuous bilinear form. Since the first variation of the potential energy is zero at the position $u_0(\lambda)$ of equilibrium we will have

$$a(u_0(\lambda), \delta u) - \langle S(u_0(\lambda)), \delta u\rangle_{\Gamma-\Gamma_1} - \int_\Omega \lambda \Delta'(u_0)\delta u d\Omega = 0, \tag{4.169}$$

and similarly at the position of equilibrium $u_0(\lambda) + v(\lambda)$ we will have

$$a(u_0(\lambda)+v(\lambda), \delta u) - \langle S(u_0(\lambda)+v(\lambda)), \delta u\rangle_{\Gamma-\Gamma_1} - \int_\Omega \lambda \Delta'(u_0(\lambda)+v(\lambda))\delta u d\Omega = 0. \tag{4.170}$$

We expand Δ' in Taylor series around $u_0(\lambda)$:

$$\begin{aligned} &a(u_0(\lambda)+v(\lambda), \delta u) - \langle S(u_0(\lambda)+v(\lambda)), \delta u\rangle_{\Gamma-\Gamma_1} \\ &- \int_\Omega \lambda[\Delta'(u_0) + \Delta''(u_0)v(\lambda)v + \frac{1}{2}\Delta'''(u_0)[v(\lambda)]^2 v^2 \ \ldots\ ]\delta u d\Omega = 0. \end{aligned} \tag{4.171}$$

Now we take into account (4.169) for $\lambda \to \lambda_c$ and we divide through $\|v(\lambda)\|$, where $\|\cdot\|$ denotes the norm in the space V of u (e.g. the $H^1(\Omega)$-norm). We use (4.167) and we set

$$u_1 =: \lim_{\lambda\to\lambda_c} \frac{v(\lambda)}{\|v(\lambda)\|}\,. \tag{4.172}$$

Here u_1 denotes the bifurcation mode [1]. Moreover we assume roughly speaking, that around λ_c, S is approximately a linear continuous function of $u(\lambda_c)$. Thus (4.171) implies that

$$a(u_1,\delta u) - \langle S(u_1),\delta u\rangle_{\Gamma-\Gamma_1} = \int_\Omega \lambda_c \Delta''(u_0)u_1\delta u d\Omega. \tag{4.173}$$

Suppose now that $\Delta(u) = \frac{1}{2}u^2 + c(x)u + d(x)$ ([1] p.46). Then (4.173) with (4.166) give rise to the eigenvalue hemivariational inequality

$$a(u_1,\delta u) + \int_{\Gamma-\Gamma_1} j^0(u_1,\delta u)d\Gamma - \int_\Omega \lambda_c u_1 \delta u d\Omega \geq 0 \tag{4.174}$$

which yields the value of $\lambda = \lambda_c$ for which the system presents bifurcation of the equilibrium path $\lambda \to u(\lambda)$.

The linearity of S around λ_c is quite restrictive. Thus we may proceed from (4.171) as it follows

$$\begin{aligned}
0 &= \limsup_{\lambda\to\lambda_c}\Big\{\frac{a(u_0(\lambda),\delta u)}{\|v\|} + a(\frac{v(\lambda)}{\|v\|},\delta u) - \frac{1}{\|v\|}\{\langle S(u_0(\lambda)+v(\lambda)),\delta u\rangle \\
&\quad -\langle S(u_0(\lambda)),\delta u\rangle\} - \frac{1}{\|v\|}\langle S(u_0(\lambda)),\delta u\rangle \\
&\quad -\frac{1}{\|v\|}\int_\Omega \lambda[\Delta'(u_0) + \Delta''(u_0)v + \frac{1}{2}\Delta'''(u_0)v^2 ...]\delta u d\Omega\Big\} \\
&\leq \lim_{\lambda\to\lambda_c}\Big\{\frac{1}{\|v\|}[a(u_0(\lambda),\delta u) - \langle S(u_0(\lambda)),\delta u\rangle - \int \lambda[\Delta'(u_0)]\delta u d\Omega]\Big\} \\
&\quad + \lim_{\lambda\to\lambda_c}\Big\{a(\frac{v(\lambda)}{\|v\|},\delta u) - \int_\Omega \lambda\Delta''(u_0)\frac{v(\lambda)}{\|v\|}\delta u d\Omega - \int_\Omega \lambda\frac{1}{2}\Delta'''(u_0)\frac{v(\lambda)}{\|v\|}v\delta u d\Omega ...\Big\} \\
&\quad + \limsup_{\lambda\to\lambda_c} -\frac{1}{\|v\|}\{\langle S(u_0(\lambda)+v(\lambda)) - S(u_0(\lambda)),\delta u\rangle\} \\
&= 0 + a(u_1,\delta u) - \int_\Omega \lambda\Delta''(u_0)u_1\delta u d\Omega \\
&\quad + \limsup_{\lambda\to\lambda_c} -\frac{1}{\|v\|}\{\langle S(u_0(\lambda)+v(\lambda)) - S(u_0(\lambda)),\delta u\rangle\}\,.
\end{aligned} \tag{4.175}$$

Let us now assume that

$$\limsup_{\lambda\to\lambda_c} -\frac{1}{\|v\|}\{\langle S(u_0(\lambda)+v(\lambda)) - S(u_0(\lambda)),\delta u\rangle\} \leq -\langle S(u_1(\lambda_c)),\delta u\rangle. \tag{4.176}$$

From (4.175) and (4.176) with (4.166) we obtain again (4.174).

This last model problem describes well the cases of bifurcation due to both geometric and physical instabilizing effects. In the case of geometric instabilizing effects $a(.,.)$ may lose its bilinearity (e.g. in the v.Kármán plate theory) and then additional assumptions must guarantee the passage from (4.168) to

(4.169). Note that, if $\Delta(u)$ has a more general form and containts the appropriate derivatives of u, one can formulate an eigenvalue problem with the inner product $(.,.)$ of the space of the problem under consideration, i.e. with the term $\lambda(u,v)$ in the right hand side.

Condition (4.176) says more explicitly that

$$\limsup_{\lambda\to\lambda_c} -\frac{1}{\|v(\lambda)\|}\langle S(u_0(\lambda)+v(\lambda))-S(u_0(\lambda)),w\rangle \le -\langle S(u_1(\lambda_c)),w\rangle, \quad (4.177)$$

where $v(\lambda)\to 0$, when $\lambda\to\lambda_c$

$$\lim_{\lambda\to\lambda_c}\frac{v(\lambda)}{\|v(\lambda)\|}=u_1(\lambda_c),$$

and $w\in V$ arbitrary.

We formulate a sufficient condition for (4.176).

Let us assume that $S: V\to H^{-1/2}(\Gamma)$ is locally Lipschitz. We denote by $\langle .,.\rangle$ the duality pairing on $H^{1/2}(\Gamma)\times H^{-1/2}(\Gamma)$. For every $h\in H^{1/2}(\Gamma)$, $\langle S(.),h\rangle$ is locally Lipschitz on V, therefore we can apply the mean-value theorem of Lebourg (Chapter 1): This theorem implies that

$$\frac{1}{\|v\|}\langle S(u_0(\lambda)+v(\lambda))-S(u_0(\lambda)),-h\rangle=\langle z^*(\lambda),\frac{v(\lambda)}{\|v\|}\rangle_{V,V^*} \quad (4.178)$$

with $z^*(\lambda)\in\partial(\langle S(.),-h\rangle)(u_0(\lambda)+t_\lambda v(\lambda))$, $t_\lambda\in(0,1)$. By definition

$$\langle z^*(\lambda),\frac{v(\lambda)}{\|v\|}\rangle_{V,V^*}\le(\langle S(.),-h\rangle)^0(u_0(\lambda)+t_\lambda v(\lambda);\frac{v(\lambda)}{\|v\|}) \quad (4.179)$$

The directional derivative of Clarke is upper semicontinuous (u.s.c.) (see Proposition 1.3). Thus it follows from (4.178),(4.179) that

$$\begin{aligned}\limsup_{\lambda\to\lambda_c}\frac{1}{\|v\|} &< S(u_0(\lambda)+v(\lambda))-S(u_0(\lambda)),-h)\\ &\le(\langle S(.),-h\rangle)^0(u_0(\lambda_c);u_1(\lambda_c)). \end{aligned}\quad (4.180)$$

Therefore, (4.176) holds if

$$(\langle S(.),-h\rangle)^0(u_0(\lambda_c);u_1(\lambda_c))\le\langle S(u_1(\lambda_c)),-h\rangle,\quad \forall\, h\in V. \quad (4.181)$$

In order to examine more carefully relation (4.181), let us assume that S is a Fréchet- differentiable function around $u_0(\lambda_c)$. Then (4.181) is equivalent with

$$\langle S'(u_0(\lambda_c)),u_1(\lambda_c)\rangle=S(u_1(\lambda_c)). \quad (4.182)$$

We note that in the case: S linear and continuous around $u_0(\lambda_c)$, (4.182) is satisfied. We give further an example, where (4.182) is fulfilled without the linearity and continuity assumptions.

Let $S : V \to V$, $S(v) = \phi(\|v\|^2)Av$, where $A \in \mathcal{L}(V, V)$ (linear continuous operator) and $\phi \in C^1(\mathbb{R})$ with $\phi(0) = \phi(1)$. Then S is Fréchet - differentiable and

$$\langle S'(0), v\rangle = S(v), \quad \forall\, v \text{ with } \|v\| = 1.$$

Indeed

$$\langle S'(0), v\rangle = \lim_{t\to 0_+} \frac{S(tv) - S(0)}{t} = \lim_{t\to 0}[\phi(t^2\|v\|^2)Av] = \phi(0)Av = \phi(1)Av.$$

But if $\|v\| = 1$, $S(v) = \phi(1)Av$. Therefore, S satisfies (4.182) with $u_0(\lambda_c)$ and any $u_1(\lambda_c)$, without being linear and continuous around $u_0(\lambda_c)$.

References

[1]B. Budiansky, Theory of Buckling and Postbuckling Behaviour of Elastic Structures. In Adv. in Appl. Mech. (ed. by Chia-Shun Yih), *Acad. Press*, London 1974, pp. 1-65.

[2]K.C. Chang, Variational Methods for Non-Differentiable Functionals and their Applications to Partial Differential Equations. *J. Math. Anal. Appl.* **80** (1981), 102-129.

[3]P.G. Ciarlet and P. Rabier, Les equations de von Kármán. Lect. Notes in Math. Vol. 826, *Springer Verlag*, Berlin 1980.

[4]F.H. Clarke, *Optimization and Nonsmooth Analysis*, J. Wiley and Sons, New York, 1983.

[5]J. Ekeland, On the Variational Principle, *J.Math. Anal. Appl.* **47** (1974), 324-353.

[6]G. Fichera, *Existence Theorems in Elasticity. In: Encyclopedia of Physics* (ed. by S. Flüge) Vol **VI** a/2. Springer-Verlag, Berlin 1972.

[7]I. Karamanlis, Buckling Problems in Composite von Kármán Plates. Doct. Thesis, Aristotle University Dept. of Civil Eng. 1991.

[8]H.N. Karamanlis, P.D. Panagiotopoulos, The Eigenvalue Problem in Hemivariational Inequalities and its Application to Composite Plates. *Journal of the Mech. Behaviour of Materials* (Freund Publ. House, Tel Aviv) **5**, No. 1, (1993), 67-76.

[9]C. Lefter and D. Motreanu, Critical Point Methods in Nonlinear Eigenvalue Problems with Discontinuities, in: International Series of Numerical Mathematics, Vol. **107**, *Birkhäuser Verlag*, Basel, Boston 1992.

[10]D. Motreanu, Existence for Minimization with Nonconvex Constraints, *J. Math. Anal. Appl.* **117** (1986), 128-137.

[11]D. Motreanu and P.D. Panagiotopoulos, Hysteresis: The Eigenvalue Problem for Hemivariational Inequalities, in "Models of Hysteresis" (ed. by A. Visintin), *Pitman Research Notes in Math.* 286, Longman Harlow, 1993, pp. 102-117.

[12]Z. Naniewicz and P.D. Panagiotopoulos, *Mathematical Theory of Hemivariational Inequalities and Applications*, Marcel Dekker, N.York, 1995.

[13]J. Naumann, H.V. Wenk, On Eigenvalue Problems for Variational Inequalites, *Rendiconti di Matem.* (3) Vol. **9** Serie VI(1976), 439-463.

[14]P.D. Panagiotopoulos, *Hemivariational Inequalities. Application to Mechanics and Engineering*, Springer Verlag, N.York, Berlin, 1993.

[15]P.D. Panagiotopoulos, Semicoercive Hemivariational inequalities. On the delamination of composite plates. *Quart. of Appl. Math.* **XLVII** (1989), 611-629.

[16]P.D. Panagiotopoulos and G. Stavroulakis, The Delamination Effect in Laminated von Kármán Plates under Unilateral Boundary Conditions. A Variational-Hemivariational Inequality Approach, *Journal of Elasticity* **23** (1990), 69-96.

[17]P.D. Panagiotopoulos,*Inequality Problems in Mechanics and Applications. Convex and Nonconvex Energy Functions*, Birkhäuser Verlag, Boston, Basel, 1985 (Russian Translation MIR Publ. Moscow 1989).

[18]P.D. Panagiotopoulos, Coercive and Semicoercive Hemivariational Inequalities. *Nonlinear Anal. T.M.A.* **16** (1991), 209-231.

[19]P.D. Panagiotopoulos, Hemivariational Inequalites and Substationarity in the Static Theory of von Kármán Plates. *ZAMM* **65** (1985), 219-229.

[20]P.D. Panagiotopoulos and G. Stavroulakis, A Variational-Hemivariational Inequality Approach to the Laminated Plate Theory under Subdifferential Boundary Conditions. *Quart. of Appl. Math.* **XLVI** (1988), 409-430.

[21]P.D. Panagiotopoulos, Hemivariational Inequalities and their Applications. In: Topics in Nonsmooth Mechanics (ed. J.J. Moreau, P.D. Panagiotopoulos and J. Strang), *Birkhäuser Verlag*, Boston, Basel, 1988.

[22]P.D. Panagiotopoulos, Non-Convex Superpotentials in the Sense of F.H. Clarke and Applications. *Mech. Res. Comm.* **8** (1981), 335-340.

[23]P.D. Panagiotopoulos, Nonconvex Energy Functions. Hemivariational Inequalities and Substationarity Principles. *Acta Mechanica* **42** (1983), 160-183.

[24]P.H. Rabinowitz, Minimax Methods in Critical Point Theory with Applications to Differential Equations. CBMS Reg. Conf. Ser. in Math. **65**, *Amer. Math. Soc.*, Providence 1986.

[25]P.H. Rabinowitz, Some Minimax Theorems and Applications to Nonlinear Partial Differential Equations, Nonlinear Analysis: A collection of papers in honor of E. Röthe, *Academic Press*, New-York, 1978, pp. 161-177.

[26]R.T. Rockafellar, La Théorie des Sous-Gradients et ses Applications à L'Optimization. Fonctions Convexes et Non-convexes, *Les Presses de l' Université de Montréal*, Montréal 1979.

[27]A. Szulkin, Minimax Principles for Lower Semicontinuous Functions and Applications to Nonlinear Boundary Value Problems, *Ann. Inst. Henri Poincaré, Analyse Non Linéaire* **3** (1986), 77-109.

[28]S. Timoshenko and J. Gere, Theory of Elastic Stability (2nd Edition) Mc Graw Hill, N. York, 1961.

5. Multiple Solutions of Eigenvalue Problems for Hemivariational Inequalities

The present chapter deals with multiplicity results for the eigenvalue problems of hemivariational inequalities. First we give a general minimax approach which permits the use of the corresponding linear eigenvalue problem for the determination of eigenvectors of a hemivariational inequality. Then the case of even nonconvex superpotential $j(x,\cdot)$ is studied by using some elements of the theory of genus of Krasnosel'skii. For the sake of completeness we recall some elements of the theory of genus and we apply them to the proof of the corresponding multiplicity results. The results of this chapter are applied to some problems of Mechanics and Engineering Sciences. Note that Section 5.1 deals with the $\lambda(V,H)$ eigenvalue problem, where Sections 5.2, 5.3, 5.4 with the $\lambda(L^2,V)$ eigenvalue problem. (Here $H \subset V$ (resp. $V \subset L^2$) are Hilbert spaces).

5.1 A Minimax Approach to the Problem

Let us consider the following eigenvalue problem (P):
Find $u \in H \subset V$ and $\lambda \in \mathbb{R}$ satisfying the inequality

$$\alpha(u,v) + \int_\Omega j^0(u,v)dx \geq \lambda(u,v)_V \qquad \text{for all } \ v \in H\ . \tag{5.1}$$

Here H and V are real Hilbert spaces with H continuously imbedded in $L^{2+\sigma}(\Omega)$ for some $\sigma > 0$, Ω is a bounded domain in an Euclidean space $\mathbb{R}^N$, $\alpha(\cdot,\cdot)$ denotes a continuous symmetric bilinear form on H and $j : \mathbb{R} \longrightarrow \mathbb{R}$ denotes the primitive

$$j(t) = \int_0^t \beta(s)ds \tag{5.2}$$

of a given function $\beta \in L^\infty_{loc}(\mathbb{R})$. The subscript of a scalar product or a norm indicates the respective vector space.

By a solution of problem (P) we mean a pair $(u,\lambda) \in H \times \mathbb{R}$ satisfying the inequality (5.1). We want to determine nontrivial solutions (u,λ) of problem (P), that is, those solutions (u,λ) with $u \neq 0$.

Problem (P) was studied for the first time by H. Karamanlis and P. D. Panagiotopoulos in [18],[19], where it is proved the existence of a solution (u,λ)

to (5.1) by using the following procedure: regularization of the problem, finite dimensional reduction and a passing to limit by means of a Galerkin basis.

In this section the study of problem (P) is based on the next two basic ideas: the use of the linear eigenvalue problem

$$\alpha(u,v) = \lambda(u,v)_V, \quad \forall\, v \in H, \tag{5.3}$$

and the application of a suitable minimax method to determine critical points of the locally Lipschitz functional

$$I_\lambda(u) = \frac{1}{2}(\alpha(u,u) - \lambda\|u\|_V^2) + \int_\Omega j(u)dx, \quad u \in H \tag{5.4}$$

for a fixed number $\lambda \in \mathbb{R}$.

For the definition of the classical Palais-Smale condition see Sect. 4.1.

The following minimax result extends to the locally Lipschitz case, Theorem 5.3 of P.H. Rabinowitz [32] holding for continuously differentiable functionals.

Theorem 5.1 Let E be a reflexive Banach space having the direct sum decomposition $E = W \oplus X$, with W a finite dimensional linear subspace of E and let $f : E \longrightarrow \mathbb{R}$ be a locally Lipschitz functional satisfying the Palais-Smale condition. Assume that the following hypotheses are satisfied :

(i) there are positive constants ρ and α such that

$$f(u) \geq \alpha \qquad \forall\; u \in X \text{ with } \; \|u\| = \rho;$$

(ii) there is a vector $e \in X$ with $\|e\| = 1$ and a number $R > \rho$ such that

$$f(u) \leq 0 \quad \forall\, u \in \partial Q,$$

where

$$Q \equiv \{w \in W |\; \|w\| < 1\} \bigoplus \{t\varepsilon | 0 < t < R\} \tag{5.5}$$

and ∂Q denotes the boundary of Q in $W \oplus$ span $\{e\}$.

Then

$$c = \inf_{h \in \Gamma} \max_{u \in \bar{Q}} f(h(u)), \tag{5.6}$$

where

$$\Gamma = \{h \in C(\bar{Q}, E) | h = id \;\; \text{(identity) on} \;\; \partial Q\} \tag{5.7}$$

is a critical value of f, i.e. there exists a critical point $u \in E$ of f for which $f(u) = c$. Moreover, one has the inequality

$$c \geq \alpha \tag{5.8}$$

Proof. For the proof we refer to Corollary 3.4 (Put $\Psi = 0$). However we can give here another direct proof. We start by showing that (5.8) holds. According to (5.5)-(5.7), it suffices to prove the following intersection property

$$h(Q) \cap \{u \in X \mid \|u\| = \rho\} \text{ is nonempty.} \tag{5.9}$$

But (5.9) is just Proposition 5.9 in Rabinowitz [32]. In order to show that the number c described by (5.6), (5.7) is a critical value of I_λ we argue by contradiction. If c were not a critical value of f, then due to the deformation theorem for locally Lipschitz functionals of Chang [5], there would exist a number $\varepsilon \in (0, \alpha/2)$ and a homeomorphism $\eta : E \to E$ such that

$$\eta(u) = u \quad \text{for} \quad u \notin \{v \in E | \ |f(v) - c| < \frac{\alpha}{2}\} \tag{5.10}$$

and

$$f(\eta(u)) \leq c - \varepsilon \text{ for } \ u \in E \ \text{ with } \ f(u) \leq c + \varepsilon \, . \tag{5.11}$$

From (5.6) one finds $h \in \Gamma$ satisfying

$$\max_{u \in \bar{Q}} f(h(u)) \leq c + \varepsilon \, . \tag{5.12}$$

Let $g = \eta o h \in C(\bar{Q}, E)$. Since, by hypothesis (ii) and equality (5.10), $\eta(u) = u$ on ∂Q, it follows $g \in \Gamma$. Then (5.11), (5.12) lead to $c \leq c - \varepsilon$ (contradiction) which completes the proof. □

From this Theorem one obtains directly the nonsmooth version due to K. C. Chang [5] of the Mountain Pass Theorem of A. Ambrosetti and P.H. Rabinowitz (cf. [32]). For the proof of the following result it is sufficient to take $W = \{0\}$ in Theorem 5.1.

Corollary 5.1 If a locally Lipschitz functional $f : E \to \mathbb{R}$ on the reflexive Banach space E satisfies the Palais-Smale condition and the hypotheses
(i) there exist positive constants ρ and α such that

$$f(u) \geq \alpha \qquad \forall \ u \in E \ \text{ with } \ \|u\| = \rho \, ;$$

(ii) $f(0) = 0$ and there is a point $e \in E$ such that

$$\|e\| > \rho \ \text{ and } f(e) \leq 0,$$

then there exists a critical value $c \geq \alpha$ of f determined by

$$c = \inf_{g \in G} \max_{t \in [0,1]} f(g(t)), \tag{5.13}$$

where

$$G = \{g \in C([0,1], E) \ |g(0) = 0, g(1) = e\} \, . \tag{5.14}$$

Note that Corollary 5.1 is a particular case of Corollary 3.2 (put $\Psi = 0$).

Further we will investigate the relationship between the critical points of the nonsmooth functional I_λ of (5.4) and the solutions of hemivariational eigenvalue problem (5.1).

We shall deal with the nonlinear part of the functional I_λ

$$J(u) = \int_\Omega j(u)dx, \tag{5.15}$$

where j given in (5.2) corresponds to a locally bounded measurable function $\beta : \mathbb{R} \to \mathbb{R}$ (cf. Proposition 1.7). To have J in (5.15) well defined in $L^{2+\sigma}(\Omega) \supset H$ we must impose upon β the growth condition

$$|\beta(t)| \leq \alpha_1 + \alpha_2 |t|^{1+\sigma}, \qquad t \in \mathbb{R}, \tag{5.16}$$

for some positive constants α_1, α_2 and σ.

Note that, if β satisfies the growth condition (5.16) with an exponent $\sigma \geq -1$, then by enlarging the constants α_1 and α_2 we may assume that (5.16) holds with an exponent $\sigma > 0$.

The next result points out a useful information regarding the generalized gradient $\partial J(u)$ of $J : L^{2+\sigma}(\Omega) \longrightarrow \mathbb{R}$. (cf. for the proof Proposition 4.3 and [4])

Proposition 5.1 Assume that the function $\beta \in L^\infty_{loc}(\mathbb{R})$ satisfies, the growth condition (5.16). Then $J : L^{2+\sigma}(\Omega) \to \mathbb{R}$ given by (5.15),(5.2) is locally Lipschitz and its generalized gradient $\partial J(u)$ for $u \in L^{2+\sigma}(\Omega)$ has the property: if $w \in \partial J(u)$, then $w \in (L^{2+\sigma}(\Omega))^*$ satisfies

$$w(x) \in [\underline{\beta}(u(x)), \overline{\beta}(u(x))] \text{ for a.e. } x \in \Omega, \tag{5.17}$$

where the functions $\underline{\beta}$ and $\overline{\beta}$ are defined by

$$\underline{\beta}(t) = \lim_{\delta \to 0} \operatorname*{essinf}_{|s-t|<\delta} \beta(s)$$

and

$$\overline{\beta}(t) = \lim_{\delta \to 0} \operatorname*{esssup}_{|s-t|<\delta} \beta(s) \tag{5.18}$$

for all $t \in \mathbb{R}$.

Recall also that if (cf. Proposition 1.7) the one-sided limits $\beta(t \pm 0)$ exist at some $t \in \mathbb{R}$, the equalities (5.18) take the form (4.18).

We give now a Theorem concerning the solutions of the hemivariational eigenvalue problem (P). This Theorem is analogous to Proposition 4.4.

Theorem 5.2 Assume that the following hypotheses are satisfied :

(H_1) Ω is a bounded domain in some Euclidean space $\mathbb{R}^N, N \geq 1$, H and V are real Hilbert spaces with H separable such that there exist the continuous linear imbeddings with a dense range

$$H \subset L^{2+\sigma}(\Omega) \subset V \tag{5.19}$$

for some $\sigma > 0$, and the first inclusion in (5.19) is compact.

(H_2) $\alpha : H \times H \rightarrow \mathbb{R}$ is a continuous symmetric bilinear form on H for which there exists a constant $C > 0$ such that

$$\alpha(u,u) \geq C\|u\|_H^2 \quad \forall \ u \in H; \tag{5.20}$$

(H_3) $\beta : \mathbb{R} \rightarrow \mathbb{R}$ is a function satisfying $\beta \in L^\infty_{loc}(\mathbb{R})$ and the growth condition (5.16).

If $u \in H$ is a critical point of the functional $I_\lambda : H \longrightarrow \mathbb{R}$ introduced in (5.4) for some $\lambda \in \mathbb{R}$, i.e. (4.3) holds, then the pair (u, λ) is a solution of hemivariational eigenvalue problem (5.1).

Proof. It is clear from Proposition 5.1 that due to $(H_1) - (H_3)$ $I_\lambda : \mathbb{R} \rightarrow \mathbb{R}$ is locally Lipschitz. Its generalized gradient $\partial I_\lambda(u)$ is equal to

$$\partial I_\lambda(u) = \alpha(u,\cdot) - \lambda(u,\cdot)_V + \partial(J|_V)(u), u \in H_1 \tag{5.21}$$

with $J : L^{2+\sigma}(\Omega) \rightarrow \mathbb{R}$ given by (5.2).
Relation (5.16) and Proposition 5.1 ensure that

$$|w| \leq \max(|\underline{\beta}(t)|, |\overline{\beta}(t)|) \leq \alpha_1 + \alpha_2|t|^{1+\sigma}, \tag{5.22}$$

for all $t \in \mathbb{R}$ and $w \in \partial j(t)$. Estimate (5.22) implies that (1.14) (cf. also Hypothesis B in F.H. Clarke [5] p.83) is verified. Thus Theorem 1.3 implies

$$\partial J(u) \subset \int_\Omega \partial j(u(x)dx, \quad u \in L^{2+\sigma}(\Omega). \tag{5.23}$$

The density of H in $L^{2+\sigma}(\Omega)$ enables us to apply Theorem 2.2 of [4] and thus we obtain

$$\partial(J|_H)(u) \subset \partial J(u), \quad u \in H\ . \tag{5.24}$$

In fact (5.24) is an equality (cf. Corollary 1.2). Combining (5.21), (5.23) and (5.24) implies that, if $u \in H$ is a critical point of I_λ, one can write

$$\lambda(u,v)_V - \alpha(u,v) \leq \int_\Omega j^0(u(x);v(x))dx \quad \forall \ v \in H\ . \tag{5.25}$$

To obtain (5.25) we used also Proposition 1.4 (ii). Relation (5.25) is just the inequality (5.1), so (u, λ) solves problem (P). Thus the proof is complete. □

Remark 5.1 In the proof of Theorem 5.2 we did not use the separability of H, the compactness of the first imbedding in (5.19) and the coercivity relation (5.20). These are needed for the proof of the next result. If the function j in (5.2) is stictly differentiable, one can interchange the integral sign and the symbol of generalized directional derivative (see [7][30]). Thus, for this situation, the problems (5.1) and (5.3) are equivalent.

Lemma 5.1 Assume that the continuous symmetric bilinear form $\alpha : H \times H \to \mathbb{R}$ satisfies hypotheses $(H_1), (H_2)$. Then there exists a sequence of eigenvalues λ_n and a sequence of corresponding eigenvectors e_n of the linear problem (5.3) such that the following properties hold

$$\text{(i)} \quad 0 < \lambda_1 \leq \lambda_2 \leq \ldots \leq \lambda_n \text{ and } \lambda_n \to \infty \text{ as } n \to \infty, \tag{5.26}$$

(ii) $\{e_n\} \geq 1$ is an orthonormal basis of H endowed with the equivalent scalar product $\alpha(u, .)$,so in particular one has

$$\alpha(e_i, e_i) = 0 \text{ for } i \neq j, \text{ and } \alpha(e_i, e_i) = 1 = \lambda_i \|e_i\|_V^2 \text{ for all } i; \tag{5.27}$$

(iii) the linear operator

$$u \in H \longrightarrow \alpha(u, \cdot) - \lambda(u, \cdot)_V \in H^* \tag{5.28}$$

is an isomorphism if $\lambda \neq \lambda_n$ for every n .

Proof. The imbeddings $H \subset V \subset H^*$ with a compact injection $H \subset V$ imply the statements (5.26), (5.27) from the Riesz-Fredholm theorem (see [4], p.236).The assertion (5.28) expresses the Fredholm alternative (cf. [4] p.242).

Lemma 5.2 Suppose that hypothses $(H_1) - (H_3)$ of Theorem 5.2 are satisfied. If a sequence $\{u_n\}$ is bounded in H and there is a sequence $w_n \in \partial(J|_H)(u_n)$ such that J is defined by (5.15), (5.2) and

$$\alpha(u_n, \cdot) - \lambda(u_n, \cdot)_V + w_n \to 0 \text{ in } H^* \text{ as } n \to \infty, \tag{5.29}$$

with λ a fixed number in $\mathbb{R}$, then $\{u_n\}$ contains a convergent subsequence in H.

Proof. The compactness of the imbedding $H \subset L^{2+\sigma}(\Omega)$ in (5.19) implies that the bounded sequence $\{u_n\}$ in H contains a convergent subsequence in $L^{2+\sigma}(\Omega)$ which we denote again by $\{u_n\}$. The fact that $J : L^{2+\sigma}(\Omega) \to \mathbb{R}$ is locally Lipschitz implies the boundedness of $\{w_n\}$ in $L^{2+\sigma}(\Omega)$. Thus the compactness of the imbedding $L^{2+\sigma}(\Omega) \subset H^*$ ensures that $\{w_n\}$ contains a convergent subsequence in H^*. The above imply the convergence of $\alpha(u_n, \cdot) - \lambda(u_n, \cdot)_V$ along a subsequence in H^*. If λ is not an element of the spectrum (5.26) of the linear eigenvalue problem (5.3), the convergence of a subsequence of $\{u_n\}$ in H is derived from property (5.28) in Lemma 5.1. Suppose now that λ belong to the sequence of eigenvalues (5.26), say $\lambda = \lambda_k$. Denote by P_k the orthogonal projection onto the eigenspace H_k associated with $\lambda = \lambda_k$, that is the linear subspace of H spanned by those vectors e_i in (5.27) corresponding to $\lambda_i = \lambda_k$. The Fredholm alternative (see Aubin [4], p. 242) and the convergence of $a(u_n, .) - \lambda(u_n, \cdot)_V$ in H^* imply that $\{u_n - P_k u_n\}$ converges along a subsequence in H_k. But H_k is finite dimensional and this ends the proof of the lemma. □

Lemma 5.3 Let $H \subset L^{2+\sigma}(\Omega)$, with $\sigma \geq 0$, be a continuous linear injection and let the function $\beta : \mathbb{R} \to \mathbb{R}$ satisfy hypothesis (H_3) of Theorem 5.2 together with the condition

(H_4) there exists constants $r > 0$ and $0 < \mu \leq 2 + \sigma$ such that

$$t\beta(t) \leq \mu j(t) < 0 \quad \forall \ |t| \geq |r|. \tag{5.30}$$

Then one can find positive constants $c_1, c_2 > 0$ such that the next estimate holds

$$J(u) \leq -c_1 \|u\|^{\mu}_{L^{\mu}(\Omega)} + c_2 \quad \forall \ u \in H\ . \tag{5.31}$$

Proof. The function $j(t)$ defined by (5.2) is absolutely continuous. Thus we may write from (5.30) the following inequality involving the derivative $j'(t)$ of $j(t)$

$$tj'(t) \leq \mu j(t) < 0 \qquad \text{for a.e. } |t| \geq |r| \tag{5.32}$$

Relation (5.32) implies by integration the inequality

$$j(t) \leq -c_1 |t|^{\mu} + c_2 \quad \forall \ t \in \mathbb{R}, \tag{5.33}$$

where c_1 and c_2 are some positive constants. Because $\mu \leq 2 + \sigma$, $L^{2+\sigma}(\Omega)$ is continuously imbedded into $L^{\mu}(\Omega)$. Thus for every $u \in H \subset L^{2+\sigma}(\Omega)$, (5.33) implies the estimate (5.31) with a different constant c_2 . □

Lemma 5.4 Assume that the hypotheses of Lemma 5.3 are fulfilled for $\sigma > 0$ and, that in addition, the following condition holds:

(H_5) $\quad \beta(t) = o(|t|)$ as $t \to 0$

Then $J(u) = o(\|u\|^2_H) \to 0$ in H.

Proof. Fix a number $\varepsilon > 0$. (H_5) implies for $\delta > 0$ that

$$|\beta(t)| \leq \varepsilon |t| \ \text{ for each } \ |t| \geq \delta. \tag{5.34}$$

On the other hand the growth condition (5.16) yields that

$$|j(t)| \leq \left(\frac{a_1}{\delta^{1+\sigma}} + \frac{a_2}{1+\sigma}\right) |t|^{2+\sigma} \ \text{ for each } \ |t| \geq \delta\ . \tag{5.35}$$

From (5.34), (5.35) we get

$$|J(u)| \leq b(\frac{\varepsilon}{2} + \|u\|_H^\sigma)\|u\|_H^2 \quad \forall \ u \in H, \tag{5.36}$$

where b denotes a positive constant which is independent of $u \in H$. In (5.36) we have used also the existence of the continuous imbedding $H \subset L^{2+\sigma}(\Omega) \subset L^2(\Omega)$. Since $\sigma > 0$, it is clear from (5.36) that keeping $\|u\|_H$ sufficiently small implies the conclusion of Lemma 5.4. □

All the above lemmata are applied in the proof of the following theorem.

Theorem 5.3 Assume that following hypotheses hold: $(H_1)-(H_3)$ of Theorem 5.2, (H_4) of Lemma 5.3 with the further conditions

$$t\beta(t) \leq 0 \quad \forall \ t \in \mathbb{R}, \tag{5.37}$$

$$\mu > 2 \ \text{ and } \ L^\mu(\Omega) \ \text{ is continuously imbedded into V}, \tag{5.38}$$

and (H_5) of Lemma 5.4. Then for every $\lambda \in \mathbb{R}$ there exists a nonzero vector $u \in H$ such that (u, λ) is a solution of the hemivariational eigenvalue problem (P) (continuous spectrum).

Proof. Let us fix the number $\lambda \in \mathbb{R}$. We show now that there are satisfied the hypotheses of Theorem 5.1 or Corollary 5.1 with $E = H$ and $f = I_\lambda$, the functional introduced in (5.4). Due to Lemma 5.1 we may consider the sequences $\{\lambda_n\}$ and $\{e_n\}$ of eigenvalues and eigenvectors (5.26), (5.27) of the linear problem (5.3), respectively. The only possible cases $\lambda_k \leq \lambda < \lambda_{k+1}$ for some $k \geq 1$ and $\lambda < \lambda_1$ will be treated separately. First let us consider the case, where there is a natural number k so that $\lambda_k \leq \lambda < \lambda_{k+1}$. For this situation we will show that Theorem 5.1 applies. The decomposition $H = E = W \bigoplus X$ of Theorem 5.1 is chosen as follows

$$W = \text{span}\,\{e_1, ..., e_k\} \text{ and } X = W^\perp \text{ (the orthogonal complement of } W \text{ in } H\text{)}. \tag{5.39}$$

Due to Lemma 5.4, the second equality in (5.39), and the assumption $\lambda < \lambda_{k+1}$ we have

$$\begin{aligned} I_\lambda(u) &= \frac{1}{2}\sum_{i\geq k+1}(1-\frac{\lambda}{\lambda_i})\alpha_i^2 + J(u) \geq \\ &\frac{1}{2}(1-\frac{\lambda}{\lambda_{k+1}})\|u\|_H^2 + o(\|u\|_H^2), \qquad \forall \ u = \sum_{i\geq k+1}\alpha_i e_i \in X \ . \end{aligned} \tag{5.40}$$

Inequality (5.40) implies clearly that the condition (i) of Theorem 5.1 is satisfied for $f = I_\lambda$, because $\lambda < \lambda_{k+1}$. Let us check condition (ii) of Theorem 5.1: we claim that there exists $R > 0$ such that

$$I(u) \leq 0 \qquad \forall\, u \in W \cup \{u \in \text{span}\,\{e_1, ..., e_{k+1}\} \mid \|u\|_H \geq R\}. \tag{5.41}$$

Indeed, suppose that $u \in W$. Then it results from (5.27), (5.37), the first equality in (5.39) and the assumption $\lambda \geq \lambda_k$ that

$$I_\lambda(u) = \frac{1}{2}\sum_{i=1}^{k}(1-\frac{\lambda}{\lambda_i})\alpha_i^2 + \int_\Omega \int_0^{u(x)} \beta(t)dtdx \leq 0 \tag{5.42}$$

$$\forall \ u = \sum_{i=1}^{k} \alpha_i e_i \in W.$$

Let now $u = \sum_{i=1}^{k+1} \alpha_i e_i$ belong to the span $\{e_1, ..., e_{k+1}\}$ with $\|u\| = 1$ and let $t > 0$ be fixed . Then (5.27), assumption $\lambda \geq \lambda_k$ and Lemma 5.3 imply that

$$\begin{aligned} I_\lambda(tu) &= \frac{1}{2}(\sum_{i=1}^{k+1}(1-\frac{\lambda}{\lambda_i})\alpha_i^2)t^2 + J(tu) \\ &\leq \frac{1}{2}(1-\frac{\lambda}{\lambda_{k+1}})\alpha_{k+1}^2 t^2 - c_1\|u\|_{L^\mu(\Omega)}^\mu t^\mu + c_2 \ , \end{aligned} \tag{5.43}$$

with $c_1 > 0$. Since the coefficients of the powers of t in the right-hand side of inequality (5.43) are bounded uniformly with respect to u (recall that u runs in the unit sphere of span $\{e_1, ...e_{k+1}\}$), and since $\mu > 2$, as stated in (5.38), one concludes that

$I_\lambda(tu) \to -\infty$ uniformly with respect to

$$u \in \{u \in \text{span } \{e_1, ..., e_{k+1}\} | \|u\|_H = 1\} \ \text{ as } \ t \to \infty. \tag{5.44}$$

Properties (5.42) and (5.44) show that (5.41) is true. Taking into account the definition of the set Q in (5.5) it results from (5.41) that assumption (ii) of Theorem 5.1 is valid when f is replaced by the functional I_λ and $e = e_{k+1}/\|e_{k+1}\|$. It remains only to verify the Palais-Smale condition for the locally Lipschitz functional $I_\lambda : H \to \mathbb{R}$. Due to Lemma 5.2 it is sufficient to verify, that, if $\{u_n\}$ is a sequence in H such that, with some constant $M > 0$,

$$|I_\lambda(u_n)| \leq M \ \text{ for all } n, \tag{5.45}$$

and there is $z_n \in \partial I_\lambda(u_n)$ satisfying

$$z_n \to 0 \ \text{ in } \ H^* \ \text{ as } \ n \to \infty \tag{5.46}$$

then $\{u_n\}$ is bounded in H. By (5.21) we can write

$$z_n = \alpha(u_n, \cdot) - \lambda(u_n, \cdot)_V + w_n, \quad w_n \in \partial(J|_H)(u_n). \tag{5.47}$$

In view of (5.38) we may fix a number $\nu \in (\mu^{-1}, \frac{1}{2})$. Then, for n sufficiently large, one obtains

$$\begin{aligned} M + \nu\|u_n\|_H &\geq I(u_n) - \nu\alpha(u_n, u_n) + \nu\lambda\|u_n\|_V^2 - \nu\int_\Omega w_n u_n dx \\ &\geq C(\frac{1}{2}-\nu)\|u_n\|_H^2 + b\lambda(\nu - \frac{1}{2})\|u_n\|_{L^\mu(\Omega)}^2 + \\ &+ \int_\Omega (j(u_n) - \nu w_n u_n)dx, \end{aligned} \tag{5.48}$$

where b is a positive constant. At this point we made use of the positivity of λ and of the continuity of the imbedding $L^\mu(\Omega) \subset V$ assumed in (5.38). Furthermore, from hypothesis (H_4), Lemma 5.3 and the inequality $\nu > \mu^{-1}$ we obtain that

$$\begin{aligned}\int_\Omega (j(u_n) - \nu w_n u_n)dx &\geq (1-\mu\nu)J(u_n) + d \\ &\geq c_1(\mu\nu - 1)\|u_n\|^\mu_{L^\mu(\Omega)} + d_1 \end{aligned} \tag{5.49}$$

where d and d_1 are real constants independent of u_n . We apply now the following method of Rabinowitz [32], p.31: corresponding to each $\varepsilon > 0$ there is a number $K(\varepsilon) > 0$ such that

$$\|u\|^2_{L^\mu(\Omega)} \leq K(\varepsilon) + \varepsilon\|u\|^{2\mu}_{L^\mu(\Omega)} \quad \text{for all} \quad u \in L^\mu(\Omega). \tag{5.50}$$

Due to the inclusions $H \subset L^{2+\sigma}(\Omega) \subset L^\mu(\Omega)$ we may put $u = u_n$ in (5.50). Since $c_1(\mu\nu - 1) > 0$ we observe that, by taking $\varepsilon > 0$ small enough in (5.50), inequalities (5.48)÷(5.50) imply the existence of a real constant A such that

$$M + \nu\|u_n\|_H \geq C(\frac{1}{2} - \nu)\|u_n\|^2_H + A \quad \text{for all} \quad n. \tag{5.51}$$

The coefficient of the quadratic term in (5.51) is positive, and thus estimate (5.51) implies the boundedness of the sequence $\{u_n\}$ in H. Therefore for the considered case $\lambda_k \leq \lambda < \lambda_{k+1}$ we may apply Theorem 5.1 to the locally Lipschitz functional $I_\lambda : H \to \mathbb{R}$. We obtain the existence of a critical value $c > 0$ fulfilling the minimax property (5.6), (5.7), where we must substitute $f = I_\lambda$, $E = H$, $W = \text{span}\ \{e_1, ..., e_k\}$ and $e = e_{k+1}/\|e_{k+1}\|$. Thus, there exists a critical point $u \in H$ of I_λ, such that $I_\lambda(u) = c$. Then Theorem 5.2 implies that the pair $(u, \lambda) \in H \times \mathbb{R}$ is a solution of problem (5.1). Since $c > 0$, one has $u \neq 0$; thus the proof is completed for the case $\lambda_k \leq \lambda < \lambda_{k+1}$.

It remains to discuss the case $\lambda < \lambda_1$, where λ_1 denotes the first eigenvalue appearing in the sequence (5.26). The assumption $\lambda < \lambda_1$ and property (5.27) guarantee that the function $|\cdot|_H : H \to \mathbb{R}$ given by

$$|u|^2_H = \frac{1}{2}(\alpha(u,u) - \lambda\|u\|^2_V), \quad u \in H, \tag{5.52}$$

is a norm on H which is equivalent to the initial norm $\|\cdot\|_H$ (cf. (5.20)). From (5.4) and (5.52) it results that

$$I_\lambda(u) = \frac{1}{2}|u|^2_H - J(u), \quad u \in H. \tag{5.53}$$

Lemma 5.4 asserts that $J(u) = o(\|u\|^2_H)$ as $u = 0$. Thus (5.52) implies that $J(u) = o(|u|^2_H)$ as $u \to 0$ in H. From this fact and from (5.53) we can deduce that for the functional I_λ, on the space H endowed with the new norm $|\cdot|_H$, hypothesis (i) of Corollary 5.1 holds. To check hypothesis (ii) of Corollary 5.1 we follow an idea of Rabinowitz [32], p.10: for a fixed vector $u \neq 0$ in H and an arbitrary number $t > 0$ we obtain from Lemma 5.3 that

$$I_\lambda(tu) \to -\infty, \quad \text{as} \quad t \to \infty. \tag{5.54}$$

This is true because

$$I_\lambda \leq \frac{t^2}{2}|u|_H^2 - c_1\|u\|_{L^\mu(\Omega)}^\mu t^\mu + c_2$$

with $c_1 > 0$ and $\mu > 2$. Then assumption (ii) of Corollary 5.1 follows directly from property (5.54). Now we will verify the Palais-Smale condition for $I_\lambda : H \to \mathbb{R}$ given by (5.53), where H is endowed with the norm $|\cdot|_H$. We note that we can apply Lemma 5.2. Thus it suffices to show that any sequence $\{u_n\}$ in H satisfying (5.45) and (5.46) for some $z_n \in \partial I_\lambda(u_n)$ is bounded in H. We write z_n in the form (5.47) and we see that, if n is sufficiently large, the estimate

$$\begin{aligned} M + \frac{1}{\mu}|u_n|_H &\geq I_\lambda(u_n) - \frac{1}{\mu}|u_n|_H^2 - \frac{1}{\mu}\int_\Omega w_n u_n dx \\ &\geq (\frac{1}{2} - \frac{1}{\mu})|u_n|_H^2 + \int_\Omega (j(u_n) - \frac{1}{\mu} w_n u_n) dx \end{aligned} \tag{5.55}$$

holds. Then hypothesis (H_4), Proposition 5.1, and (5.55) imply the relation

$$M + \frac{1}{\mu}|u_n|_H \geq (\frac{1}{2} - \frac{1}{\mu})|u_n|_H^2 + B \text{ for } n \text{ sufficiently large} \tag{5.56}$$

where B denotes, a constant independent of n. From (5.56) we obtain the boundedness of $\{u_n\}$ in H.

Now we can apply Corollary 5.1, from which it results the existence of a positive critical value c of the functional I_λ fulfilling the minimax properties (5.13), (5.14). Thus there exists a critical point $u \in H$ of I_λ with $u \neq 0$. From Theorem 5.2 it follows that the pair (u, λ) is a (nontrivial) solution of the hemivariational eigenvalue problem (5.1). This completes the proof of Theorem 5.3.□

In this Section any $w \in \partial(J|_H)(u)$ is regarded, because of (5.24), as an element of $(L^{2+\sigma}(\Omega))^* = L^{(2+\sigma)/(1+\sigma)}(\Omega)$.

In the proof of Theorem 5.3 we have obtained the formula $I_\lambda(u) = c \geq \alpha > 0$ for an eigenvector $u \in H$, corresponding to $\lambda \in \mathbb{R}$ in problem (P). The constant α is given, namely $\alpha = (1/2)(1 - \lambda\lambda_{k+1}^{-1}) - \varepsilon$ or $(1/2) - \varepsilon$, with $\varepsilon > 0$ sufficiently small, such that α remains positive. Thus an a priori estimate for the eigenfunction $u \in H$ is obtained. A numerical method for the search of the eigenvectors $u \in H$ of the problem (P) could be based on the minimax characterizations (5.6), (5.7) with $f = I_\lambda$.

It is also worth noting that the results of this section hold also for $V = L^2(\Omega)$ (cf. (5.19) and (5.38)). Then the L^2-inner product will appear at the right hand side of (5.2).

5.2 Multiplicity Results for Even Nonlinearities. Prerequisites

In this Section and in the following ones we use the following notations for a locally Lipschitz function f on a Banach space X:

$$\begin{aligned} K &:= \{u \in X : 0 \in \partial f(u)\}, \\ f_c &:= \{u \in X : f(u) \leq c\}, \\ K_c &:= \{u \in K : f(u) = c\}, \end{aligned} \tag{5.57}$$

where $c \in \mathbf{R}$. The two following results from [5] express basic facts in the non-smooth critical point theory. The second one extends the classical deformation lemma to locally Lipschitz functions. For the proofs we refer to Chapter 2 and 3. Moreover we denote by (PS) the Palais-Smale Condition.

Lemma 5.5 If f satisfies (PS), then K_c is compact.

Lemma 5.6 Let X be a reflexive Banach space, and let f be a locally Lipschitz function satisfying (PS). If c is a real number and N is any neighborhood of K_c then for any $\varepsilon_0 > 0$ there exists $\varepsilon \in (0, \varepsilon_0)$ and a homeomorphism $\eta : X \to X$ such that

1) $\eta(x) = x$ for $x \notin f_{c+\varepsilon_o} \setminus f_{c-\varepsilon_o}$,

2) $\eta(f_{c+\varepsilon} \setminus N) \subset f_{c-\varepsilon}$,

3) if $K_c = \emptyset$ then $\eta(f_{c+\varepsilon}) \subset f_{c-\varepsilon}$,

4) if f is even then η can be chosen to be odd.

In what follows we shall need the notion of Krasonel'skii genus and thus we give some elements of the corresponding theory in the form of Lemma 5.7.

Let X be a real Banach space and Σ the collection of all symmetric (with respect to the origin) subsets of $X \setminus \{0\}$ which are closed in X. A nonempty set $A \in \Sigma$ is said to have genus k (we put $\gamma(A) = k$), if k is the smallest integer with the property that there exists an odd continuous mapping $\eta : A \to \mathbf{R}^k \setminus \{0\}$. If there is no such k, then we say that $\gamma(A) = +\infty$ and if $A = \emptyset$, we set $\gamma(A) = 0$. Denote by $d(u, A)$ the distance from u to the set A and let $N_\delta(A) := \{u \in X : d(u, A) \leq \delta\}$. Some properties of the genus are summarized below.

Lemma 5.7 Let $A, B \in \Sigma$. The following hold:

1) If there exists an odd continuous mapping $f : A \to B$, then $\gamma(A) \leq \gamma(B)$.

2) If $A \subset B$, then $\gamma(A) \leq \gamma(B)$.

3) $\gamma(A \cup B) \leq \gamma(A) + \gamma(B)$.

4) If $\gamma(B) < +\infty, \gamma(\overline{A \backslash B}) \geq \gamma(A) - \gamma(B)$.

5) If N is an open symmetric and bounded neighbourhood of the origin in $\mathbb{R}^k$ and if A is homeomorphic to the boundary of N, denoted ∂N, by an odd homeomorphism, then $\gamma(A) = k$.

6) If A is compact, then $\gamma(A) < +\infty$ and $\gamma(N_\delta(A)) = \gamma(A)$ for all sufficiently small $\delta > 0$.

7) Let Y be a closed subspace of finite codimension of X. If $A \cap Y = \emptyset$, then $\gamma(A) \leq \mathrm{codim} Y$.

8) If $\gamma(A) > 1$, then A contains infinitely many distinct points.

For the proofs we refer to [10].

By means of standard arguments of the critical point theory [10], [34], [6], [14], [20], [28] it is possible to prove the following multiplicity result for locally Lipschitz, even, functions.

Let $B_\rho := \{x \in X : \|x\| < \rho\}$ be the open ball of radius ρ centered at 0 and $\partial B_\rho = \{x \in X : \|x\| = \rho\}$, the boundary of B_ρ.

Theorem 5.4 Let X be a reflexive Banach space. Suppose that $f : X \to \mathbb{R}$ is an even locally Lipschitz function satisfying (PS) and the conditions

(i) $f(0) = 0$,

(ii) there exists a subspace X_1 of X of finite codimension and numbers $\beta, \rho > 0$ such that $f(x) \geq \beta$ on $X_1 \cap \partial B_\rho$,

(iii) there is a finite dimensional subspace X_2 of X, with $\mathrm{dim} X_2 > \mathrm{codim} X_1$, such that $f(u) \to -\infty$ as $\|u\| \to \infty$, $u \in X_2$.

Then K contains at least $\mathrm{dim} X_2 - \mathrm{codim} X_1$ pairs of nontrivial elements $(u_1, -u)$.

Proof. We can assume that f has no critical points in f_{-d} for some $d > 0$, otherwise, since d is arbitrary, there are infinitely many critical points and there is nothing to prove. Set $m = \mathrm{codim} X_1, k = \mathrm{dim} X_2, Q = \{x \in X_2 : \|x\| \leq R\}$, where $R > \rho$ is chosen so that $f \leq -d$ on ∂Q. Define, for $1 \leq j \leq k$,
$F := \{\eta \in C(Q, X) : \eta$ is odd and $\eta|_{\partial Q}$ is homotopic to $\mathrm{id}|_{\partial Q}$ in f_{-d} by an odd homotopy$\}$,
$\Gamma_j := \{\eta(Q \backslash V) : \eta \in F, V$ is open in Q and symmetric, $V \cap \partial Q = \emptyset$ and for each $Y \subset V$ such that $Y \in \Sigma, \gamma(Y) \leq k - j\}$,
$\Delta_j := \{A \subset X : A$ is compact, $A = -A$ and for each open set $U \supset A$ there is $A_o \in \Gamma_j$ such that $A_o \subset U\}$.

It is clear that $Q \in \Delta_j$ with $A_o = Q, V = \emptyset$ and $\eta = \mathrm{id}|_Q$ so that $\Delta_j \neq \emptyset$. We set

$$c_j := \inf_{A \in \Delta j} \sup_{u \in A} f(u). \tag{5.58}$$

The following properties are proved in [34 ; Lemma 4.5 and 4.6] by means of standard topological arguments. The results remain true in the locally Lipschitz context.

P$_1$. For $m+1 \leq j \leq k, c_j \geq \beta$.

P$_2$. $\Delta_{j+1} \subset \Delta_j$.

P$_3$. If $A \in \Delta_j, W$ is a closed and symmetric set containing A in its interior and $\alpha : W \to X$ is an odd mapping such that $\alpha_{|W \cap f_{-d}}$ is homotopic to $\mathrm{id}|_{W \cap f_{-d}}$ in f_{-d} by an odd homotopy, then $\alpha(A) \in \Delta_j$.

P$_4$. If $Z \in \Sigma$ is compact, $\gamma(Z) \leq p$ and $f|_Z > -d$, then there exists a $\delta > 0$ such that for each $A \in \Delta_{j+p}$, $A \backslash \mathrm{int}\{N_\delta(Z)\} \in \Delta_j$.

By P$_1$ and P$_2$, we have $\beta \leq c_{m+1} \leq ... \leq c_k$. Suppose that $c_j = ... = c_{j+p} = c$ for some $j \in [m+1, k] \cap \mathbb{N}$, and $p \geq 0$. Since f is even, K_c is symmetric and since f satisfies the (PS) condition, K_c is compact. Moreover, $c > 0$ and thus $0 \notin K_c$, so that $K_c \in \Sigma$. We shall prove that $\gamma(K_c) \geq p+1$.

Suppose on the contrary that $\gamma(K_c) \leq p$. By Lemma 5.7 point (6), we can choose $\delta > 0$ so that $\gamma(N_\delta(K_c)) = \gamma(K_c)$. Let $N = \mathrm{int}\{N_\delta(K_c)\}$ and choose an $\varepsilon_o > 0$. By Lemma 5.5, there exists an odd homeomorphism η such that

$$\eta(x) = x, \qquad \forall x \notin f_{c+\varepsilon_o} \backslash f_{c-\varepsilon_o},$$

and $\eta(f_{c+\varepsilon} \backslash N) \subset f_{c-\varepsilon}$, for some $\varepsilon \in (0, \varepsilon_o)$. There exists $A \in \Delta_{j+p}$ such that

$$\sup\{f(u) : u \in A\} \leq c + \varepsilon.$$

Since K_c is compact, $\gamma(K_c) \leq p$ and $f|_{K_c} > -d$, by P$_4$ and for δ sufficiently small we have $A \backslash N \in \Delta_j$. Moreover, $A \backslash N \subset f_{c+\varepsilon} \backslash N$ and thus $\eta(A \backslash N) \subset f_{c-\varepsilon}$. Thus, if ε_o is sufficiently small, $\eta(x) = x, \forall x \in f_{-d}$, and, by P$_3$, $\eta(A \backslash N) \in \Delta_j$. We obtain

$$\sup\{f(u) : u \in \eta(A \backslash N)\} \leq c - \varepsilon.$$

This is a contradiction, because

$$\sup\{f(u) : u \in \eta(A \backslash N)\} \geq c.$$

Thus $\gamma(K_c) \geq p+1$. In particular $\gamma(K_{c_j}) \geq 1$, so that each K_{c_j} has at least two points u_j and $-u_j$. This gives the required number of critical points, if all c_j are distinct. If they are not, then $p > 0$ for some j, so that $\gamma(K_{c_j}) > 1$ and we conclude by means of Lemma 5.7 point 8. □

We point out here that the results of Theorem 5.4 are a special case of Corollary 3.6.

Remark 5.2 Besides the "deformation" result given in Lemma 5.6, there exist also other types for nonsmooth deformation results (see [34],[8] and Chapters 2 and 3).

Further the results of Theorem 5.4 will be applied to the investigation of the multiplicity of solutions in eigenvalue problems for hemivariational inequalities.

Let $\langle V, V^*\rangle$ be a dual sytem of real Hilbert spaces such that

$$V \subset L^2(\Omega) \subset V^*,$$

where Ω is an open bounded subset of $\mathbb{R}^n$ with a smooth boundary. We assume that the injection $V \subset L^2(\Omega)$ is dense and compact. We denote by $\|\cdot\|$ the norm of V and by $\|\cdot\|_{L^2}$ the $L^2(\Omega)$–norm.

Let $a : V\times V \to \mathbb{R}$ be a continuous bilinear form. We introduce the mapping $A \in L(V, V^*)$ (where $L(V, V^*)$ denotes the space of bounded linear mappings from V into V^*) defined by

$$\langle Au, v\rangle = a(u, v), \qquad \forall u, v \in V. \tag{5.59}$$

Let $j : \Omega \times \mathbb{R} \to \mathbb{R}, (x, y) \to j(x, y)$, be a function such that $j(x, .)$ is locally Lipschitz, a.e. in Ω. We assume that

(H_1) $a : V \times V \to \mathbb{R}$ is symmetric and there exists $\alpha > 0$ such that

$$a(u, u) \geq \alpha\|u\|^2, \qquad \forall u \in V.$$

(H_2) the functionals

$$\psi(u) = \int_\Omega j(x, u(x))dx \tag{5.60}$$

and

$$\int_\Omega j^0(x, u; v)dx, \qquad \forall v \in V,$$

are well-defined on V and satisfy

$$\int_\Omega j^0(x, u; v)dx \geq \psi^0(u; v), \qquad \forall v \in V. \tag{5.61}$$

We formulate the following eigenvalue problem:
(P) Find $u \in V$ and $\lambda \in \mathbb{R}$ satisfying

$$a(u, v) - \lambda \int_\Omega uv dx + \int_\Omega j^0(x, u; v)dx \geq 0, \qquad \forall v \in V. \tag{5.62}$$

For a fixed $\lambda \in \mathbb{R}$ we introduce the functionals Φ_λ and J_λ on V by

$$\Phi_\lambda(u) := \frac{1}{2}a(u, u) - \frac{1}{2}\lambda\|u\|^2_{L^2},$$

$$J_\lambda := \Phi_\lambda + \psi_{|V}.$$

If the inequality (5.61) holds and if $\lambda \in \mathbb{R}$ and $u = u(\lambda) \in V$ solves the problem

(P') $\qquad 0 \in \partial J_\lambda(u),$

then the pair $(\lambda, u(\lambda))$ is an eigensolution for problem (P).

Since $\Phi \in C^1(X, \mathbb{R})$, we have (cf. Chapter 1)

$$\partial J_\lambda(u) = \Phi'_\lambda(u) + \partial\psi_{|V}(u),$$

with

$$\langle \Phi'_\lambda(u), h\rangle = \langle Au, h\rangle - \lambda(u,h)_{L^2}, \qquad \forall u, h \in V.$$

Further we will study the nontrivial solutions for Problem (P). This will be done by finding nontrivial solutions of problem (P'). Note that if $j^0(x,0,v) \geq 0, \forall x \in \Omega, v \in \mathbb{R}$, then, for each $\lambda \in \mathbb{R}$, Problem (P) has the trivial solution $u = 0$.

Denote by λ_j the j-th eigenvalue (counted according to its multiplicity) of the problem

$$a(u,v) = \lambda(u,v)_{L^2}, \qquad \forall v \in V. \tag{5.63}$$

Due to the assumption (H_1), $\lambda_j > 0$ and we can assume that $\lambda_j < \lambda_{j+1}$. The corresponding eigenfunctions $e_i \in V$, i.e.,

$$a(e_i, v) = \lambda_i(e_i, v)_{L^2} \qquad \forall v \in V, \tag{5.64}$$

can be chosen, such as to fulfill

$$a(e_i, e_i) = 1$$

and

$$a(e_i, e_j) = 0 \text{ if } j \neq i.$$

We shall prove that, if $\lambda_k < \lambda < \lambda_{k+1}$, then Problem (P) has, under suitable assumptions, at least k distinct pairs $(+u(\lambda), -u(\lambda))$ of nontrivial solutions. To this end it is sufficient to prove this assertion for Problem (P'). Thus the study of Problem (P) is reduced to finding $2k$ critical points pairwise symmetric of the functional J_λ on V. To prove their existence we apply Theorem 5.4.

But first let us introduce a compactness condition which is directly associated to the eigenvalue problem of hemivariational inequalities, i.e. to Problem (P) defined in (5.62). Throughout the rest of this section we assume that assumptions (H_1) and

(H'_2) (i) for each $y \in \mathbb{R}$, the map $x \to j(x,y)$ is measurable;

(ii) there exists $k \in L^2(\Omega, \mathbb{R}^+)$ such that, for all y and y' in $\mathbb{R}$ and x in Ω, one has

$$|j(x,y) - j(x,y')| \leq k(x)|y - y'|,$$

(iii) $\int_\Omega j(x,0)dx = 0$,

are satisfied.

Fix $\lambda \in \mathbb{R}$. We say that the functional J_λ on V satisfies the (HPS) condition (H-Palais Smale condition) if every sequence $\{u_n; n \in \mathbb{N}\}$ in V such that $J_\lambda(u_n) \to c \in \mathbb{R}$ and

$$\langle \Phi'_\lambda(u_n), v\rangle + \int_\Omega j^0(x, u_n; v)dx \geq -\varepsilon_n\|v\|, \qquad \forall v \epsilon V, \tag{5.65}$$

where $\varepsilon_n \to 0$, possesses a convergent subsequence.

We prove that (HPS) is equivalent to the following (HPS′) condition: if $\{u_n; n \in \mathbb{N}\}$ is a sequence such that $J_\lambda(u_n) \to c \in \mathbb{R}$ and

$$\langle \Phi'_\lambda(u_n), v\rangle + \int_\Omega j^0(x, u_n; v)dx \geq \langle z_n, v\rangle, \qquad \forall v \in V, \tag{5.66}$$

where $z_n \to 0$ in V^*, then $\{u_n; n \in \mathbb{N}\}$ possesses a convergent subsequence.

In order to prove the equivalence between (HPS) and (HPS′) we use the following lemma due to Szulkin [34 Lemma 1.3].

Lemma 5.8 Let X be a real Banach space and $\chi : X \to (-\infty, +\infty]$ a convex lower semicontinuous function such that

(i) $\chi(0) = 0$,

(ii) $\chi(y) \geq -\|y\|, \forall y \in X$.

Then there exists $w \in X^*$ such that $\|w\| \leq 1$ and

$$\chi(y) \geq \langle w, y\rangle, \qquad \forall y \in X.$$

Proposition 5.1 Conditions (HPS) and (HPS′) are equivalent.

Proof. Obviously (5.66) implies (5.65). Suppose now that (5.65) is satisfied. If $\varepsilon_n \leq 0$ then we take $z_n = 0$. If $\varepsilon_n > 0$ then we put

$$\chi_n(y) := \frac{1}{\varepsilon_n}\left[\langle \Phi'_\lambda(u_n), y\rangle + \int_\Omega j^0(x, u_n; y)dx\right], \qquad \forall y \in V,$$

and (5.65) becomes $\chi_n(y) \geq -\|y\|$, $y \in V$. Notice that $\chi(0) = 0$ and that the function

$$y \to \int_\Omega j^0(x; u_n, y)dx$$

is convex on V. On the other hand, assumption (H_2), implies the validity of

$$\left|\int_\Omega j^0(x, u_n; y)dx - \int_\Omega j^0(x, u_n; y')dx\right| \leq \|k\|_{L^2}\, \|y - y'\|_{L^2}, \qquad \forall y, y' \in V.$$

From this property follows then the strong continuity of the map $y \to \int_\Omega j^0(x,u_n; y)dx$ using the compact imbedding $V \subset L^2(\Omega)$. Thus, all assumptions of Lemma 5.8 are satisfied and thus there exists $w_n \in V^*$ such that $\|w_n\|_{V^*} \leq 1$ and

$$\chi_n(y) \geq \langle w_n, y\rangle, \qquad \forall y \in V.$$

Setting $z_n = \varepsilon_n w_n$, we see that

$$\langle \Phi'_\lambda(u_n), y\rangle + \int_\Omega j^0(x, u_n; y)dx \geq \langle z_n, y\rangle, \qquad \forall y \in V,$$

and $z_n \to 0$ in V^*. Hence (5.66) holds. □

It is easy to verify that (HPS) is stronger than (PS).

Proposition 5.2 If J_λ satisfies (HPS), then J_λ satisfies (PS).

Proof. Let $\{u_n; n \in \mathbb{N}\}$ be a sequence such that $J_\lambda(u_n)$ is bounded and $\mu(u_n) \to 0$ (see (2.1) for the definition of $\mu(u_n)$-replace in (2.1) λ by μ). There exists a subsequence of $\{u_n\}$ (denoted again by $\{u_n\}$) and a sequence $\{z_n\}$ in V^* such that $J_\lambda(u_n) \to c \in \mathbf{R}$, $z_n \in \Phi'_\lambda(u_n) + \partial\psi|_V(u_n)$ and $z_n \to 0$. Thus (5.61) and Proposition 5.1, imply that there exists a convergent subsequence of $\{u_n\}$. □

Note that condition (HPS) is a generalization of the compactness condition introduced by Szulkin [34]. Szulkin wanted to establish minimax "principles" applicable to variational inequalities, here we are interested in minimax "principles" applicable to hemivariational inequalities [15] (cf. also [22]-[25] and Chapters 2 and 3).

5.3 Multiplicity Results: The Lipschitz Case

First we deal with the multiplicity in the Lipschitz case.To this end we add a new hypothesis for the function $j : \Omega \times \mathbf{R} \to \mathbf{R}$.

(H_3) (i) For each $x \in \Omega, j(x, .)$ is even,

(ii) $\frac{\int_\Omega j(x,u)dx}{\|u\|^2} \to +\infty$ as $\|u\| \to 0, u \in V$.

Assumption (H_2') and Theorem 1.3 imply that the inequality (5.61) holds true.

Now we can prove the first multiplicity result (for the resonant case $\lambda = \lambda_k$ cf. Chapter 7).

Theorem 5.5 Assume that assumptions (H_1), (H_2') and (H_3) hold and that $\lambda_k < \lambda < \lambda_{k+1}$. Then Problem (P) has at least k distincts pairs $(-u(\lambda), +u(\lambda))$ of nontrivial solutions.

Proof: Let $\lambda_k < \lambda < \lambda_{k+1}$. In order to obtain our result we apply Theorem 5.4 for $f = J_\lambda$ (see Section 5.2). The assumptions of Theorem 5.4 are fulfilled:

i) J_λ satisfies (HPS) and thus (PS), due to Proposition 5.2. Indeed, let $\{u_n\} \subset V$ be a sequence such that $J_\lambda(u_n) \to c$ and

$$\langle \Phi'_\lambda(u_n), v\rangle + \int_\Omega j^0(x, u_n; v)dx \geq -\varepsilon_n\|v\|, \qquad \forall v \epsilon V. \tag{5.67}$$

First we show that $\{u_n\}$ is bounded. If not, $\|u_n\| \to \infty$ and passing to a subsequence, we can assume that

$$z_n := u_n/\|u_n\| \to z \text{ (weakly) in } V,$$

and thus $z_n \to z$ in $L^2(\Omega)$.

We have

$$\|u_n\|^{-2}J_\lambda(u_n) = \frac{1}{2}a(z_n, z_n) - \frac{1}{2}\lambda\|z_n\|_{L^2}^2 + \int_\Omega \|u_n\|^{-2}j(x, u_n)dx.$$

Since the injection $V \subset L^2(\Omega)$ is continuous, there exists a constant $c > 0$ such that $\|x\|_{L^2} \leq c\|x\|, \forall x \in V$, and thus

$$\|u_n\|^{-2}J_\lambda(u_n) \geq \frac{1}{2}\alpha - \frac{1}{2}\lambda\|z_n\|_{L^2}^2 - c\|u_n\|^{-1}\|k\|_{L^2}.$$

Taking the limit as $n \to \infty$, implies that

$$\|z\|_{L^2}^2 \geq \alpha/\lambda,$$

and thus $z \neq 0$. Since j satisfies (H_2'), we see that

$$\begin{aligned}\int_\Omega j^0(x, u_n; v)dx &\leq \int_\Omega |j^0(x, u_n; v)|dx \\ &\leq \int_\Omega |k(x)|\ |v(x)|dx \\ &\leq \|k\|_{L^2}\|v\|_{L^2}, \qquad \forall v \in V.\end{aligned}$$

Thus

$$\langle \Phi_\lambda'(u_n), v\rangle + \|k\|_{L^2}\|v\|_{L^2} \geq -\varepsilon_n\|v\|, \qquad \forall v \in V.$$

Dividing by $\|u_n\|$ and taking the limit as $n \to \infty$, we obtain

$$a(z, v) - \lambda\langle z, v\rangle \geq 0.$$

Since v is arbitrary, we have

$$a(z, v) - \lambda\langle z, v\rangle = 0, \qquad \forall v \in V.$$

We know that $z \neq 0$ and thus λ is an eigenvalue of the linear problem (5.3). This is a contradiction since $\lambda \in (\lambda_k, \lambda_{k+1})$. Accordingly, the sequence $\{u_n\}$ is bounded and there exists a subsequence (again denoted by $\{u_n\}$) such that $u_n \to u$ weakly in V. Set $v = u - u_n$ in (5.67). We obtain

$$\langle \Phi_\lambda'(u_n), u - u_n\rangle + \int_\Omega j^0(x, u_n; u - u_n)dx \geq -\varepsilon_n\|u - u_n\|,$$

and thus

$$a(u_n, u_n - u) \leq \varepsilon_n\|u - u_n\| + \int_\Omega j^0(x, u_n; u - u_n)dx + \lambda\langle u_n, u_n - u\rangle.$$

Then it follows

$$\begin{aligned} a(u_n - u, u_n - u) \leq\ & \varepsilon_n \|u - u_n\| + \int_\Omega j^0(x, u_n; u - u_n)dx \\ & + \lambda \langle u_n, u_n - u \rangle + a(u, u - u_n), \end{aligned}$$

and thus

$$\begin{aligned} \alpha \|u_n - u\|^2 \leq\ & \varepsilon_n \|u - u_n\| + \|k\|_{L^2} \|u - u_n\|_{L^2} \\ & + \lambda [\|u_n\|_{L^2} - \langle u_n, u \rangle] + \langle Au, u - u_n \rangle. \end{aligned}$$

Letting $n \to +\infty$ and using the facts that $u_n \to u$ weakly in V and $u_n \to u$ in $L^2(\Omega)$, we obtain $u_n \to u$ in V.

ii) By assumption (H_3), J_λ is even and $J_\lambda(0) = 0$.

iii) Set $X_1 = V$. We have $\operatorname{codim} X_1 = 0$, and we prove now the existence of $\beta, \rho > 0$ such that $J_\lambda(x) \geq \beta$ on ∂B_ρ. If not, for all $\rho > 0$ and $n \in \mathbb{N}$, there exists $u_n \in V$ such that $\|u_n\| = \rho$ and

$$J_\lambda(u_n) \leq \rho^2 / n.$$

We may then find a sequence $\{u_n\} \subset V$ such that $\|u_n\| \neq 0, u_n \to 0$ and

$$\|u_n\|^{-2} J_\lambda(u_n) \leq 1/n.$$

Let us put $z_n = u_n / \|u_n\|$. After passing to a subsequence, we can assume that $z_n \to z$ in $L^2(\Omega)$.

We have

$$1/n \geq \|u_n\|^{-2} J_\lambda(u_n) \geq \frac{1}{2}\alpha - \frac{1}{2}\lambda \|z_n\|^2_{L^2} + \int_\Omega j(x, u_n) \|u_n\|^{-2} dx.$$

Taking the limit as $n \to \infty$, we arrive at a contradiction to (H_3).

vi) Let $X_2 = \operatorname{span}\{e_1, ..., e_k\}$. For $u = \alpha_1 e_1 + \ldots + \alpha_k e_k \in X_2$, we have

$$\begin{aligned} J_\lambda(u) &= \int_\Omega j(x, u)dx + \frac{1}{2} \sum_{i=1}^k \alpha_i^2 \{a(e_i, e_i) - \lambda \|e_i\|^2_{L^2}\} \\ &\leq c\|k\|_{L^2} \|u\| + \frac{1}{2}(1 - \lambda/\lambda_k) \|u\|^2. \end{aligned}$$

Since $\lambda > \lambda_k$, $J_\lambda(u) \to -\infty$ as $\|u\| \to +\infty, u \in X_2$.

Since all the assumptions of Theorem 5.4 are satisfied, J_λ has at least $\dim X_2 - \operatorname{codim} X_1 = k$ pairs of nontrivial critical points. Moreover, due to (5.61) , each of these critical points is also a solution of Problem (P), and the proof is completed. □

Remark 5.3 Note that Theorem 5.5 remains true for the problem: Find $u \in V$ and $\lambda \in \mathbb{R}$ satisfying

$$a(u,v) - \lambda \langle Lu, v \rangle + \int_\Omega j^0(x, u; v)dx \geq 0, \qquad \forall v \in V,$$

where $L : V \to V^*$ is a general linear selfadjoint compact operator.

Remark 5.4 Let us denote by Γ the boundary of Ω and let Γ_1 be a nonempty part of Γ. If we suppose that $\gamma : V \to L^2(\Gamma_1)$ is a compact and linear operator then all our previous considerations can be adapted without difficulty for the study of the problem: Find $u \in V$ and $\lambda \in \mathbb{R}$ satisfying

$$a(u,v) - \lambda \int_\Omega uv dx + \int_{\Gamma_1} j^0(x, \gamma(u); \gamma(v))d\Gamma \geq 0, \qquad \forall v \in V,$$

where $d\Gamma$ denotes the $(n-1)$-dimensional Hausdorff measure.

5.4 Multiplicity Results: The Locally Lipschitz Case

In this section we replace the global Lipschitz assumption (assumption (H_2') (ii)), which allows only a sublinear growth condition for $j(x,.)$, by a polynomial growth condition. Here we assume that V is densely and compactly imbedded in a space $L^p(\Omega)$ with $p > 2$. We suppose that the following assumptions hold:

(H_2'') (i) $|z| \leq c(1+|t|^{p-1}), \forall t \in \mathbb{R}$, a.e. $x \in \Omega, \forall z \in \partial j(x,t)$ for some constant $c > 0$.

(ii) $j(x,0)$ is (essentially) bounded for a.e. $x \in \Omega$,

(iii) $j(x,t) \leq c_1 + c_2|t|^\sigma, \forall t \in \mathbb{R}$, a.e. $x \in \Omega$, with constants $c_1 \geq 0, c_2 \geq 0$ and $0 \leq \sigma < 2$;

(H_3') (i) $j(x,.)$ is an even function for each $x \in \Omega$,

(ii) $\int_\Omega j(x,0)dx = 0$;

(H_4) for each unbounded sequence $\{v_n\}$ in V, with $\int_\Omega j(x, v_n(x)\|v_n\|^{-1})dx$ convergent, and each sequence j_n in $L^{p/(p-1)}(\Omega)$ with

$$j_n(x) \in \partial j(x, v_n(x)) \text{ for a.e. } x \in \Omega,$$

there exists a subsequence of $\{v_n\}$, denoted also by $\{v_n\}$, satisfying

$$j_n/\|v_n\| \to 0 \text{ in } L^{p/(p-1)}(\Omega) \text{ as } n \to \infty;$$

(H_5) $\liminf_{t\to 0} t^{-2} j(x,t) \geq \frac{1}{2}\lambda_{k+1}$ uniformly with respect to $x \in \Omega$.

Hypothesis (H_5) is related to the linear eigenvalue problem (5.64). Assumption (H_2'')(i), and the Theorem 1.3 imply that (5.61) holds true.

We state now our result.

Theorem 5.6 Assume that assumptions (H_1), (H_2''), (H_3'), (H_4) and (H_5) are satisfied and that $\lambda_k < \lambda < \lambda_{k+1}$. Then Problem (P) has at least k distincts pairs $(-u(\lambda), +u(\lambda))$ of nontrivial solutions.

Proof. By (H_2'') and Theorem 1.1 we obtain the estimate

$$\begin{aligned} |j(x,t)| &\leq |j(x,0)| + |j(x,t) - j(x,0)| \\ &\leq M + |\zeta(x,s)||t| \\ &\leq M + c(1+|t|^{p-1})|t| \\ &\leq a_1 + a_2|t|^p, \qquad \forall\, t \in \mathbb{R}, \quad \text{a.e. } x \in \Omega, \end{aligned} \tag{5.68}$$

where a_1, a_2, M are positive constants and $\zeta(x,s) \in \partial j(x,s)$ with $0 < s < t$.

We show that condition (PS) is satisfied. Let $\{v_n\}$ be a sequence in V sush that $J_\lambda(v_n)$ (where J_λ is the functional introduced in Section 5.2) is convergent in $\mathbb{R}$ and

$$Av_n - \lambda i(v_n) + j_n \to 0 \text{ in } V^* \text{ as } n \to \infty, \tag{5.69}$$

where $j_n \in L^{p/(p-1)}(\Omega)$ has the property

$$j_n(x) \in \partial j(x, v_n(x)) \text{ for a.e. } x \in \Omega. \tag{5.70}$$

Here $i : V \to V^*$ denotes the operator defined by

$$\langle i(u), v\rangle = \int_\Omega u(x)v(x)dx, \qquad \forall u, v \in V.$$

In order to write (5.69) we have used the fact that (cf. Theorem 1.3)

$$\partial\psi(v) \subset \int_\Omega \partial j(x, v(x))dx, \qquad v \in L^p(\Omega),$$

where Ψ is defined in (5.60). We show first that the sequence $\{v_n\}$ is bounded. We argue by contradiction: suppose that

$$\|v_n\| \to +\infty \text{ as } n \to +\infty.$$

Then for $z_n = v_n/\|v_n\|$, it follows that $a(z_n, z_n)$ is bounded, and thus convergent along a subsequence denoted again by z_n. The boundedness of $\{z_n\}$ in V implies that for a subsequence of $\{z_n\}$ (denoted again by $\{z_n\}$) we may assume that $z_n \to z$ in $L^2(\Omega)$. Hence

$$\int_\Omega j(x, z_n(x))dx$$

is convergent. From assumption (H_4) and (5.70) we obtain that

$$j_n/\|v_n\| \to 0 \text{ in } V^*. \tag{5.71}$$

From (5.69) and (5.70) the relation

$$Az_n - \lambda i(z_n) \to 0 \text{ in } V^*.$$

results. The mapping $v \to Av - \lambda i(v)$ is a topological isomorphism from V onto V^* and thus λ is not an eigenvalue of A in the weak sense. Therefore $z_n \to z$, $\|z\| = 1$ and

$$a(z, v) - \lambda\langle z, v\rangle = 0, \qquad \forall v \in V.$$

That means that λ must be an eigenvalue whose corresponding eigenfunction is z. This contradicts the fact that $\lambda_k < \lambda < \lambda_{k+1}$. We thus deduced that the sequence $\{v_n\}$ is bounded. Then we can assume that $\{v_n\}$ converges in $L^p(\Omega)$. Because $j_n(x) \in \partial j(x, v_n(x))$ it follows that $\{j_n\}$ is bounded in $L^{p/(p-1)}(\Omega)$. Consequently, we may suppose that there exists a subsequence again denoted by j_n such that j_n is convergent in V^*. Then (5.69) implies that $Av_n - \lambda i(v_n)$ is convergent in V^*. Using the fact that $A - \lambda i$ is an isomorphism, we deduce that $\{v_n\}$ contains a convergent subsequence in V. Thus the (PS) holds for J_λ.

Note that J_λ is an even functional and the requirement (i) in Theorem 5.4 is true due to (H_3').

In order to prove (ii) of Theorem 5.4, we choose $X_1 = V$. Let us fix some $\varepsilon > 0$. By (H_5) we can find $\delta > 0$ such that

$$j(x,t) \geq \frac{1}{2}(\lambda_{k+1} - \varepsilon)t^2, \qquad \forall x \in \Omega \text{ and } |t| \leq \delta.$$

Thus, due to (5.68) we may write that,

$$-j(x,t) \leq \frac{1}{2}(-\lambda_{k+1} + \varepsilon)t^2 + (a_1\delta^{-p} + a_2)|t|^p, \qquad \forall t \in \mathbf{R}, \text{ a.e. } x \in \Omega$$

and thus

$$J_\lambda(v) \geq \frac{1}{2}a(v,v) + \frac{1}{2}(\lambda_{k+1} - \lambda - \varepsilon)\|v\|_{L^2}^2 - (a_1\delta^{-p} + a_2)\|v\|_{L^p}^p, \quad \forall v \in V.$$

By choosing $\varepsilon < \lambda_{k+1} - \lambda$ we can write

$$J_\lambda(v) \geq \frac{1}{2}\alpha\|v\|^2 - (a_1\delta^{-p} + a_2)\|v\|_{L^p}^p, \qquad \forall v \in V.$$

The continuity of the injection $V \subset L^p(\Omega)$ yields that a constant $C > 0$ exists such that

$$\|v\|_{L^p} \leq C\|v\|, \qquad \forall v \in V.$$

Thus

$$\begin{aligned} I(v) &\geq \frac{1}{2}\alpha\|v\|^2 - C^p(a_1\delta^{-p} + a_2)\|v\|^p \\ &= (\frac{1}{2}\alpha - C^p(a_1\delta^{-p} + a_2)\|v\|^{p-2})\|v\|^2, \forall v \in V. \end{aligned} \tag{5.72}$$

Taking into account that $p > 2$, we have for $\rho > 0$ small enough

$$C^p(a_1\delta^{-p} + a_2)\rho^{p-2} < \frac{1}{2}\alpha.$$

Then

$$I(v) \geq b := (\frac{1}{2}\alpha - C^p(a_1\delta^{-p} + a_2)\rho^{p-2})\rho^2, \qquad \forall \|u\| = \rho$$

and assumption (ii) in Theorem 5.4 is verified.

Let us consider now point (iii) of Theorem 5.4 and let us choose $X_2 := \text{span}\{e_1, ..., e_k\}$. Then for every $v = \alpha_1 e_1 + \ldots + \alpha_k e_k$

$$\begin{aligned} J_\lambda(v) &= \int_\Omega j(x,u)dx + \frac{1}{2}\sum_{i=1}^k \alpha_i^2\{a(e_i,e_i) - \lambda\|e_i\|_{L^2}^2\} \\ &\leq \frac{1}{2}\sum_{i=1}^k (1 - \lambda/\lambda_i)\alpha_i^2 + c_1 \operatorname{meas}(\Omega) + c_2\|v\|_{L^\sigma}^\sigma \\ &\leq \frac{1}{2}(1 - \lambda/\lambda_k)a(v,v) + d_2\|v\|^\sigma + d_1, \end{aligned}$$

and thus

$$J_\lambda(v) \leq \frac{1}{2}(1 - \lambda/\lambda_k)\alpha\|v\|^2 + d_2\|v\|^\sigma + d_1, \qquad \forall v \in X_2, \tag{5.73}$$

where d_1 and d_2 are some positive constants. Because $\lambda_k > 0, \lambda > \lambda_k$ and $\sigma < 2$, it results from (5.73) that $J_\lambda(v) \to -\infty$ as $\|v\| \to \infty$ for $v \in X_2$. Therefore assertion (iii) of Theorem 5.4 is verified and thus the proof is complete. □

Note that Assumption (H_3) is satisfied if j is globally Lipschitz and that assumption (H_5) expresses an interaction between the behaviour of function j near $t = 0$ and the spectrum of the corresponding linear eigenvalue problem.

Remark 5.5 Let Γ be the boundary of Ω and let Γ_1 be a nonempty part of Γ. If we suppose that $\gamma : V \to L^p(\Gamma_1)$ is a compact and linear operator, then all our previous considerations can also be adapted without difficulty to the problem: Find $u \in V$ and $\lambda \in \mathbb{R}$ satisfying

$$a(u,v) - \lambda \int_\Omega uv dx + \int_{\Gamma_1} j^0(x, \gamma(u); \gamma(v)) ds \geq 0, \qquad \forall v \in V.$$

Remark 5.6 Hemivariational inequalties in subspaces of $L^p(\Omega)$ have been studied by Fundo [12][13] by means of Ekeland's variational principle.

Remark 5.7 Theorem 5.6 holds true for the eigenvalue problem of Remark 5.3.

5.5 Buckling of Beams on Adhesive Supports

We consider an elastic beam (linear elasticity) of length l measured along the axis Ox of the coordinate system Oxy. The beam is simply supported at its

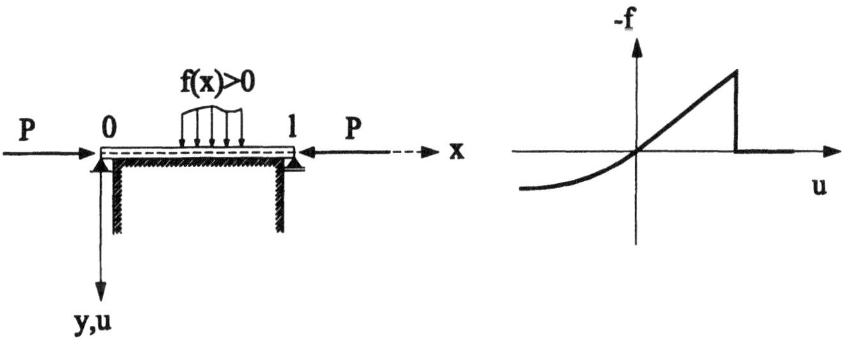

Fig. 5.1 Adhesively supported beam.

ends $x = 0$ and $x = l$ and on the interval $(0, l)$ is adhesively connected with a support (Fig. 5.1). The displacements of the beam are denoted by $x \to u(x)$ and the behaviour of the adhesive support is described by a nonmonotone possibly multivalued law between $-f(x)$ and $u(x)$, where $x \to f(x)$ denotes the reaction force per unit length vertical to the Ox axis, due to the adhesive material (cf. [30] p.51). The adhesive material can sustain a small tensile force before rupture (debonding). In Fig. 5.1 a rupture of brittle type is depicted in the $(-f, u)$ diagramm. The beam is assumed to have the modulus of elasticity E and let I be the moment of inertia of it. The beam is subjected to the compressive force P and we want to determine the buckling loading of the beam. This problem is yet an open problem in Engineering Sciences. From the large deflection theory of beams (cf. [21] p. 14), we may write the following relations which describe the behaviour of the beam:

$$a^2 u''''(x) + u''(x) = a^2 f(x) \qquad \text{on } (0,l); \tag{5.74}$$

$$u(0) = u(l) = 0, \tag{5.75}$$

$$u''(0) = u''(l) = 0. \tag{5.76}$$

Here $a^2 := IE/P$. We assume that the $(-f, u)$ graph results from a $L^\infty_{Loc}(\mathbf{R})$ function by filling- in the jumps, i.e. according to Proposition 1.7 there exists a locally Lipschitz function $j : \mathbf{R} \to \mathbf{R}$ such that

$$-f(x) \in \partial j(u(x)), \quad \forall x \in (0,l). \tag{5.77}$$

In order to be able to apply the results of Theorem 5.6, we assume that

(i) $j(0) = 0$

(ii) $|j(x) - j(y)| \le k|x-y|, \forall x, y \in \mathbf{R} (k > 0)$

(iii) $j(x)/|x|^2 \to +\infty$ as $|x| \to 0$

(iv) j even,

or more generally all the conditions in Theorem 5.5 and 5.6.

We set

$$V := H^2(\Omega) \cap H^1_0(\Omega).$$

It is a Hilbert space with the inner product (cf. [9] p.216, Lemma 4.2)

$$a(u,v) := \int_0^l u''(x) v''(x) dx.$$

Let $L : V \to V^*$ be the linear compact operator defined by

$$\langle Lu, v \rangle := \int_0^l u'(x) v'(x) dx, \qquad \forall u, v \in V.$$

The superpotential law (5.77) implies that

$$j^0(u(x); y) \ge -f(x) y, \qquad \forall x \in (0,l), \forall y \in \mathbf{R}.$$

Therefore, multiplying (5.74) by $v(x)$ and integrating over $(0,l)$ implies, by taking into account the boundary conditions (5.75) and (5.76), the weak formulation

$$a^2 \int_0^l u''(x)v''(x)dx - \int_0^l u'(x)v'(x)dx + \\ + a^2 \int_0^l j^0(u(x); v(x))dx \geq 0, \qquad \forall v \in V.$$

Thus buckling of the beam occurs if $\lambda := 1/a^2$ is an eigenvalue for the following hemivariational inequality

$$a(u,v) - \lambda \langle Lu, v\rangle + \int_0^l j^0(u(x); v(x))dx \geq 0, \qquad \forall v \in V. \tag{5.78}$$

Here $\lambda_k = \frac{k^2\pi^2}{l^2}$ $(k = 1, 2, 3...)$ and thus if $\lambda_k < \lambda < \lambda_{k+1}$, then problem (5.78) has at least $2k$ solutions which are pairwise symmetric.

Note that in this application we have applied Theorem 5.6, when λ stays before $\langle Lu, v\rangle$ and not before the L^2-inner product. The proof remains the same with minor modifications, as well as the content of Theorems 5.5 and 5.6, as we have noticed in Section 5.3 in Remark 5.3.

Suppose further that we have instead of a simply supported beam a beam with one end free and the other end built-in. Then $\lambda_k = \frac{(2k)^2\pi^2}{l^2}$ $(k = 1, 2, 3...)$. Finally in the case of a beam with built-in ends $\lambda_k = \frac{(2k-1)^2\pi^2}{4l^2}$ $(k = 1, 2, 3...)$.

Note that the results of Theorems 5.5 and 5.6 can be applied to the theory of buckling of Timoshenko plates and of cylindrical shells (Sections 4.6 and 4.7). Recall that all the above hold for even nonconvex superpotentials a fact fulfilled in most cases (cf. [27] p.27 ff). We leave as an exercise for the reader the nontrivial extension of Theorems 5.5 and 5.6 for $a(.,.)$ replaced by $a(.,.) + [C(.),.]$, where C is a compact nonlinear operator (cf. also Chapter 6).

5.6 Eigenvalue Problems for Linear Elastic Bodies Subjected to Nonmonotone Multivalued Boundary Conditions

Due to the physical nonlinearity of the boundary conditions which are assumed to hold we may have bifurcation of the solution. Thus we are led to the consideration of the corresponding eigenvalue problem (cf. Chapter 4 and [15],[26]).

We recall from [27],[30],[31] that such boundary conditions may describe adhesive contact effects, unilateral contact with a granular support, nonmonotone frictional behaviour etc.

We denote by Ω an open bounded subset of R^3 which is occupied by the body. The boundary of Ω is denoted by Γ and is assumed to be appropriately regular ($C^{0,1}$, i.e. a Lipschitzian boundary, is sufficient). The points $x \in \Omega, x = \{x_i\}, i = 1, 2, 3$, are referred to a Cartesian orthogonal coordinate system. We denote by $S = \{S_i\}$ the stress vector on Γ. We recall that $S_i = \sigma_{ij}n_j$, where $\sigma = \{\sigma_{ij}\}$ is an appropriately defined symmetric stress tensor and $n = \{n_i\}$ is the outward unit normal vector on Γ. The vector S may be decomposed into a normal component S_N and a tangential component S_T with respect to Γ, i.e.,

$$S_N = \sigma_{ij}n_jn_i \quad \text{and} \quad S_{T_i} = \sigma_{ij}n_j - (\sigma_{ij}n_in_j)n_i.$$

Analogously to S_N and S_T, u_N and u_T denote the normal and the tangential components of the displacement vector u with respect to Γ. S_N and u_N are considered as positive if they are parallel to n and have the same direction.

We denote further by $H^1(\Omega)$ the classical Sobolev space and by $H^{1/2}(\Gamma)$ the Sobolev space of the traces on Γ of the functions $v_i \in H^1(\Omega)$. Let us denote further by $\langle\cdot,\cdot\rangle$ the duality pairing between $[H^{1/2}(\Gamma)]^3$ and its dual space.

We denote by H_T the space

$$H_T = \{v | v \in [H^{1/2}(\Gamma)]^3, \ v_in_i = 0 \text{ a.e. on } \Gamma\}$$

and we recall ([31] p.32) that if $a = \{a_i\} \in [H^{1/2}(\Gamma)]^3$, and $a_N = a_in_i$, $a_T = \{a_{T_i}\}$ where $a_{T_i} = a_i - a_Nn_i$, then the mapping $a \to \{a_N, a_T\}$ is an isomorphism from $[H^{1/2}(\Gamma)]^3$ onto $H^{1/2}(\Gamma) \times H_T$. In the dual spaces, a'_N and a'_T are uniquely determined by the relation

$$\langle a', a\rangle = \langle a'_N, a_N\rangle_{1/2} + \langle a'_T, a_T\rangle_{H_T} \quad \forall a \in [H^{1/2}(\Gamma)]^3,$$

where $\langle\cdot,\cdot\rangle_{1/2}$ and $\langle\cdot,\cdot\rangle_{H_T}$ denote the duality pairings on $H^{1/2}(\Gamma)\times H^{-1/2}(\Gamma)$ and $H'_T \times H_T$. Obviously $a' \to \{a'_N, a'_T\}$ is again an isomorphism from $([H^{1/2}(\Gamma)]^3)'$ onto $H^{-1/2}(\Gamma) \times H'_T$. For all the above it is sufficient that Γ be $C^{1,1}$-regular ($C^{0,1}$-regularity, i.e. a Lipschitz boundary is also possible, cf. [16]).

Let $\Omega \subset R^3$ be an open bounded subset occupied by a deformable body in its undeformed state. On the assumption of small strains $\varepsilon = \{\varepsilon_{ij}\}$ we can write the relation

$$\int_\Omega \sigma_{ij}(u)\varepsilon_{ij}(v-u)d\Omega = (f, v-u) + \langle\sigma_{ij}n_j, (v_i - u_i)\rangle, \quad \forall v \in [H^1(\Omega)]^3, \quad (5.79)$$

where $u \in [H^1(\Omega)]^3$, $\sigma_{ij} \in L^2(\Omega)$, $f_i \in L^2(\Omega)$, and $(f,v) = \int_\Omega f_iv_id\Omega$ $i,j = 1,2,3$. Relation (5.79), which is obtained from the "operator equations of the problem" by applying the Green-Gauss theorem, is the expression of the "principle" of virtual work for the body when it is considered to be free (without boundary conditions). Note that for the derivation of (5.79) we have multiplied the equilibrium equation

$$\sigma_{ij,j} + f_i = 0,$$

where the $f_i \in L^2(\Omega)$ is the volume force vector, by $v_i - u_i$ and then we have integrated over Ω. On the assumption of "appropriately smooth" functions, we have applied the Green-Gauss theorem by taking into account the strain-displacement relationship

$$\varepsilon_{ij} = \frac{1}{2}(u_{i,j} + u_{j,i}).$$

An extension by density leads to the variational equality (5.79).

Let us assume further that the body is linear elastic, i.e. that

$$\sigma_{ij} = C_{ijhk}\varepsilon_{hk},$$

where $C = \{C_{ijhk}\}$, $i, j, h, k = 1, 2, 3$, is the elasticity tensor which has the well-known symmetry and ellipticity properties

$$C_{ijhk} = C_{jihk} = C_{khij}$$

$$C_{ijhk}\varepsilon_{ij}\varepsilon_{hk} \geq c\varepsilon_{ij}\varepsilon_{ij} \quad \forall \varepsilon = \{\varepsilon_{ij}\} \in R^6, \quad c = \text{const} > 0.$$

We denote the bilinear form of linear elasticity by $a(\cdot, \cdot)$, i.e.

$$a(u, v) = \int_{\Omega} C_{ijhk}\varepsilon_{ij}(u)\varepsilon_{hk}(v)d\Omega, \tag{5.80}$$

where $C_{ijhk} \in L^\infty(\Omega)$. Note also that instead of (5.79) we can write the relation

$$\begin{aligned}\int_{\Omega} \sigma_{ij}\varepsilon_{ij}(v - u)d\Omega &= (f, v - u) + \langle S_N, v_N - u_N\rangle_{1/2} + \\ &+ \langle S_T, v_T - u_T\rangle_{H_T}, \qquad \forall v \in [H^1(\Omega))]^3. \quad (5.81)\end{aligned}$$

Let us assume now that on Γ the classical boundary conditions $S_{T_i} = 0$, $i = 1, 2, 3$, and $u_N = 0$ hold. Then (5.81) with (5.80) leads to the following variational equality: Find $u \in V_0 = \{v \mid v \in [H^1(\Omega)]^3,\ v_N = 0 \text{ on } \Gamma\}$ such that

$$a(u, v) = (f, v) \quad \forall v \in V_0.$$

We assume that on a subset Γ_S of the boundary Γ of $\Omega \subset R^3$ a nonmonotone multivalued law holds between the displacements u and the reactions $-S$, that is

$$-S \in \partial j(u) \quad \text{on } \Gamma_S,$$

where j is a locally Lipschitz function on $\mathbb{R}^3$. Let on $\Gamma_U = \Gamma - \Gamma_S$, mes$\Gamma_U > 0$, the displacements be zero. Accordingly we can formulate from (5.81) a hemivariational eigenvalue problem of the form : Find $u \in V$, $\lambda \in \mathbb{R}$ such that

$$a(u, v - u) + \int_{\Gamma_S} j^0(\gamma u, \gamma v - \gamma u)d\Gamma \geq \lambda(u, v) \quad \forall v \in V. \tag{5.82}$$

Here $a(.,.)$ denotes the elastic energy of the body, $V = \{v | v \in [H^1(\Omega)]^3,\ v = 0 \text{ on } \Gamma_U\}$, γ denotes the trace mapping $v \to \gamma v : V \to [L^2(\Gamma)]^3$ which is compact (recall that $v \to \gamma v : V \to [H^{1/2}(\Gamma)]^3$ is continuous and $H^{1/2}(\Gamma) \subset L^2(\Gamma)$ is compact) and $L : V \to V^*$ is an appropriately defined linear compact operator. Note that γ can be omitted for the sake of simplicity in (5.82). Then we can apply Theorems 5.5 and 5.6 (cf. Remarks 5.3,5.4,5.5) and we obtain the existence of nontrivial solutions. Note that due to mes$\Gamma_U > 0$, $a(.,.)$ is coercive and thus only the assumptions (H_2') and (H_3), or $(H_2'')(H_3')(H_4)(H_5)$ for the superpotential j must be fulfilled.

5.7 Hemivariational Eigenvalue Problems for Boundary Conditions of Fuzzy-type

It is well known (cf. e.g. [30] p.55 and [27] p.35) that using a nonconvex superpotential introduced by Rockafellar we can describe mathematically stress-strain laws and displacement-reaction relations including regions of "fuzzy" type.

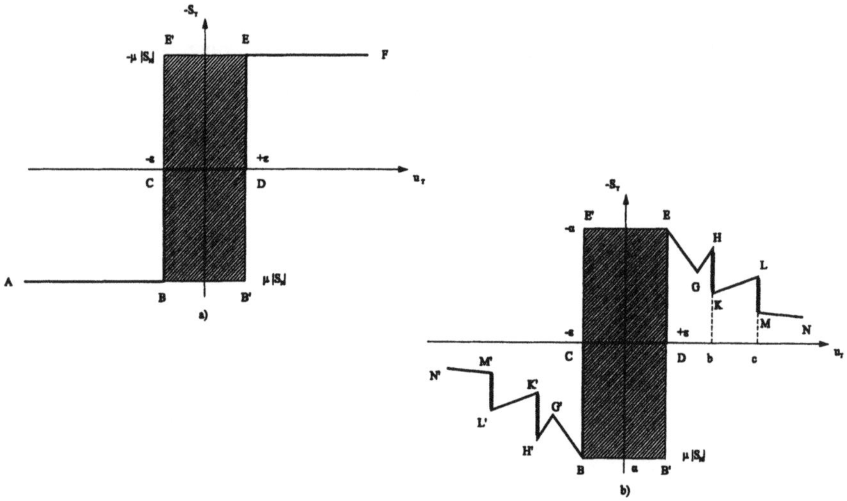

Fig. 5.2 Nonmonotone laws with nonfully determined values (fuzzy laws).
a) a friction law b) adhesive contact law (tangential)

In Fig. 5.2 reaction-displacement laws with nonfully determined values (fuzzy laws) in some regions are depicted. For instance, in Fig. 5.2 we give a friction law where $|S_T| \leq \mu|S_N|$ is verified as a strict inequality, not only for $u_T = 0$ but for $-\varepsilon \leq u_T < \varepsilon$, where ε is a small number, i.e. the reaction S_T may take for $-\varepsilon \leq u_T \leq \varepsilon$ any value in the interval $[-\mu|S_N|, \mu|S_N|]$. Thus the limit between adhesive frictional behaviour and sliding frictional behaviour is not clear. Therefore we have called this type of laws "fuzzy laws" in [30] where they have been described for the first time. In order to study such a law it is necessary to introduce the following nonconvex superpotential (see [33]):

Let l be an open subset of the real line R and let M be a measurable subset of l such that for every open and nonempty subset I of l, $\text{mes}\,(I \cap M)$ and $\text{mes}\,(I \cap (l - M))$ are positive. Let (b_1, b_2 are positive constants)

$$g(u) = \begin{cases} +b_1 & \text{if} \quad u \in M \\ -b_2 & \text{if} \quad u \notin M \end{cases}$$

and

$$F(u) = \int_0^u g(t)dt.$$

Then F is Lipschitzian and it can be verified that

$$\partial F(u) = [-b_2, b_1] \tag{5.83}$$

for every $u \in l$, i.e., we obtain an uncountable number of jumps in l (Fig. 5.3).

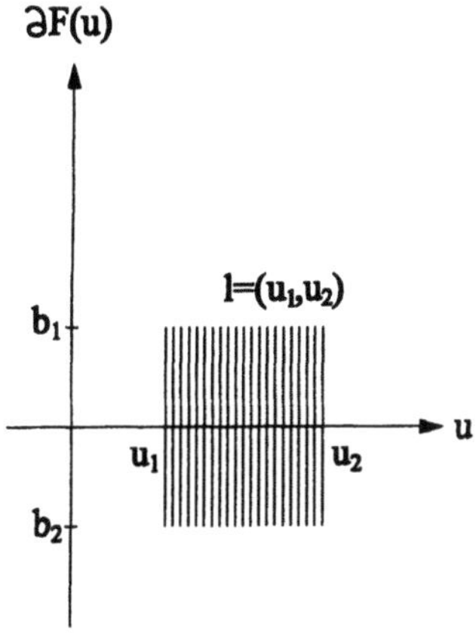

Fig. 5.3 The graph of (5.83).

Using the nonconvex superpotential F in (5.83) one may write the one-dimensional law of Fig. 5.2a in the form

$$-S_T \in \partial j(u_T) + \partial F(u_T).$$

Here j is the convex superpotential resulting by "integration" from the monotone law $ABCDEF$ and F results from (5.83) where $b_1 = b_2 = \mu|S_N|$ (μ denotes the friction coefficient) and $l = (-\varepsilon, \varepsilon)$. Analogously we proceed for the law of Fig. 5.2b (see [27] p.36).

Let us assume that we have the three-dimensional reaction-displacement law

$$-S \in \partial j(x, u) \qquad S(x) \in \mathbb{R}^3, \;\; u(x) \in \mathbb{R}^3 \qquad x \in \Gamma_S$$

holding on $\Gamma_S \subset \Gamma$. For $u_i \in [-\varepsilon, +\varepsilon]$, $\; i = 1, 2, 3$, where ε is small and given, we can localize a fuzzy behaviour. Thus we may write this law in the form

$$-S \in \partial j(x, u) + \partial F(u_1) + \partial F(u_2) + \partial F(u_3). \tag{5.84}$$

Here F is the same as in (5.83) and ∂ denotes as usual the generalized gradient. Thus the corresponding hemivariational eigenvalue problem reads: Find $u \in V$ such as to satisfy

$$a(u, v - u) + \int_{\Gamma_S} [j_T^0(x, \gamma u, \gamma v - \gamma u) +$$

$$+\sum_{i=1}^{3} F^0(\gamma u_i, \gamma v_i - \gamma u_i)]d\Gamma \geq \lambda(u, v-u) \quad \forall v \in V. \tag{5.85}$$

Here we assume that on $\Gamma_U = \Gamma - \Gamma_S$, $u_i = 0$, $i = 1, 2, 3$. Then V and all the notions are the same as in Section 5.6. In (5.85) we can omit γ for the sake of simplicity. Theorems 5.5 and 5.6 imply (with minor modifications in its proof) the existence of solutions of (5.85).

5.8 Nonmonotone Multivalued Relations in Structural Analysis and the Corresponding Eigenvalue Problem

Structural analysis deals with discretized models of structures. These models are obtained by means of the Finite Element Method (F.E.M.). Suppose that a structure is discretized by n-finite elements and that the method of the natural generalized stresses and strains are applied for the description of the discretized stress and strain fields. This means that the strains of each element are not affected by rigid body displacements and the stresses within each element are selfequilibrated [1],[2],[3]. Moreover, we assume that the geometric nonlinearity (assumption of large displacements) is treated by means of an incremental procedure, i.e. the loading is simulated by a sequence of small increments. A dot on a quantity denotes the increment of this quantity. We denote the stress (resp. the strain) vector of the i-element by s_i (resp. e_i) and the displacement (resp. the load) vector of the k-node by u_k (resp. p_k). Let m be the total number of nodes. We assume that the behaviour of each element contains a linear part and a strongly nonlinear part with nonmonotonicity and multivaluedness. Thus we may write the incremental stress-strain law, with respect to a reference configuration in the form

$$\dot{s}_i \in \partial j_i(\dot{e}_i) + \partial \frac{1}{2} \dot{e}_i^T F_{0i}^{-1} \dot{e}_i, \quad i = 1, \ldots, n, \tag{5.86}$$

where j_i is a locally Lipshitz superpotential, ∂ is the generalized gradient, and F_{0i} is the symmetric tangential natural flexibility matrix of the step under consideration. The nonconvex superpotential relation in (5.86) may describe the zig-zag behaviour of reinforced concrete in tension (Scanlon's diagram [11]), or of a composite material [29],[17], the semirigid connections in steel structures [30] etc. For the whole structure (5.86) takes the form

$$\dot{s} - F_0^{-1} \dot{e} \in \Lambda \dot{e}, \tag{5.87}$$

where $\dot{e} = [\dot{e}_1^T, \ldots, \dot{e}_n^T]$ (T denotes the transpose of a vector or a matrix), $F_0 = \text{diag}[F_{01}, \ldots, F_{0n}]$, and $\Lambda(\dot{e}) = [\partial j_1(\dot{e}_1)^T, \partial j_2(\dot{e}_2)^T, \ldots, \partial j_n(\dot{e}_n)^T]^T$. In order to complete the equations of the problem we must take into account the conditions of equilibrium and the compatibility conditions appropriately written for the stability problem as

$$G\dot{s} + K_G \dot{u} = \dot{p} \tag{5.88}$$

and

$$\dot{e} = G^T \dot{u} \tag{5.89}$$

respectively. Here the first equation (5.88) is linearized equation of equilibrium whithin the step under consideration. K_G is the geometric stiffness matrix and for the stability question we replace $\dot{p}$ by $\lambda\dot{u}$ (cf. Chapter 4). Note that K_G depends on the stresses at the end of the preceding step and is a symmetric, positive semidefinite (assumption) matrix. We obtain that

$$\dot{s}^T(\dot{e}^\star - \dot{e}) = \dot{p}^T(\dot{u}^\star - \dot{u}) \quad \forall \dot{u}^\star \in \mathbb{R}^l,$$

where $\dot{e}^\star$ and $\dot{u}^\star$ satisfy (5.89) and l is the dimension of the displacement vector u. By the definition of the generalized gradient, (5.87) implies that the structure obeys to the following hemivariational inequality: Find $u \in \mathbb{R}^l$ such as to satisfy the inequality

$$\sum_{i=1}^{n} j_i^0([G^T\dot{u}]_i\,,\ [G^T(\dot{u}^\star - \dot{u})]_i) + \dot{u}^T K(\dot{u}^\star - \dot{u}) + \dot{u}^T K_G(\dot{u}^\star - \dot{u}) \geq$$

$$\geq \lambda\dot{u}^T(\dot{u}^\star - \dot{u}) \quad \forall \dot{u}^\star \in R^l \quad \text{where} \quad K = GF_0^{-1}G^T. \tag{5.90}$$

Since the lines of matrix G are by definition linearly independent, the total stiffness matrix $K = GF_0^{-1}G^T$ is with F_0^{-1} (strictly) positive definite.

Theorems 5.2, 5.5 and 5.6 can be adapted for the above finite dimensional problem, if the respective assumptions for j_i $i = 1, \ldots, n$ are fulfilled.

References

[1]J.H. Argyris, Three-Dimensional Anisotropic and Inhomogeneous Elastic Media. Matrix Analysis for Small and Large Displacements. *Ing. Archiv* **34** (1965), 33-55.

[2]J.H. Argyris, Continua and Discontinua, Proc. 1st Conf. Matrix Meth. Struct. Mech. Wright Patterson Air Force Base, Dayton, Ohio 1965, AFFDL TR (1966), 66-80.

[3]J.H. Argyris, D.W. Scharpf: Some General Considerations on the Natural Mode Technique. Aeron. *J. Royal Aeron. Soc.* **73**(1969), 218-226 and 361-368.

[4]J. P. Aubin, *Applied Functional Analysis,* John Wiley & Sons, New York, (1979).

[5]K. C. Chang, Variational Methods for Non-differentiable Functionals and their Applications to Partial Differential Equations, *J.Math. Anal. Appl.* **80**(1981), 102-129.

[6]D.C. Clark, A variant of the Lusternik-Schnirelman theory, *Indiana Univ. Math. J.* **22**(1972), 65-74.

[7]F. H. Clarke, *Optimization and Nonsmooth Analysis,* John Wiley and Sons, New York, (1983).

[8]J.N. Corvellec, M. Degiovanni and M. Marzocchi, Deformation Properties for Continuous Functionals and Critical Point Theory, *Topol. Methods Nonlinear Analysis* **1**(1993), 151-171.

[9]G. Duvaut and J.L. Lions, *Les Inéquations en Mécanique et en Physique,* Dunod, Paris (1972).

[10]I. Ekeland and A. Szulkin, Minimax Results of Lusternik-Schnirelman type and Applications. Séminaire de Mathématiques Supérieures, Vol. 107, Université de Montréal, Montréal (1989).

[11]H. Floegl, H.A. Mang: Tension Stiffening Concept Based on Bond Slip. *ASCE (ST 12)* **108**(1982), 2681-2701.

[12]M. Fundo, Hemivariational Inequalities in Subspaces of $L^p(\Omega)$ $(p \geq 3)$, *J. Nonlin. Anal. T.M.A.* (accepted).

[13]M. Fundo, An Existence Result on a General Type of Hemivariational Inequalites, *Applic. Anal.* **65**(1997), 373-394.

[14]N. Ghoussoub, Duality and Perturbation Methods in Critical Point Theory, Cambridge University Press, Cambridge (1993).

[15]D. Goeleven, D. Motreanu and P.D. Panagiotopoulos, Multiple Solutions for a Class of Eigenvalue Problems in Hemivariational Inequalities, *J. Nonlin. Anal. T.M.A.* **29**(1997), 9-26.

[16]J. Haslinger, I. Hlavaček, Approximation of the Signorini Problem with Friction by a Mixed Finite Element Method, *J. Math. Anal.,* **86**(1982), 99-122.

[17]J. Hult, L. Travniček, Carrying Capacity of Fibre Bundles with Varying Strength and Stiffness. *J. Méc. Théor. et Appl.* **2**(1983), 643-657.

[18]H. N. Karamanlis, *Buckling problems in composite von Kármán Plates*, Doct. Thesis, Aristotle University (1991).

[19]H. N. Karamanlis and P. D. Panagiotopoulos, The Eigenvalue Problem in Hemivariational Inequalities and its Application to Composite Plates. *Journal of the Mech. Behaviour of Materials* (Freund Publ., Tel Aviv), **5**(1993)25-38.

[20]J. Mawhin and M. Willem, *Critical Point Theory and Hamiltonian Systems*. Springer, New York (1989).

[21]P. Moliotis, *Theory of Elastic Stability*. Thessaloniki, 1969 (in greek).

[22]D. Motreanu and P.D. Panagiotopoulos, A Minimax Approach to the Eigenvalue Problem of Hemivariational Inequalities and Applications, *Applic. Anal.* **58**(1995), 53-76.

[23]D. Motreanu and P.D. Panagiotopoulos, On the Eigenvalue Problem for Hemivariational Inequalities: Existence and Multiplicity of Solutions, *J. Math. Anal. Appl.* **197**(1996), 75-89.

[24]D. Motreanu and P.D. Panagiotopoulos, An Eigenvalue Problem for a Hemivariational Inequality involving a Nonlinear Compact Operator, *Set-Valued Anal.* **3**(1995), 1-9.

[25]D. Motreanu and P.D. Panagiotopoulos, Nonconvex Energy Functions, Related Eigenvalue Hemivariational Inequalities on the Sphere and Applications, *J. Global Optimization* **6**(1995), 163-177.

[26]D. Motreanu and P. D. Panagiotopoulos, *Hysteresis the Eigenvalue Problem for Hemivariational Inequalities*, in "Models of Hysteresis" (ed. by A.Visintin) Pitman Res. Notes in Mathematics Nr. 286, Longman, Harlow (1993), pp. 102-117.

[27]Z. Naniewicz and P. D. Panagiotopoulos, *Mathematical Theory of Hemivariational Inequalities*, Marcel Dekker, N.York, (1995).

[28]L. Nirenberg, Variational and Topological Methods in Nonlinear Problems, *Bull. Am. Math. Soc.* **4**(1981), 267-302.

[29]G. Ondracek (ed.): Verbundwerkstoffe. Phasenverbindung und mechanische Eigenschaften, Deutsche Gesellschaft für Metallkunde, Band 1, 1985.

[30]P. D. Panagiotopoulos, *Hemivariational inequalities. Application to Mechanics and Engineering*, Springer Verlag, Berlin, N.York (1993).

[31]P.D. Panagiotopoulos, *Inequality Problems in Mechanics and Applications. Convex and Nonconvex Energy Functionals*, Birkhäuser Verlag, Boston, Basel, (1985) (Russian Transl. MIR Publ. Moscow, 1989).

[32]P. H. Rabinowitz, Minimax Methods in Critical Point Theory with Applications to Differential Equations, CBMS Regional Conf. Sec. in Math., No. 65, *Amer. Math. Soc.*, Providence, R. I., (1986).

[33]R.T. Rockafellar, La théorie des Sous-Gradients et ses Applications à l'Optimization. Fonctions Convexes et Non-convexes, Les Presses de l' Université de Montréal, Montréal, (1979).

[34]A. Szulkin, Minimax Principles for Lower Semicontinuous Functions and Applications to Nonlinear Boundary Value Problems, *Ann. Inst. Henri Poincaré Anal. Non Linéaire* **3**(1986), 77-109.

6. Eigenvalue Problems for Hemivariational Inequalities on the Sphere

The theory of hemivariational inequalities is closely related to the derivation of variational "principles" for problems involving nonconvex, nonsmooth energy functions. We refer to [10]-[12] for all related references and to [13],[14] concerning the area of Nonsmooth Mechanics. Mathematical questions concerning the existence of solutions of hemivariational inequalities have been treated by means of compactness arguments [11][12], or by means of pseudomonotonicity arguments for multivalued mappings [7]-[9].

In this chapter we continue our study on eigenvalue problems for hemivariational inequalites arising in problems involving nonconvex, nonsmooth energy functions. They lead to the search for critical points of nonconvex, nonsmooth potential functions, which in turn give rise to hemivariational inequalities. In this chapter we focus our attention on eigenvalue problems for hemivariational inequalites whose solutions must belong to a sphere. Having here in mind problems of Elasticity, or generally of Engineering (adhesively connected plates), and problems of Economics (network flow problems, cf.[9]) leading to hemivariational inequalities, for which, either the solution remains a priori bounded, or the cost, or the weight of the structure are prescribed, we study the corresponding eigenvalue problem on a sphere

$$S_r = \{u \in V : \|u\|_V = r\}, \quad r > 0$$

in a real Hilbert space V. In Section 6.1 the existence of the solution is investigated and in Section 6.2 the multiplicity of solutions. In Section 6.3 another type of eigenvalue problem is studied and Section 6.4 describes certain applications of the method concerning constant weight or cost problems.

6.1 An Existence Result

Let V be a real Hilbert space, with the scalar product $(\cdot,\cdot)_V$ and the associated norm $\|\cdot\|_V$. V is densely and compactly imbedded in $L^p(\Omega;\mathbb{R}^N)$ for some $p \geq 2$, an integer $N \geq 1$ and a bounded domain Ω in $\mathbb{R}^m$, $m \geq 1$. The pairing on $V \times V^*$ is denoted by $\langle\cdot\rangle_V$. The Euclidean norm in the Euclidean space $\mathbb{R}^N$ is denoted by $|\cdot|$ and the pairing over $\mathbb{R}^N \times (\mathbb{R}^N)^*$ by $\langle\cdot,\cdot\rangle_{\mathbb{R}^N}$.

In the space V we consider the sphere S_r of center 0 and radius $r > 0$, i.e. , $\|u\|_V = r$. It is a Riemannian manifold with the Riemannian structure induced by the Hilbert space V. The (geodesic) distance on S_r is denoted by $d(\cdot,\cdot)$, that is, for the points $u, v \in S_r$, $d(u,v)$ is equal to the length of the minimal arc of the great circle on S_r joining u and v.

Since V is continuously imbedded into $L^p(\Omega;\mathbf{R}^N)$, there is a constant $c_0 > 0$ such that

$$\|v\|_{L^p} \leq c_0\|v\|_V, \quad v \in V. \tag{6.1}$$

Let $\alpha : V\times V \to \mathbf{R}$ be a continuous symmetric bilinear form, let $C : S_r \to V^*$ a compact (nonlinear) mapping (in the sense that the closure of the range $C(S_r)$ of C is compact in V^*) and let $j : \Omega \times \mathbf{R}^N \to \mathbf{R}^N$ be a Caratheodory function, locally Lipschitz in the second variable and with $j(\cdot,0)$ bounded on Ω. We denote by $A : V \to V^*$ the operator which corresponds to the bilinear form α ,

$$\langle Au, v\rangle_V = \alpha(u,v), \quad u, v \in V.$$

For the sake of simplicity we denote Clarke's [4] generalized gradient $\partial_y j(x,y)$ of j with respect to the second variable simply by $\partial j(x,y)$. The same convention will be used relative to Clarke's generalized directional derivative $j^0(x,y;z)$. For any other locally Lipschitz real-valued functional G, the notations ∂G and G^0 denote the generalized gradient and the generalized directional derivative, respectively.

This Section deals with the following eigenvalue problem with constraints: Find $u \in V$ and $\lambda \in \mathbf{R}$ such that the two relations below (6.2), (6.3) hold

$$\alpha(u,v) + \langle C(u), v\rangle_V + \int_\Omega j^0(x,u(x);v(x))dx \geq \lambda(u,v)_V, \quad v \in V, \tag{6.2}$$

$$\|u\|_V = r. \tag{6.3}$$

The following hypotheses are assumed to hold.

(H_1) The growth condition

$$|w| \leq c(1+|y|^{p-1}) \text{ for all } \quad w \in \partial j(x,y), \quad x \in \Omega, y \in \mathbf{R}^N,$$

where $c > 0$ is a constant, holds for j.

(H_2) A (Fréchet) differentiable function $g : V \to \mathbf{R}$ and a lower semicontinuous (l.s.c.) on S_r function $h : V \to \mathbf{R}$ can be determined, such that

$$\langle C(u), v\rangle_V \;\geq\; \langle g'(u), v\rangle_V + h(u+v) - h(u) \tag{6.4}$$

for all $u \in S_r$ and $v \in T_uS_r$;

$$g(u) + h(u) \geq c_1, \quad u \in S_r, \tag{6.5}$$

for a constant c_1 ;

$$h(\exp_u(tv)) \leq (1-t)h(u) + th(u+v), \tag{6.6}$$

for $u \in S_r$, $v \in T_uS_r$ and $0 < t < 1$.

(H_3) For every sequence $\{u_n\} \subset S_r$ with $u_n \to u$ weakly in V, $\alpha(u_n, u_n) + {}$ $+\langle C(u_n), u_n\rangle_V \longrightarrow \alpha_0 \in \mathbb{R}$, and for every $w \in L^{p/(p-1)}(\Omega; \mathbb{R}^N)$ with

$$w(x) \in \partial j(x, u(x)) \quad \text{for a.e.} \quad x \in \Omega \tag{6.7}$$

such that $(A - \lambda_0 \Lambda)u_n$ converges in V^* , where

$$\lambda_0 := r^{-2}(\alpha_0 + \int_\Omega \langle w(x), u(x)\rangle_{\mathbb{R}^N} dx),$$

there exists a strongly convergent subsequence of $\{u_n\}$ in V (thus in S_r).

In hypothesis (H_2) , the notation $T_u S_r$ means the tangent space of S_r at $u \in S_r$, thus

$$T_u S_r = \{v \in V : (v, u)_V = 0\}. \tag{6.8}$$

The notation $\exp_u$ in (6.6) denotes the exponential mapping (in the sense of Riemannian manifolds) of S_r at $u \in S_r$, $\exp_u : T_u S_r \longrightarrow S_r$, defined by

$$\exp_u(v) = \gamma_v(1), \quad v \in T_u S_r, \tag{6.9}$$

where $t \to \gamma_v(t)$ in (6.9) denotes the unique geodesic (arc of a great circle) of S_r satisfying

$$\gamma_v(0) = u \quad (\text{if } \ v \in T_u S_r) \ \text{ and } \ \gamma_v'(0) = v.$$

The mapping $\Lambda : V \to V^*$ of (H_3) represents the duality mapping

$$\langle \Lambda u, v\rangle_V = (u, v)_V, \quad u, v \in V.$$

Now we are in the position to formulate our existence result for the eigenvalue problem (6.2), (6.3).

Theorem 6.1 On the assumptions (H_1) – (H_3) the eigenvalue problem (6.2), (6.3) has a solution $(u, \lambda) \in V \times \mathbb{R}$ $(u \in S_r)$ with

$$\lambda = r^{-2}\left(\alpha(u, u) + \langle C(u), u\rangle_V + \int_\Omega \langle w(x), u(x)\rangle_{\mathbb{R}^N} dx\right)$$

for certain $w \in L^{p/(p-1)}(\Omega; \mathbb{R}^N)$ satisfying (6.7).

Proof. By (H_1) and Lebourg's mean value theorem (see Clarke [4], p.41) one obtains that j verifies the following growth condition

$$\begin{aligned} |j(x,y)| &\leq |j(x,0)| + |j(x,y) - j(x,0)| \\ &\leq |j(x,0)| + (sup\{|w| : w \in \partial j(x, \bar{y}), \bar{y} \in [0, y]\})|y| \\ &\leq |j(x,0)| + c(|y| + |y|^p) \leq c_2 + c_3|y|^p, \quad \forall x \in \Omega, y \in \mathbb{R}^N, \end{aligned} \tag{6.10}$$

with positive constants c_2 and c_3 .

We introduce the functional $J : L^p(\Omega; \mathbb{R}^N) \to \mathbb{R}$ by setting

$$J(v) = \int_\Omega j(x, v(x))dx, \quad v \in L^p(\Omega; \mathbb{R}^N), \tag{6.11}$$

the functional $E : V \to \mathbb{R}$ by setting

$$E(v) = (1/2)\alpha(v, v) + g(v), \quad v \in V \tag{6.12}$$

and $I : V \to \mathbb{R}$ by setting

$$I = E + J|_V + h. \tag{6.13}$$

Relations (6.1),(6.5) and (6.10) imply that I is bounded from below on S_r

$$\begin{aligned} I(u) &\geq -(1/2)\|\alpha\|\|u\|_V^2 + g(u) + h(u) - c_2|\Omega| - c_3\|u\|_{L^p}^p \\ &\geq -(1/2)\|\alpha\|r^2 + c_1 - c_2|\Omega| - c_3 c_0^p r^p, \quad u \in S_r \ . \end{aligned} \tag{6.14}$$

Due to (6.14) and because functional I defined in (6.13) is l.s.c. we can apply to I the variational principle of Ekeland on the complete metric space S_r (see Ekeland [5]). Then there exists a sequence $\{u_n\} \subset S_r$ such that

$$I(u_n) \leq \inf_{S_r} I + \frac{1}{n} \tag{6.15}$$

and

$$I(w) \geq I(u) - (1/n)d(w, u_n) \ \text{ for each } \ w \in S_r. \tag{6.16}$$

If we set

$$w = \exp_{u_n}(tv) \ \text{ with } \ t > 0 \ \text{ and } \ v \in T_{u_n}S_r,$$

inequality (6.16) becomes

$$E(\exp_{u_n}(tv)) - E(u_n) + J(\exp_{u_n}(tv)) - J(u_n) +$$

$$+h(\exp_{u_n}(tv)) - h(u_n) \geq -(1/n)d(\exp_{u_n}(tv), u_n),$$

or by virtue of (6.6),

$$E(\exp_{u_n}(tv)) - E(u_n) + J(\exp_{u_n}(tv)) - J(u_n) +$$

$$+t(h(u_n + v) - h(u_n)) \geq -(1/n)d(\exp_{u_n}(tv), u_n) \tag{6.17}$$

for $v \in T_{u_n}S_r$ and $t > 0$.

The exponential mapping $\exp_{u_n} : T_{u_n}S_r \to S_r$ has the following properties:

$$\frac{d}{dt}(\exp_{u_n}(tv)) \mid_{t=0} = v, \quad v \in T_{u_n}S_r, \tag{6.18}$$

$$d(\exp_{u_n}(tv), u_n) = \|tv\|_V = t\|v\|_V \tag{6.19}$$

for $v \in T_{u_n}S_r$ and $t > 0$ sufficiently small.

Dividing by $t > 0$ and letting $t \to 0$ in (6.17) we obtain that

$$\alpha(u_n, v) + \langle g'(u_n), v\rangle_V + h(u_n + v) - h(u_n) + J^0(u_n; v) \geq -(1/n)\|v\|_V$$

$$\text{for all} \quad v \in T_{u_n} S_r. \tag{6.20}$$

To obtain (6.20) we use the differentiability of the functional E in (6.12), the definition of $J^0(u_n; v)$ and the formulae (6.18), (6.19). From (6.20) and (6.4) we obtain that

$$\alpha(u_n, v) + \langle C(u_n), v\rangle_V + J^0(u_n; v) \geq -(1/n)\|v\|_V, \quad v \in T_{u_n} S_r. \tag{6.21}$$

Notice that the left-hand side of (6.21) is continuous,convex with respect to $v \in T_{u_n} S_r$ and vanishes at 0. Consequently, Lemma 1.3 in Szulkin [18] (i.e. Lemma 5.8) can be applied in the tangent space $T_{u_n} S_r$. It yields an element $z_n \in (T_{u_n} S_r)^*$ of norm ≤ 1 verifying

$$\alpha(u_n, v) + \langle C(u_n), v\rangle_V + J^0(u_n; v) \geq (1/n)\langle z_n, v\rangle_V \ , \ v \in T_{u_n} S_r \ . \tag{6.22}$$

Applying Hahn-Banach theorem one obtains an element (denoted again by z_n) in V^* with

$$\|z_n\|_{V^*} \leq 1 \tag{6.23}$$

so that (6.22) remains true (on $T_{u_n} S_r$).

The density of V in $L^p(\Omega; \mathbb{R}^N)$ implies (cf. Corollary 1.2)

$$\partial(J|_V)(u) \subset \partial J(u), \qquad u \in V \ . \tag{6.24}$$

Then (6.22) and (6.24) imply that an element

$$w_n \in \partial J(u_n), \qquad n \geq 1 \tag{6.25}$$

can be determined satisfying the equality

$$\langle Au_n + C(u_n) + w_n - (1/n)z_n, v\rangle_V = 0, \quad v \in T_{u_n} S_r \ . \tag{6.26}$$

From (6.26) and the characterization of the tangent space $T_{u_n} S_r$ (cf. (6.8)) we obtain that there exist numbers $\lambda_n \in \mathbb{R}$ such that

$$Au_n + C(u_n) + w_n - (1/n)z_n = \lambda_n \Lambda u_n, \quad n \geq 1 \ . \tag{6.27}$$

The boundedness of $\{u_n\} \subset S_r$ ensures the existence of a subsequence again denoted by $\{u_n\}$ such that

$$u_n \to u \quad \text{weakly in} \ V \ , \tag{6.28}$$

$$u_n \to u \quad \text{strongly in} \ L^p(\Omega; \mathbb{R}^N), \tag{6.29}$$

$$\{C(u_n)\} \ \text{is strongly convergent in} \ V^*, \tag{6.30}$$

$$\alpha(u_n, u_n) + \langle C(u_n), u_n\rangle_V \to \alpha_0 \in \mathbb{R} \ . \tag{6.31}$$

Here we use also the compactness of the imbedding $V \subset L^p(\Omega; \mathbb{R}^N)$ and of the mapping $C : S_r \to V^*$. Since the functional J in (6.11) is locally Lipschitz on $L^p(\Omega; \mathbb{R}^N)$, relations (6.25) and (6.29) imply that the sequence $\{w_n\}$ is bounded in $L^p(\Omega;\mathbb{R}^N)$. According to the compact imbedding $V^* \subset L^{p/(p-1)}(\Omega;\mathbb{R}^N)$ we may suppose that for some $w \in V^*$

$$w_n \to w \quad \text{strongly in} \quad V^* . \tag{6.32}$$

The upper semicontinuity of the generalized gradient ∂J (cf. Proposition 1.4) and relations (6.29), (6.25) imply that

$$w \in \partial J(u) . \tag{6.33}$$

We note that the growth condition (H_1) is just relation (1.14), so it is allowed to apply Theorem 1.3. It follows that (6.33) implies (6.7).

By scalar product multiplication of (6.27) with u_n one obtains

$$\alpha(u_n, u_n) + \langle C(u_n) + w_n - (1/n)z_n, u_n\rangle_V = \lambda_n r^2 .$$

This equality and relations (6.31)-(6.33),(6.23),(6.28) imply that

$$\lambda_n \to \lambda_0 := r^{-2}(\alpha_0 + \int_\Omega \langle w(x), u(x)\rangle_{\mathbb{R}^N} dx), \quad \text{as } n \to \infty . \tag{6.34}$$

Furthermore, we get from (6.27),(6.30),(6.32) and (6.23) that

$$\{Au_n - \lambda_n \Lambda u_n\} \ \text{converges in} \quad V^* . \tag{6.35}$$

Combining (6.34),(6.35) we conclude that $\{(A - \lambda_0\Lambda)u_n\}$ converges in V^*. This fact together with (6.31),(6.33),(6.7) permits us to apply hypothesis (H_3). We arrive at the conclusion that along a subsequence

$$u_n \to u \quad \text{strongly in} \quad V , \tag{6.36}$$

with $u \in S_r$.

Passing to the limit in (6.15) implies that

$$\lim_{n\to\infty} I(u_n) = \inf_{S_r} I . \tag{6.37}$$

The lower semicontinuity of h on S_r and relations (6.36),(6.37) lead to

$$h(u) \le \liminf_{n\to\infty} h(u_n) = \lim_{n\to\infty}(I(u_n) - E(u_n) - J(u_n)) = \inf_{S_r} I - E(u) - J(u) ,$$

and therefore

$$I(u) = \inf_{S_r} I . \tag{6.38}$$

Due to (6.38)

$$\begin{aligned} 0 \le I(\exp_u(tv)) - I(u) \ &= \ E(\exp_u(tv)) - E(u) + J(\exp_u(tv)) - J(u)) \\ &\quad + h(\exp_u(tv)) - h(u) , \quad v \in T_u S_r \ \text{ and } \ t > 0 , \end{aligned}$$

or, by (6.6),

$$t^{-1}(E(\exp_u(tv))-E(u))+t^{-1}(J(\exp_u(tv))-J(u))+h(u+v)-h(u) \geq 0\,. \quad (6.38a)$$

For $t \to 0$ we get from the foregoing inequality, by means of (6.18), that

$$\alpha(u,v) + \langle C(u), v\rangle_V + J^0(u;v) \geq 0\,, \quad v \in T_u S_r\,, \tag{6.39}$$

which can be written in the inclusion form

$$-(Au + C(u))|_{T_u S_r} \in \partial(J|_{S_r})(u)\,.$$

This is equivalent to the existence of some $w \in V^*$ satisfying (6.33) and

$$\langle Au + C(u) + w, v\rangle_V = 0\,, \quad v \in T_u S_r\,. \tag{6.40}$$

Comparing (6.8) and (6.40) we obtain that a $\lambda \in \mathbb{R}$ exists with the property

$$Au + C(u) + w = \lambda u\,. \tag{6.41}$$

From (6.33) and hypothesis (H_1) we obtain (cf. Section 1.3 and [4] p.83)

$$w \in \partial J(u) \subset \int_\Omega \partial j(x, u(x))dx$$

in the sense that $w \in L^{p/(p-1)}(\Omega; \mathbb{R}^N)$ and (6.7) holds. Then (6.41) and Proposition 1.4 imply that

$$\begin{aligned}\lambda(u,v)_V &= \alpha(u,v) + \langle C(u), v\rangle_V + \int_\Omega \langle w(x), v(x)\rangle_{\mathbb{R}^N} dx \\ &\leq \alpha(u,v) + \langle C(u), v\rangle_V + \int_\Omega j^0(x, u(x); v(x))dx\,, \quad v \in V\,,\end{aligned}$$

which is the relation (6.2). Equality (6.3) was already proved (see (6.36)). Thus the eigenvalue λ can be calculated directly from (6.41). □

Corollary 6.1 Under assumptions $(H_1)-(H_3)$ the variational hemivariational inequality

$$\alpha(u,v) + \langle g'(u), v\rangle_V + h(u+v) - h(u) + \int_\Omega j^0(x, u(x); v(x))dx \geq 0$$

$$\text{for all } \; v \in V \; \text{ with } \; (u,v)_V = 0,$$

has a solution $u \in V$ with $\|u\|_V = r$.

Proof. This result has been obtained within the proof of Theorem 6.1. More precisely, it is deduced from the inequality (6.38a). □

Let us now give some aplications of the foregoing general theory.

(a) We assume first that on a Hilbert space V, $\alpha : V \times V \to \mathbb{R}$ is coercive, i.e.

$$\alpha(v,v) \geq \bar{\alpha}\|v\|_V^2 , \quad v \in V$$

where $\bar{\alpha}$ is a constant > 0, that $C = 0$ and that $j : \Omega \times \mathbb{R}^N \to \mathbb{R}$ verifies (H_1) and the following generalized sign condition

$$\int_\Omega \langle w(x), u(x)\rangle_{\mathbb{R}^N} dx < 0 , \tag{6.42}$$

for all $u \in S_r$ and for $w \in L^{p/(p-1)}(\Omega; \mathbb{R}^N)$ satisfying (6.7). Then (H_2) holds with $g = h = 0$ and (H_3) is also valid, because, due to (6.42), λ_0 in (H_3) satisfies $\lambda_0 < \bar{\alpha}$ for all $u \in S_r$. Consequently, λ_0 is an element of the resolvent set of the operator $A : V \to V^*$ and therefore (H_3) is true.

Several other situations can be treated analogously, for example, when λ_0 is an eigenvalue of A, but with a finite dimensional eigenspace corresponding to it. This results from the a priori boundedness $\|u_n\|_V = r$ in (H_3).

(b) Concerning hypothesis (H_2) the convexity condition (6.6) relative to the sphere S_r is in fact a fairly weak requirement for the function $h : V \to \mathbb{R}$. For instance, let us put $h \equiv$ constant $= k_0$ on S_r and $h =$ any function $\geq k_0$ on $V \setminus S_r$. Then (6.6) is verified and h is l.s.c. on S_r.

6.2 Multiplicity of Solutions for a Special Case

In this section we study under certain additional symmetry assumptions the multiplicity of solutions $(u, \lambda) \in V \times \mathbb{R}$, with $\|u\|_V = r$, for the following eigenvalue hemivariational inequality on the sphere S_r:

Find $u \in V$ and $\lambda \in \mathbb{R}$ such that

$$\begin{cases} \alpha(u,v) + C(u,v) + \int_\Omega j^0(x, u(x); v(x))dx \geq \lambda(u,v)_V, & v \in V, \\ \|u\|_V = r . \end{cases} \tag{6.43}$$

Clearly the problem is related to the case of Section 6.1. The data α and j in (6.43) have the same meaning as in Section 6.1, but here C denotes a real function $C : S_r \times V \to \mathbb{R}$. A kind of compactness assumption for C is hidden in the hypothesis below (due to the compact inclusion of V into $L^p(\Omega; \mathbb{R}^N)$). No continuity assumption is necessary.

(H_2') There exists a locally Lipschitz function $f : L^p(\Omega; \mathbb{R}^N) \to \mathbb{R}$, bounded on S_r, satisfying

$$C(u,v) \geq f^0(u;v) \text{ for all } u \in S_r \text{ and } v \in V \text{ with } (u,v)_V = 0.$$

Hypothesis (H_3) is replaced by

(H_3') For every sequence $\{u_n\} \subset S_r$ with $u_n \to u$ weakly in V, for every $z_n \in \partial f(u_n)$ with

$$\alpha(u_n, u_n) + \langle z_n, u_n \rangle_V \to \alpha_0 \in \mathbb{R} ,$$

where α_0 is some real number, and for every $w \in L^{p/(p-1)}(\Omega; \mathbb{R}^N)$ verifying (6.7) such that $(A - \lambda_0 \Lambda)u_n$ converges in V^*, where λ_0 is as in (H_3), there exists a (strongly) convergent subsequence of $\{u_n\}$ in V (thus in S_r).

In addition, we suppose a symmetry condition.

(H_4) j is even with respect to the second variable $y \in \mathbb{R}^N$, i.e.,

$$j(x, y) = j(x, -y) \quad \text{for all} \quad x \in \Omega,\ y \in \mathbb{R}^N;$$

f is even on the sphere S_r, i.e.,

$$f(u) = f(-u) \quad \text{for all} \quad u \in V \text{ with } \|u\|_V = r .$$

Our multiplicity result concerning the eigenvalue problem (6.43) is formulated as follows.

Theorem 6.2 Assume that the assumptions (H_1),(H_2'),(H_3'),(H_4) are fulfilled. Then the constrained eigenvalue problem (6.43) admits infinitely many pairs of solutions $\{(\pm u_n, \lambda_n)\}_{n \geq 1} \subset S_r \times \mathbb{R}$ with

$$\lambda_n = r^{-2}(\alpha(u_n, u_n) + \langle z_n, v_n \rangle_V + \int_\Omega \langle w_n(x), u_n(x) \rangle_{\mathbb{R}^N} dx)$$

for some $z_n \in V^*$ and $w_n \in L^{p/(p-1)}(\Omega; \mathbb{R}^N)$ satisfying

$$z_n \in \partial f(\pm u_n), \quad n \geq 1 \tag{6.44}$$

and

$$w_n(x) \in \partial j(x,\ \pm u_n(x)) \quad \text{for a.e.} \quad x \in \Omega, \quad n \geq 1 . \tag{6.45}$$

Proof. Consider the locally Lipschitz functional $F : V \to \mathbb{R}$ given by

$$F(u) = (1/2)\alpha(u, u) + f(u) + J(u), \quad u \in V , \tag{6.46}$$

with $J : L^p(\Omega; \mathbb{R}^N) \to \mathbb{R}$ given by (6.11). From (H_4) it follows that F is even on S_r, i.e.

$$F(u) = F(-u) \quad \text{for} \quad u \in V \quad \text{with} \quad \|u\|_V = r . \tag{6.47}$$

As in (6.14) we check that F is bounded below on S_r

$$F(u) \geq C_0, \quad u \in S_r \tag{6.48}$$

for some constant C_0.

Let us now show that the Palais-Smale condition holds on S_r (in the sense of Chang [2]). To this end we recall that the generalized gradient $\partial(F|_{S_r})(u)$ at $u \in S_r$ is expressed by

$$\partial(F|_{S_r})(u) = \{J - r^{-2}\langle J, u \rangle_V \Lambda u; J \in \partial F(u)\}. \tag{6.49}$$

According to (6.49) it suffices to show, that if $\{v_n\} \subset S_r$ is a sequence for which there exists a sequence $\{J_n\}$ in V^* with

$$J_n \in \partial F(v_n) \tag{6.50}$$

and

$$J_n - r^2 \langle J_n, v_n \rangle_V \Lambda v_n \to 0 \quad \text{strongly in} \quad V^* , \tag{6.51}$$

then $\{v_n\}$ contains a strongly convergent subsequence in V.

From (6.50),(6.51) and (6.46) and the formula for the generalized gradient of a sum (Proposition 1.6) one deduces that two sequences $\{z_n\}$ and $\{w_n\}$ in V^* can be determined such that

$$z_n \in \partial f(v_n) \tag{6.52}$$

$$w_n \in \partial(J|_V)(v_n), \tag{6.53}$$

$$Av_n + z_n + w_n - r^2 \langle Au_n + z_n + w_n, v_n \rangle_V \Lambda v_n \to 0 \; . \tag{6.54}$$

Since $\|v_n\|_V = r$ we can extract a subsequence again denoted by $\{v_n\}$ such that

$$v_n \to u \quad \text{weakly in} \quad V , \tag{6.55}$$

for some $u \in V$. As in the proof of Theorem 6.1 we can choose subsequences of $\{z_n\}$ (cf. (H_3')) and of $\{w_n\}$ for which one has

$$z_n \to z \quad \text{strongly in} \quad V^* , \tag{6.56}$$

$$w_n \to w \quad \text{strongly in} \quad V^* , \tag{6.57}$$

with $z, w \in V^*$. Additionally, we can suppose that

$$\{\alpha(v_n, v_n)\} \quad \text{is convergent in} \quad \mathbb{R} \tag{6.58}$$

$$\langle z_n + w_n, v_n \rangle_V \to \langle z + w, v \rangle_V \; . \tag{6.59}$$

Due to the upper semicontinuity property of the generalized gradient (Proposition 1.4) relations (6.52)-(6.54),(6.59) and hypothesis (H_2') yield

$$z \in \partial f(u) , \tag{6.60}$$

$$w \in \partial(J|_V)(u) \; . \tag{6.61}$$

From (6.61) we deduce as in the proof of Theorem 6.1 that w satisfies (6.7). Relations (6.54)-(6.59) allow us to derive that

$$(A - \lambda_0 \Lambda) v_n \quad \text{converges strongly in} \quad V^* , \tag{6.62}$$

with λ_0 obtained from (6.58),(6.59) as required in (H_3'). Now we apply hypothesis (H_3') with v_n instead of u_n. This justifies that a strongly convergent subsequence of $\{v_n\}$ exists and thus the Palais-Smale condition for the function F on S_r is verified.

For any closed, symmetric with respect to the origin, subset S of S_r, let us denote by $\gamma(S)$ the Krosnoselski's genus of S. We recall that $\gamma(S)$ is the smallest integer $k > 0$ for which there exists an odd continuous mapping from S into $\mathbb{R}^k \setminus \{0\}$ (see Section 5.2 and Rabinowitz [17] for more details). We consider the following class of subsets of the sphere S_r

$$\begin{aligned} \Gamma_n = \{S \subset S_r : S \text{ closed, symmetric with respect to } 0, \\ \text{with } \gamma(S) \geq n\}, \quad n \geq 1 \,, \end{aligned} \tag{6.63}$$

and we form the corresponding minimax value of F over Γ_n

$$\beta_n = \inf_{S \in \Gamma_n} \max_{u \in S} F(u), \quad n \geq 1 \,. \tag{6.64}$$

It is clear that each class Γ_n contains compact sets, for instance $S_r \cap V_{n+1}$ with V_{n+1} an $(n+1)$-dimensional linear subspace of V. From (6.48) it follows then that each β_n is a real number.

The Palais-Smale condition and (6.48) are the only requirements which are necessary to apply Theorem 3.2 of Chang [2] (cf. Chapters 2 and 3). This ensures that $\beta_n \in \mathbb{R}$ given by (6.63),(6.64) are critical values of F on S_r. Hence there exists a critical point u_n (in fact $\pm u_n$ by (6.47)) of F, i.e.,

$$0 \in \partial F(\pm u_n) \tag{6.65}$$

with $F(\pm u_n) = \beta_n$. Recalling now (6.49) we can express (6.65) in the form

$$\alpha(\pm u_n, v) + \langle z_n + w_n, v\rangle_V = \lambda_n(\pm u_n, v), \quad v \in V \,, \tag{6.66}$$

for $\lambda \in \mathbb{R}$ and for z_n, w_n satisfying (6.60),(6.61) where u, z, w are replaced by $\pm u_n, z_n, w_n$. Arguing as in the final part of Theorem 6.1 we are led to the inequality

$$\begin{aligned} \alpha(\pm u_n, v) + \langle z_n, v\rangle_V + \int_\Omega j^0(x, \pm u_n; v(x))dx \geq \\ \geq \lambda_n(\pm u_n, v)_V \,, \quad v \in V, \ n \geq 1 \,. \end{aligned} \tag{6.67}$$

Hypothesis (H_2') ensures that the first relation in (6.43) holds for $u = \pm u_n, n \geq 1$, where λ_n is obtained from (6.66). The proof is thus complete. □

Remark 6.1 Hypotheses (H_1),(H_4) are obviously necessary for this approach. Hypothesis (H_3') is of the same nature as (H_3). Hypothesis (H_2') is verified, for instance, for f being the restriction to V of a locally Lipschitz function on $L^p(\Omega)$ bounded below on S_r and with $f^0(.,.)$ less or equal to $C(.,.)$.

6.3 Another Type of Eigenvalue Problem on the Sphere

Let V be an infinite dimensional real Hilbert space with the scalar product $(.,.)_V$ and the norm $\|\cdot\|_V$ which is densely and compactly imbedded in $L^p(\Omega; \mathbb{R}^N)$

for some $p \geq 2, N \geq 1$ and a bounded domain Ω in $\mathbb{R}^m$, $m \geq 1$. Consider a continuous symmetric bilinear form $a : V \times V \to \mathbb{R}$ on V whose corresponding selfadjoint bounded linear operator is $A : V \to V$, i.e.,

$$(Au, v)_V = a(u, v) \qquad \text{for all } u, v \in V. \tag{6.68}$$

It is given also a function $j : \Omega \times \mathbb{R}^N \to \mathbb{R}$ of Carathéodory type, i.e.

(i) for every $y \in \mathbb{R}^N$, the function $j(\cdot, y) : \Omega \to \mathbb{R}$ is measurable;

(ii) for every $x \in \Omega$, the function $j(x, \cdot) : \mathbb{R}^N \to \mathbb{R}$ is locally Lipschitz;

(iii) $j(\cdot, 0) : \Omega \to \mathbb{R}$ is bounded.

Assume further that the growth condition

$$\begin{aligned} | w | \leq c(1+ | y |^{p-1}) \qquad &\text{for all} \quad x \in \Omega,\ y \in \mathbb{R}^N \\ &\text{and} \quad w \in \partial_y j(x, y), \end{aligned} \tag{6.69}$$

holds, where c is a positive constant. As in the previous sections of this chapter we denote by $| \cdot |$ the Euclidean norm in $\mathbb{R}^N$ and by $\partial_y j(x, y)$ Clarke's generalized gradient of j at $(x, y) \in \Omega \times \mathbb{R}^N$ with respect to the second variable y. The notation ∂F without any subscript means, as usual, the (total) generalized gradient of the functional F.

Corresponding to each $f \in V^*$ and each $r > 0$ we formulate the following eigenvalue problem in the form of a hemivariational inequality:
$(P_{f,r})$ Find $u \in V$ with $\| u \|_V = r$ and $\lambda \in \mathbb{R}$ such that

$$a(u, v) - \langle f, v \rangle_V + \int_\Omega j^o(x, u; v) dx \geq \lambda (u, v)_V \qquad \text{for all } v \in V. \tag{6.70}$$

Note that the eigensolution $u \in V$ must belong to a sphere S_r in V,

$$S_r = \{v \in V : \|v\|_V = r\}. \tag{6.71}$$

Moreover, the presence of the linear term $\langle f, . \rangle_V$ does not allow the application of the Mountain Pass Theorem of Ambrosetti and Rabinowitz [1][17] in the variant of Chang [2] for locally Lipschitz functionals (cf. Chapters 2 and 3).

In (6.70) it is given an arbitrary locally Lipschitz function j on $\Omega \times \mathbb{R}^N$. The function j does not need to have for $N > 1$ the form of a primitive

$$j(x, t) = \int_0^t \Theta(x, \tau) d\tau, \quad (x, t) \in \Omega \times \mathbb{R} \tag{6.72}$$

Notice that we do not impose any coercivity hypothesis on the bilinear form a or on the operator A in (6.68).
In order to apply to the problem $(P_{f,r})$ methods from the critical point theory we assume that the data in (6.70) verify the hypothesis below:

$(H_{f,r})$ For every sequence $(v_n) \subset V$ with $\|v_n\|_V = r$, for every number $\alpha \in [-r \parallel A \parallel, r \parallel A \parallel]$ and for every measurable map $z : \Omega \to (\mathbb{R}^N)^*$ such that $v_n \to v$ strongly in $L^p(\Omega; \mathbb{R}^N)$ for some $v \in V$,

$$z(x) \in \partial_y j(x, v(x)) \quad \text{for a.e. } x \in \Omega$$

and

$$a(v_n, v_n) \to \alpha,$$

we have that

$$\inf_{\|w\|_V=1} \{a(w,w)\} - \frac{1}{r^2}(\alpha - \langle f, v\rangle_V + \int_\Omega \langle z(x), v(x)\rangle_{\mathbb{R}^N} dx) > 0. \quad (6.73)$$

Hypothesis $(H_{f,r})$ is not standard in the critical point theory, because it involves the sphere S_r. It is closely related to the verification of the Palais-Smale condition on the sphere S_r for a functional associated to problem $(P_{f,r})$.

Our results are formulated further in Theorems 6.3, 6.4. Theorem 6.3 establishes the existence of a solution $(u, \lambda) \in V \times \mathbb{R}$ to problem $(P_{f,r})$. Theorem 6.4 concerns the existence of a countable set of solutions $(u_n, \lambda_n)_{u \geq 1}$ in $V \times \mathbb{R}$ of the homogeneous eigenvalue problem $(P_{o,r})$ (i.e., $f = 0$) under the additional assumption that $j : \Omega \times \mathbb{R}^N \to \mathbb{R}$ is even in the second variable, i.e. that

$$j(x, -y) = j(x, y) \qquad (x, y) \in \Omega \times \mathbb{R}^N. \quad (6.74)$$

If j is equal to the primitive in (6.72) of the function $\Theta : \Omega \times \mathbb{R} \to \mathbb{R}$, condition (6.74) is satisfied, if Θ is odd in the second variable, i.e. if

$$\Theta(x, -t) = -\Theta(x, t) \qquad (x, t) \in \Omega \times \mathbb{R}. \quad (6.75)$$

Theorem 6.3 Assume that $a : V \times V \to \mathbb{R}$ is a continuous, symmetric, bilinear form, $f \in V^*$ and the function $j : \Omega \times \mathbb{R}^N \to \mathbb{R}$ satisfies the assumptions (i), (ii), (iii), (6.69) and $(H_{f,r})$ for a fixed $r > 0$. Then the problem $(P_{f,r})$ admits a solution $(u, \lambda) \in V \times \mathbb{R}$. Moreover, there exists a mapping $z : \Omega \to (\mathbb{R}^N)^*$ such that the function $x \in \Omega \to \langle z(x), u(x)\rangle_{R^N}$ is integrable,

$$z(x) \in \partial_y j(x, u(x)) \quad \text{for a.e. } x \in \Omega$$

and

$$\lambda = \frac{1}{r^2}(\langle Au - f, u\rangle_V + \int_\Omega \langle z(x), u(x)\rangle_{\mathbb{R}^N} dx).$$

Proof. We introduce the functional $J : L^p(\Omega; \mathbb{R}^N) \to \mathbb{R}$ by

$$J(v) = \int_\Omega j(x, v(x))dx, \qquad v \in L^p(\Omega; \mathbb{R}^N). \quad (6.76)$$

Under the hypotheses (i), (ii) and the growth condition (6.69) the functional J is well-defined on $L^p(\Omega; \mathbb{R}^N)$ and is Lipschitz continuous on the bounded subsets of $L^p(\Omega; \mathbb{R}^N)$ (cf. Chapter 1). Moreover, for Clarke's generalized gradient $\partial J(v) \subset L^p(\Omega; \mathbb{R}^N)^*$ one has the inclusion (cf. (1.16))

$$\partial J(v) \subset \int_\Omega \partial_y j(x, v(x))dx, \quad v \in L^p(\Omega; \mathbb{R}^N). \tag{6.77}$$

Since the Hilbert space V is densely and continuously imbedded in $L^p(\Omega; \mathbb{R}^N)$ the generalized gradient $\partial(J|_V)(v) \subset V^*$ of the restriction $J|_V : V \to \mathbb{R}$ has the property that

$$\partial(J|_V)(v) \subset \partial J(v), \quad v \in V. \tag{6.78}$$

Our argument concerns the locally Lipschitz functional $I : V \to \mathbb{R}$ defined by

$$I(v) = \frac{1}{2}a(v, v) - \langle f, v\rangle_V + J(v), \quad v \in V, \tag{6.79}$$

with J given in (6.76). Denoting by $\Lambda : V \to V^*$ the duality isomorphism

$$\langle \Lambda v, u\rangle_V = (v, u)_V, \quad u, v \in V, \tag{6.80}$$

we see from (6.68), (6.79) and (6.80) that the generalized gradient $\partial I(v) \subset V^*$ may be expressed as

$$\partial I(v) = \Lambda A v - f + \partial(J|_V)(v), \quad v \in V. \tag{6.81}$$

By the growth condition (6.69) and hypothesis (iii) one deduces the existence of positive constants a_1, a_2 such that the estimate

$$|j(x, y)| \leq a_1 + a_2|y|^p \quad \text{for all } (x, y) \in \Omega \times R^N. \tag{6.82}$$

holds. Indeed, estimate (6.82) is obtained from the inequalities

$$\begin{aligned} |j(x,y)| &\leq |j(x,y) - j(x,0)| + |j(x,0)| \\ &\leq |j(x,0)| + \max_{\substack{w \in \partial_y j(x,Y) \\ y \in [0,Y]}} |\langle w, y\rangle|_{R^N} \\ &\leq |j(x,0)| + \max_{y \in [0,Y]} c(1 + |y|^{p-1})|y| \\ &\leq |j(x,0)| + c(1 + |y|^{p-1})|y| \quad \text{for all } (x,y) \in \Omega \times R^N. \end{aligned}$$

We used above Lebourg's mean value theorem for locally Lipschitz functions (see in Chapter 1, Theorem 1.1). Then we obtain from (6.79) and (6.82) that

$$\begin{aligned} |I(u)| &\leq \frac{1}{2}\|A\|\|v\|_V^2 + \|f\|_{V^*}\|v\|_V + \int_\Omega |j(x, v(x))|dx \qquad (6.83) \\ &\leq \frac{1}{2}\|A\|\|v\|_V^2 + \|f\|_{V^*}\|v\|_V + a_1|\Omega| + a_2\|v\|_{L^p}^p, \quad v \in V, \end{aligned}$$

where $|\Omega|$ denotes the Lebesgue measure of the domain $\Omega \subset \mathbb{R}^n$. The continuity of the imbedding $V \subset L^p(\Omega; \mathbb{R}^N)$ ensures the existence of a positive constant $C_p(\Omega)$ such that

$$\|v\|_{L^p} \leq C_p(\Omega)\|v\|_V \quad \text{for all } v \in V. \tag{6.84}$$

Then (6.83) and (6.84) yield that

$$|I(v)| \leq \frac{1}{2}\|A\|\|v\|_V^2 + \|f\|_{V^*}\|v\|_V + a_1|\Omega| + a_2(C_p(\Omega))^p\|v\|_V^p$$
$$\text{for all } v \in V. \tag{6.85}$$

Relation (6.85) implies that functional I is bounded on the sphere S_r. Thus there exists a constant $M > 0$ such that

$$|I(v)| \leq M \quad \text{for all } v \in V \quad \text{with } \|v\|_V = r. \tag{6.86}$$

We consider now the restriction $I|_{S_r}$ of the functional I to the sphere $S_r \subset V$. The generalized gradient $\partial(I|_{S_r})(v)$ is equal to

$$\partial(I|_{S_r})(v) = \{w - \frac{1}{r^2}\langle w, v\rangle_V \Lambda v, \text{ where } w \in \partial I(v)\} \tag{6.87}$$
$$\text{for all } v \in S_r.$$

Let us now check that $I|_{S_r}$ satisfies the Palais-Smale condition on the sphere S_r in the sense of Chang [2]. Precisely, we have to show that if $\{v_n\}$ is a sequence on the sphere S_r such that

$$\min\{\|\zeta\|_{V^*}; \zeta \in \partial(I|_{S_r})(v_n)\} \to 0 \quad \text{as } n \to \infty, \tag{6.88}$$

then $\{v_n\}$ possesses a strongly convergent subsequence in V.
Taking into account (6.87), we see from (6.88) that there exists a sequence $\{w_n\} \subset V^*$ satisfying

$$w_n \in \partial I(v_n) \quad \text{for all } n \tag{6.89}$$

and

$$w_n - \frac{1}{r^2}\langle w_n, v_n\rangle_V \Lambda v_n \to 0 \quad \text{in } V^* \quad \text{as } n \to \infty. \tag{6.90}$$

From (6.81), (6.89) and (6.90) we get the existence of a sequence $\{z_n\}$ in V^* with the properties

$$z_n \in \partial(J|_V)(v_n) \qquad \text{for all } n \tag{6.91}$$

and

$$\Lambda A v_n - f + z_n - \frac{1}{r^2}\langle \Lambda A v_n - f + z_n, v_n\rangle_V \Lambda v_n \to 0 \quad \text{in } V^* \quad \text{as } n \to \infty. \tag{6.92}$$

Since $\|u_n\|_V = r$ we can extract a subsequence again denoted by $\{v_n\}$ converging weakly in V to some $v \in V$

$$v_n \to v \quad \text{weakly in } V \quad \text{as } n \to \infty. \tag{6.93}$$

The compactness of the imbedding $V \subset L^p(\Omega; \mathbf{R}^N)$ implies that

$$v_n \to v \quad \text{strongly in } L^p(\Omega; \mathbf{R}^N) \quad \text{as } n \to \infty. \tag{6.94}$$

The functional J is locally Lipschitz on $L^p(\Omega; \mathbf{R}^N)$. Thus we find from (6.94), (6.91) and (6.78) that $\{z_n\}$ is bounded in $L^p(\Omega; \mathbf{R}^N)^*$. Thus a subsequence can be determined such that

$$z_n \to z \quad \text{weakly in } L^p(\Omega; \mathbf{R}^N)^* \quad \text{as } n \to \infty. \tag{6.95}$$

Due to the compactness of the imbedding $L^p(\Omega; \mathbf{R}^N)^* \subset V^*$ relation (6.95) implies

$$z_n \to z \quad \text{strongly in } V^* \quad \text{as } n \to \infty. \tag{6.96}$$

Combining (6.93) and (6.96) one obtains that

$$\langle z_n, v_n \rangle \to \langle z, v \rangle_V \quad \text{as } n \to \infty. \tag{6.97}$$

Due to the boundedness of $\{v_n\}$ in V we may suppose that $a(u_n, u_n) \to \alpha$ as $n \to \infty$ for some $\alpha \in R$. Then from (6.93), (6.96) and (6.97) it turns out that (6.92) implies that

$$Av_n - \frac{1}{r^2}(\alpha - \langle f, v \rangle_V + \langle z, v \rangle) v_n \text{ converges in } V \text{as } n \to \infty. \tag{6.98}$$

From (6.91), (6.78), (6.94) and (6.95) we deduce that

$$z \in \partial J(v) \tag{6.99}$$

(see Proposition 1.4 and Clarke [4], p.29). Thus (6.96) and (6.99) ensure the existence of a measurable mapping $z : \Omega \to (\mathbf{R}^N)^*$ which fulfills the conditions

$$z(x) \in \partial_y j(x, v(x)) \quad \text{for a.e. } x \in \Omega \tag{6.100}$$

and

$$\langle z, v \rangle_V = \langle z, v \rangle_{L^p(\Omega; \mathbf{R}^N)} = \int_\Omega \langle z(x), v(x) \rangle_{\mathbf{R}^N} dx. \tag{6.101}$$

Note that due to (6.68), (6.100) and $v \in L^p(\Omega; \mathbf{R}^N)$, one has $\langle z(x), v(x) \rangle_{\mathbf{R}^N} \in L^1(\Omega; \mathbf{R})$. Replacing (6.101) in (6.98) one gets that

$$Av_n - \frac{1}{r^2}(\alpha - \langle f, v \rangle_V + \int_\Omega \langle z(x), v(x) \rangle_{\mathbf{R}^N} dx) v_n \tag{6.102}$$

$$\text{converges in } V \quad \text{as } n \to \infty,$$

with z verifying (6.100).
Consequently, we can apply hypothesis $(H_{f,r})$, and thus inequality (6.73) is valid. Now we can write the following inequality

$$\begin{aligned}
\inf_{\|w\|_V = 1} a(w, w) \quad &- \quad \frac{1}{r^2}(\alpha - \langle f, v \rangle_V + \int_\Omega \langle z(x), v(x) \rangle_{\mathbf{R}^N} dx) \|v_n - v_k\|_V^2 \\
&\leq a(v_n - v_k, v_n - v_k) - \frac{1}{r^2}(\alpha - \langle f, v \rangle_V \\
&\quad + \int_\Omega \langle z(x), v(x) \rangle_{\mathbf{R}^N} dx (v_n - v_k), v_n - v_k)_V \qquad (6.103) \\
&\leq \|A(v_n - v_k) - \frac{1}{r^2}(\alpha - \langle f, v \rangle_V \\
&\quad + \int_\Omega \langle z(x), v(x) \rangle_{\mathbf{R}^N} dx)(v_n - v_k)\|_V \|v_n - v_k\|_V \quad \text{for all } n, k.
\end{aligned}$$

The convergence in (6.102) and the relations (6.73) and (6.103) show that $\{v_n\}$ contains a Cauchy subsequence in V, thus v_n converges along a subsequence in V to v. This completes the verification of the Palais-Smale condition for $I|_{S_r}$.

The boundedness property in (6.86) and the Palais-Smale condition for $I|_{S_r}$ that we just verified, are the only requirements to apply the Palais-Smale Minimization Theorem (Chang's version for locally Lipschitz functions [2], Theorem 3.5) to the functional $I|_{S_r}$, with I given by (6.79). We point out that the above mentioned result is given in [2] for a locally Lipschitz functional defined on a whole reflexive Banach space, but its variant for a function on a sphere S_r can be easily derived (or, more generally, for locally Lipschitz functionals on Banach manifolds as noted by Chang [2], Remark 3.3). In this respect we refer also to Chapters 2 and 3. Thus the infimum of I on S_r is achieved. Therefore

$$c = \inf_{S_r} I \tag{6.104}$$

is a critical value of $I|_{S_r}$. Thus a $u \in S_r$ exists with

$$0 \in \partial(I|_{S_r})(u). \tag{6.105}$$

On the other hand one has

$$\partial(I|_{S_r})(u) = \{w - \frac{1}{r^2}\langle w, u\rangle_V \Lambda u;\ w \in \partial I(u)\}. \tag{6.106}$$

From (6.105), (6.106) and (6.81) it is clear that there exists

$$z \in \partial(J|_V)(u) \subset \partial J(u) \tag{6.107}$$

such that u solves the equation

$$Au - f + z = \frac{1}{r^2}\langle Au - f + z, u\rangle_V \Lambda u. \tag{6.108}$$

Due to (6.107) and (6.77) we can find a mapping $z : \Omega \to (\mathbf{R}^N)^*$ having the properties

$$z(x) \in \partial_y j(x, u(x)) \quad \text{for a.e. } x \in \Omega \tag{6.109}$$

and

$$\langle z, v\rangle_V = \int_\Omega \langle z(x), v(x)\rangle_{\mathbf{R}^N} dx \quad \text{for all } v \in V. \tag{6.110}$$

Now we set

$$\lambda = \frac{1}{r^2}(\langle Au - f, u\rangle_V + \int_\Omega \langle z(x), u(x)\rangle_{\mathbf{R}^N} dx). \tag{6.111}$$

and we obtain from (6.108) $\div$ (6.111) that

$$\lambda(u, v)_V - a(u, v) + \langle f, v\rangle_V = \int_\Omega \langle z(x), v(x)\rangle_{\mathbf{R}^N} dx \leq$$
$$\leq \int_\Omega \max_{\bar{z} \in \partial_y j(x, u(x))} \langle \bar{z}, v(x)\rangle_{\mathbf{R}^N} dx = \int_\Omega j^0(x, u(x); v(x) dx \tag{6.112}$$

for all $v \in V$. (cf. also Proposition 1.4).

Relation (6.112) implies, that $u \in S_r$ is an eigensolution corresponding to the eigenvalue λ given in (6.111) for the eigenvalue problem (6.70). This completes the proof of Theorem 6.3. □

Subsequently, from the proof of Theorem 6.3 the following minimization result is obtained.

Corollary 6.2 Assume that the conditions of Theorem 6.3 are verified. Then the locally Lipschitz functional $I : V \to \mathbb{R}$ given by (6.79) assumes its infimum on the sphere S_r.

We turn now to the problem of multiplicity of solutions of the hemivariational eigenvalue problem. Since we want to apply Clark's Theorem (see [3] and [17], p.53) in Chang's variant ([2], Theorem 3.2) in the locally Lipschitz setting, we have to assume that the functional $I : V \to \mathbb{R}$ defined in (6.79) is even. This occurs if the integrand $j : \Omega \times \mathbb{R}^N \to \mathbb{R}$ is even in the second variable, i.e. (6.74) holds, and $f = 0$.

Theorem 6.4 In addition to the hypotheses of Theorem 6.3 suppose that (6.74) is satisfied and $f = 0$. Then the eigenvalue problem $(P_{o,r})$ (i.e., $(P_{f,r})$ with $f = 0$) admits infinitely many distinct pairs of solutions $(\pm u_n, \lambda_n)_{n\geq 1} \subset V \times \mathbb{R}$ with

$$\lambda_n = \frac{1}{r^2}(a(u_n, u_n) + \int_\Omega \langle z_n(x), u_n(x)\rangle_{\mathbb{R}^N} dx), \quad n \geq 1, \tag{6.113}$$

Here each $z_n : \Omega \to (\mathbb{R}^N)^*$ denotes a mapping such that $\langle z_n, u_n\rangle_{\mathbb{R}^N} \in L^1(\Omega; \mathbb{R})$ and

$$z_n(x) \in \partial_y j(x, u_n(x)) \quad \text{for a.e. } x \in \Omega.$$

Proof. First we check that if $(u, \lambda) \in V \times \mathbb{R}$ is a solution of problem $(P_{o,r})$, then $(-u, \lambda)$ is also a solution of $(P_{o,r})$. Indeed, by putting $-v$ in place of v in (6.70) with $f = 0$ one obtains

$$a(-u, v) + \int_\Omega j^0(x, u(x); -v(x))dx \geq \lambda(-u, v)_V \quad \text{for all } v \in V.$$

Taking Clarke's generalized directional derivative for even function (cf. (6.74)) we find that

$$j^0(x, -u(x); v(x)) = j^0(x, u(x); -v(x)) \quad \text{for a.e. } x \in \Omega.$$

Therefore $(-u, \lambda)$ solves also $(P_{o,r})$. Consider the locally Lipschitz functional $I : V \to \mathbb{R}$ in (6.79) with $f = 0$, i.e.

$$I(v) = \frac{1}{2}a(v, u) + J(v), \quad v \in V, \tag{6.114}$$

where J is given by (6.76). Assumption (6.74) implies that I in (6.114) is even, i.e.

$$I(-v) = I(v), \quad v \in V. \tag{6.115}$$

In the proof of Theorem 6.3 we have shown that the functional I is bounded on the sphere S_r defined by (6.71), and thus (6.86) holds. Hence all the following minmax values

$$b_n = \inf_{S \in \gamma_n} \max_{v \in S} I(v), \quad n \geq 1, \tag{6.116}$$

are real numbers. We recall that here γ_n represents the following class of subsets of the sphere S_r

$$\gamma_n = \{S \subset S_r; S \text{ closed and symmetric } \gamma(S) \geq n\}, \tag{6.117}$$

where $\gamma(S)$ means the genus of Krasnoselski of the set S (cf. Chapter 5 and [16],[17]).

It was also shown in the proof of Theorem 6.3 that the functional $I|_{S_r}$ satisfies the Palais-Smale condition on the sphere S_r. Therefore we can apply Theorem 3.2 of [2] to conclude that the infinitely many numbers b_n of (6.116) are critical values of $I|_{S_r}$. Moreover the properties of the genus guarantee that even if the numbers b_n are not distinct, infinitely many critical points $u_n \subset S_r$ of I exist. Arguing as in the final part of the proof of Theorem 6.3 we see that each critical point $u_n \in S_r$ of $I|_{S_r}$ is an eigenfunction of problem $(P_{o,r})$ whose associated eigenvalue λ_n is expressed as in (6.113). The proof is thus complete.□

Remark 6.2 If $N = 1$ and the function $j : \Omega \times \mathbf{R} \to \mathbf{R}$ is given by the primitive in (6.72), then Theorems 6.3 and 6.4 generalize the Theorems 5.5 and 5.6 of Chang [2] (cf. in this respect also [6]).

6.4 Applications

6.4.1 Generalities

In Mechanics and Engineering the cost or the weight of a structure is expressed usually as a linear function of the norm of the unknown function. Thus the constraint $\|u\|_V = r$ means that we have a system with prescribed cost or weight, or in some cases, energy consumption. The stability analysis, or the parametric analysis for $f = \lambda u$, $\lambda \in \mathbf{R}, u \in V$, where f is the loading and u is the displacement (or f is a force -i.e. the cause- and u is a flux -i.e. the result- in the terminology of Onsager), of such a system involving nonconvex nonsmooth energy functions, leads to an eigenvalue problem for a hemivariational inequality on a sphere. It is obvious that for almost all eigenvalue problems which were applications of the theory of the Chapters 4 and 5 the corresponding "given cost" or "weight" problems can be studied by means of the theory of the Sections 6.1÷6.3. The corresponding existence and multiplicity results are obtained from the Theorems 6.1÷6.4. Further we give two applications leading to a different mathematical theory.

6.4.2 Buckling of Beams and Plates with Prescribed Weight on Adhesive Supports

We put ourselves in the framework of Section 5.5 and we assume that the beam must have a prescribed weight. This constraint is written in the form $\|u\| = r$, or also in the form $(Lu, u) = r$, where $\|\cdot\|$ denotes the norm in V and $r \in \mathbb{R}$ is given. Thus instead of (5.78) we have to solve the problem:
Find $(u, \lambda) \in V \times \mathbb{R}$ such as to satisfy

$$a(u, v-u) + \int_0^l j^0(u; v-u)dx \geq \lambda(\tilde{L}u, v)_V = \lambda \int_0^l u'(v'-u')dx \qquad (6.118)$$

$$(\tilde{L}u, u) = r. \qquad (6.119)$$

We recall that $(.,.)_V$ is the inner product in V and $\tilde{L}$ is a linear compact operator. Here due to (5.77) we have $N = 1$ and

$$\partial j(t) = [\underline{\beta}(t), \overline{\beta}(t)] \qquad \forall\, t \in \mathbb{R} \qquad (6.120)$$

(cf. Proposition 1.7). Then by assuming that

$$\underline{\beta}(t) > 0 \text{ for } t < 0, \quad \overline{\beta}(t) < 0 \text{ for } t > 0 \qquad (6.121)$$

we can verify that

$$\int_0^l z(x)u(x)dx < 0 \text{ for any } z(x) \in \partial j(u(x)) \text{ a.e. on } (0, l). \qquad (6.122)$$

Thus for $r > 0$ sufficiently small, the inequality (6.73) (for $f = 0$) is fulfilled. Then we can apply to (6.118)(6.119) the Theorems 6.3 and 6.4, which hold (see Section 6.5) for the present problem with a slight modification in their proofs due to the appearence of the operator $\tilde{L}$. Analogous applications can be given for the Timoshenko plate (Section 4.6) and the shell buckling problem (Section 4.7).

The more general case of the buckling of an adhesively supported von Kármán plate having prescribed weight or cost can also be studied by means of the theory of this chapter on the additional assumptions of thickness deformability (4.147). More specifically, because of the appearence of the compact nonlinear operator C in the von Kármán plate theory (cf. Section 4.5.2 and Fig. 4.1) we have to apply the theory of Sections 6.1 and 6.2. The adhesive bond between the plate and the support is expressed by means of a relation of the type (4.95). The boundary conditions of the plate may permit a rigid body displacement, because the Theorems 6.1 and 6.2 do not need any coercivity assumption. Thus the buckling problem of an adhesively connected (cf. Fig. 4.1) von Kármán plate with prescribed weight or cost (in the form of a norm $\|\cdot\| = r$) can be put in the form (6.2),(6.3), where (4.130) replaces (6.2). Here we assume that the plate Ω is fixed with a hinge, along its boundary Γ, i.e. (4.106) hold, and thus V is identified with the space Z given in (4.107). Due to the presence of B, the proofs of Theorems 6.1 and 6.2 must be modified.

6.4.3 The Optimum Weight for the Buckling Shell problem Subjected to Loading and to Adhesive Contact Conditions

Here we consider the same problem as in Section 4.7 with the only difference that in (4.157) we split q into two parts i.e. $q = \overline{q} + q_0$ where $\overline{q}$ is the adhesive reaction (i.e. it satisfies (4.161)) and $q_0 \in L^2(0,l)$ is the given loading. Thus we may formulate the following problem:
Find $w \in V$, $\lambda \in \mathbb{R}$ with $(Lw, w) = r$ such that

$$a(w, v-w) + \frac{Eh}{a^2}\int_0^l w(v-w)dx + \int_{l_1}^{l_2} j^0(w, v-w)dx \geq$$

$$\geq \lambda(Lw, v-w) + \int_0^l q_0(v-w)dx. \tag{6.123}$$

Here L is compact operator defined by (4.163). For this problem Theorem 6.3 holds (cf. Sobolev imbedding theorem of H^2 into L^p and Section 6.5) on the assumption that (6.69) holds together with (6.73). We recall that $a(.,.)$ is an equivalent norm on V (cf. (4.159)).

6.5 The case of the Constraint $\langle Lu, u\rangle = r$

The applications in Sections 6.4.2 and 6.4.3 lead to the present mathematical problem. Let V be an infinite dimensional real Hilbert space with the scalar product $(.,.)_V$ and the norm $\|\cdot\|_V$ which is densely and compactly imbedded in $L^p(\Omega; \mathbb{R}^N)$ for some $p \geq 2, N \geq 1$ and a bounded domain Ω in $\mathbb{R}^m$, $m \geq 1$. We denote by $\langle\cdot,\cdot\rangle_V$ the pairing of V^* and V. Consider a continuous symmetric bilinear form $a : V \times V \to \mathbb{R}$ on V whose corresponding self-adjoint bounded linear operator is $A : V \to V^*$, i.e.,

$$< Au, v >_V = a(u,v) \text{ for all } u, v \in V. \tag{6.124}$$

We assume that a (or A) satisfies the coercivity condition

$$a(v,v) \geq \alpha\|v\|_V^2, \quad \forall v \in V, \tag{6.125}$$

with a constant $\alpha > 0$. Let $j : \Omega \times \mathbb{R}^N \to \mathbb{R}$ be a function of Carathéodory type, verifying

(i) for every $y \in \mathbb{R}^N$, the function $j(\cdot, y) : \Omega \to \mathbb{R}$ is measurable;

(ii) for every $x \in \Omega$, the function $j(x,\cdot) : \mathbb{R}^N \to \mathbb{R}$ is locally Lipschitz;

(iii) $j(\cdot,0) : \Omega \to \mathbb{R}$ is essentially bounded.

Assume further that j satisfies the growth conditions

$$| w | \leq c(1+ | y |^{p-1}), \quad \forall x \in \Omega,\ y \in \mathbb{R}^N,\ w \in \partial_y j(x,y), \tag{6.126}$$

and

$$j(x,y) \geq -c_0(1+|y|^{\sigma}), \forall x \in \Omega,\ y \in \mathbb{R}^N, \tag{6.127}$$

where $c > 0$, $c_o > 0$, $\sigma \in [1,2)$ are constants. We denoted by $|\cdot|$ the Euclidean norm in $\mathbb{R}^N$ and by $\partial_y j(x,y)$ Clarke's generalized gradient of j at $(x,y) \in \Omega \times \mathbb{R}^N$ with respect to the second variable y.

Corresponding to $f \in V^*$, $r > 0$ and a linear compact selfadjoint operator $L : V \to V^*$ we formulate the following eigenvalue problem in the form of a hemivariational inequality:

$(P_{f,r,L})$ Find $u \in V$ with $\langle Lu, u\rangle_V = r$ and $\lambda \in \mathbb{R}$ such that

$$a(u,v) - \langle f, v\rangle_V + \int_\Omega j^o(x,u;v)dx \geq \lambda\langle Lu, v\rangle_V \qquad \text{for all } v \in V. \tag{6.128}$$

Note that the eigensolution $u \in V$ must belong to the following set of constraints

$$S_{f,r,L} := \{v \in V : \langle Lv, v\rangle_V = r\} \tag{6.129}$$

that we suppose to be nonempty.

Moreover, the presence of the linear term $\langle f, .\rangle_V$ does not allow the application of the Mountain Pass Theorem of Ambrosetti and Rabinowitz in the variant of Chang for locally Lipschitz functionals.

Our result for the existence of solutions to problem $(P_{f,r,L})$ is stated below.

Theorem 6.5 Under the assumptions above problem $(P_{f,r,L})$ admits a solution $(u, \lambda) \in V \times \mathbb{R}$. Moreover, there exists a mapping $z : \Omega \to (\mathbb{R}^N)^*$ such that the function $x \in \Omega \to \langle z(x), u(x)\rangle_{\mathbb{R}^N}$ is integrable,

$$z(x) \in \partial_y j(x, u(x)) \quad \text{for a.e. } x \in \Omega$$

and

$$\lambda = \frac{1}{\langle Lu, \Lambda^{-1}Lu\rangle_V}(\langle Au - f, u\rangle_V + \int_\Omega \langle z(x), u(x)\rangle_{\mathbb{R}^N} dx).$$

Proof. We introduce the functional $J : L^p(\Omega; \mathbb{R}^N) \to \mathbb{R}$ by

$$J(v) = \int_\Omega j(x, v(x))dx, \quad \forall v \in L^p(\Omega; \mathbb{R}^N). \tag{6.130}$$

Under hypotheses (i), (ii) and the growth condition (6.126) the functional J is well-defined on $L^p(\Omega; \mathbb{R}^N)$ and is Lipschitz continuous on the bounded subsets of $L^p(\Omega; \mathbb{R}^N)$. Moreover, for Clarke's generalized gradient $\partial J(v) \subset L^p(\Omega; \mathbb{R}^N)^*$ one has the inclusion

$$\partial J(v) \subset \int_\Omega \partial_y j(x, v(x))dx, \ \forall v \in L^p(\Omega; \mathbb{R}^N). \tag{6.131}$$

Since the Hilbert space V is densely and continuously imbedded in $L^p(\Omega; \mathbb{R}^N)$ the generalized gradient $\partial(J|_V)(v) \subset V^*$ of the restriction $J|_V : V \to \mathbb{R}$ has the property that

$$\partial(J|_V)(v) \subset \partial J(v), \quad \forall v \in V. \tag{6.132}$$

Our argument deals with the locally Lipschitz functional $I : V \to \mathbb{R}$ defined by

$$I(v) = \frac{1}{2}a(v,v) - \langle f, v\rangle_V + J(v), \quad \forall v \in V, \tag{6.133}$$

with J given in (6.130). For a later use we denote by $\Lambda : V \to V^*$ the duality isomorphism

$$\langle \Lambda v, u\rangle_V = (v,u)_V, \ \forall u, v \in V. \tag{6.134}$$

We see from (6.124) and (6.133) that the generalized gradient $\partial I(v) \subset V^*$ may be expressed as

$$\partial I(v) = Av - f + \partial(J|_V)(v), \quad \forall v \in V. \tag{6.135}$$

By (6.125), (6.127) and hypothesis (iii) one deduces that the estimate below holds

$$I(v) \geq \alpha\|v\|_V^2 - \|f\|_{V^*}\|v\|_V - a_1\|v\|_V^\sigma - a_2, \quad \forall v \in V, \tag{6.136}$$

for $a_1, \ a_2 \in \mathbb{R}$.

Relation (6.136) implies that the functional I is bounded from below on V, thus there exists a constant $M > 0$ such that

$$I(v) \geq M, \quad \forall v \in V. \tag{6.137}$$

We consider now the restriction $I|_{S_{f,r,L}}$ (see (6.129)) of the functional I to the set $S_{f,r,L} \subset V$.

Firstly, let us note that the set $S_{f,r,L}$ introduced in (6.129) is a smooth submanifold of the Hilbert space V. This can be seen as follows:

$$S_{f,r,L} = g^{-1}(\frac{r}{2}),$$

where $g : V \to \mathbb{R}$ is the smooth function

$$g(v) = \frac{1}{2}\langle Lv, v\rangle_V, \forall v \in V$$

whose gradient is

$$\nabla g(u) = \Lambda^{-1}Lu \neq 0, \quad \forall u \in S_{f,r,L}.$$

Therefore the tangent space $T_u(S_{f,r,L})$ at any $u \in S_{f,r,L}$ is expressed by

$$T_u(S_{f,r,L}) = \{v \in V : \ \langle Lu, v\rangle_V = 0\}. \tag{6.138}$$

The generalized gradient $\partial(I|_{S_{f,r,L}}(u)$ for any $u \in S_{f,r,L}$ is equal to

$$\partial(I|_{S_{f,r,L}})(u) = \{w - \frac{\langle w, \Lambda^{-1}Lu\rangle_V}{\langle Lu, \Lambda^{-1}Lu\rangle_V}Lu; \ w \in \partial I(u)\}, \tag{6.139}$$

where (6.132), (6.134) and (6.138) have been used.

Let us now check that $I|_{S_{f,r,L}}$ satisfies the Palais-Smale condition on the manifold $S_{f,r,L}$ in the sense of Chang. Precisely, we have to show that if $\{v_n\}$ is a sequence on the manifold $S_{f,r,L}$ such that $I(v_n)$ is bounded and there exists a sequence $\{w_n\} \subset V^*$ satisfying

$$w_n \in \partial I(v_n) \text{ for all } n \tag{6.140}$$

and

$$w_n - \frac{\langle w_n, \Lambda^{-1}Lv_n\rangle_V}{\langle Lv_n, \Lambda^{-1}Lv_n\rangle_V} Lv_n \to 0 \text{ in } V^* \text{ as } n \to \infty, \tag{6.141}$$

then $\{v_n\}$ possesses a strongly convergent subsequence in V. In the preceding formulation of Palais-Smale condition we have used essentially formula (6.139) for the generalized gradient on the corresponding manifold.

From (6.135), (6.140) and (6.141) we get the existence of a sequence $\{z_n\}$ in V^* with the properties

$$z_n \in \partial(J|_V)(v_n) \text{ for all } n \tag{6.142}$$

and

$$Av_n - f + z_n - \frac{\langle Av_n - f + z_n, \Lambda^{-1}Lv_n\rangle_V}{\langle Lv_n, \Lambda^{-1}Lv_n\rangle_V} Lv_n \to 0 \quad \text{in } V^* \text{ as } n \to \infty. \tag{6.143}$$

Since relation (6.136) holds, we deduce that the sequence $\{v_n\}$ is bounded. Therefore we can extract a subsequence again denoted by $\{v_n\}$ converging weakly in V to some $v \in V$, and on the basis of the compactness of the operator L,

$$Lv_n \to Lv \text{ strongly in } V^* \text{ as } n \to \infty. \tag{6.144}$$

The compactness of the imbedding $V \subset L^p(\Omega; \mathbb{R}^N)$ implies that

$$v_n \to v \quad \text{strongly in } L^p(\Omega; \mathbb{R}^N) \quad \text{as } n \to \infty. \tag{6.145}$$

Since the functional J is locally Lipschitz on $L^p(\Omega; \mathbb{R}^N)$, we find from (6.145), (6.142) and (6.132) that $\{z_n\}$ is bounded in $L^p(\Omega; \mathbb{R}^N)^*$. Thus a subsequence can be determined such that

$$z_n \to z \quad \text{weakly in } L^p(\Omega; \mathbb{R}^N)^* \quad \text{as } n \to \infty. \tag{6.146}$$

Due to the compactness of the imbedding $L^p(\Omega; \mathbb{R}^N)^* \subset V^*$ relation (6.146) implies that

$$z_n \to z \quad \text{strongly in } V^* \quad \text{as } n \to \infty. \tag{6.147}$$

Combining (6.144) and (6.147) one obtains that

$$\langle z_n, \Lambda^{-1}Lv_n\rangle_V \to \langle z, \Lambda^{-1}Lv\rangle_V \quad \text{as } n \to \infty. \tag{6.148}$$

Then from (6.143), (6.144), (6.147) and (6.148) it turns out that Av_n converges in V^* as $n \to \infty$. Consequently, in view of (6.124), (6.125), we deduce that a subsequence of v_n converges in V to $v \in S_{f,r,L}$. This completes the verification of the Palais-Smale condition for $I|_{S_{f,r,L}}$.

The boundedness property in (6.137) and the Palais-Smale condition for $I|_{S_{f,r,L}}$, that we just verified, are the only requirements to apply the Palais-Smale Minimization Theorem in Chang's version for locally Lipschitz functions to the functional $I|_{S_{f,r,L}}$, with I given by (6.133). Thus the infimum of I on $S_{f,r,L}$ is achieved. Therefore

$$c = \inf_{S_{f,r,L}} I \tag{6.149}$$

is a critical value of $I|_{S_{f,r,L}}$. This means that a point $u \in S_{f,r,L}$ exists with $I(u) = c$ and

$$0 \in \partial(I|_{S_{f,r,L}})(u). \tag{6.150}$$

Taking into account (6.150), (6.139) and (6.135) it is clear that there exists

$$z \in \partial(J|_V)(u) \subset \partial J(u), \tag{6.151}$$

such that u solves the equation

$$Au - f + z = \frac{\langle Au - f + z, \Lambda^{-1}Lu\rangle_V}{\langle Lu, \Lambda^{-1}Lu\rangle_V} Lu. \tag{6.152}$$

Due to (6.151) and (6.131) we can find a mapping $z : \Omega \to (I\!R^N)^*$ having the properties

$$z(x) \in \partial_y j(x, u(x)) \quad \text{for a.e. } x \in \Omega \tag{6.153}$$

and

$$\langle z, v\rangle_V = \int_\Omega \langle z(x), v(x)\rangle_{I\!R^N} dx \quad \text{for all } v \in V. \tag{6.154}$$

Now we set

$$\lambda = \frac{\langle Au - f + z, \Lambda^{-1}Lu\rangle_V}{\langle Lu, \Lambda^{-1}Lu\rangle_V}, \tag{6.155}$$

With z satisfying (6.153), (6.154), we obtain from (6.152) $\div$ (6.155) that

$$\lambda\langle Lu, v\rangle_V - a(u, v) + \langle f, v\rangle_V = \int_\Omega \langle z(x), v(x)\rangle_{I\!R^N} dx \leq$$

$$\int_\Omega \max_{\bar{z} \in \partial_y j(x,u(x))} \langle \bar{z}, v(x)\rangle_{I\!R^N} dx = \int_\Omega j^0(x, u(x); v(x)) dx, \quad \forall v \in V. \tag{6.156}$$

This relation implies that $u \in S_{f,r,L}$ is an eigensolution corresponding to the eigenvalue λ given in (6.155) for the eigenvalue problem $(P_{f,r,L})$) (see (6.128)). This completes the proof of Theorem 6.5. □

Subsequently, from the proof of Theorem 6.5 (precisely relation (6.149)), the following minimization result is obtained.

Corollary 6.3 Assume that the conditions of Theorem 6.5 are verified. Then the locally Lipschitz functional $I : V \to I\!R$ given by (6.133) assumes its infimum on the smooth submanifold $S_{f,r,L}$ of V.

In addition, we derive the following global result.

Corollary 6.4 Assume that the conditions of Theorem 6.5 are verified. Then the locally Lipschitz functional $I : V \to I\!R$ given by (6.133) assumes its global infimum on V, i.e, there exists $u \in V$ such that $I(u) = \inf_V I$. Moreover, $u \in V$ solves the hemivariational inequality

$$a(u, v) - \langle f, v\rangle_V + \int_\Omega j^o(x, u; v)dx \geq 0, \ \forall v \in V.$$

Proof. Let us apply the Palais-Smale Minimization Theorem in Chang's version for locally Lipschitz functions to the functional $I : V \to V$ given by (6.133). It is known from (6.137) that I is bounded from below. We need only to verify that I satisfies the Palais-Smale condition. To this end let $\{v_n\}$ be a sequence in V such that $I(v_n)$ is bounded and there exists a sequence $\{w_n\} \subset V^*$ satisfying (6.140) with $w_n \to 0$ in V^* as $n \to \infty$. Then a sequence $\{z_n\}$ in V^* can be found so that (6.142) holds and

$$Av_n - f + z_n \to 0 \text{ in } V^* \text{ as } n \to \infty.$$

From relation (6.136) we see that the sequence $\{v_n\}$ is bounded. Passing to a subsequence we may suppose that (6.145) is valid. Then we may infer that property (6.147) is verified. The equality above allows then to conclude that a subsequence of $\{v_n\}$ converges strongly in V. Thus the Palais-Smale condition is checked. The rest of the proof proceeds as in the proof of Theorem 6.5. □

Remark 6.3 Due to Riesz theorem Theorem 6.5 holds for $(L_1u, v)_V$ instead of $\langle Lu, v\rangle_V$ where L_1 is a linear compact symmetric operator.

References

[1]A. Ambrosetti and P.H. Rabinowitz. Dual Variational Methods in Initial Point Theory and Applications, *J. Funct. Anal.* **14** (1973), 349-381.

[2]K.C. Chang, Variational Methods for Non-Differentiable Functionals and their Applications to Partial Differential Equations, *J. Math. Anal. Appl.* **80** (1981), 102-129.

[3]D.C. Clark, A variant of the Ljusternik-Schnirelmann theory, *Indiana Univ. Math. J.* **22** (1972), 65-74.

[4]F.H. Clarke, *Optimization and Nonsmooth Analysis* , John Wiley and Sons, New York (1983).

[5]I. Ekeland, On the Variational Principle , *J. Math. Anal. Appl.* **47** (1974), 324-353.

[6]D. Motreanu and P.D. Panagiotopoulos, On the Eigenvalue Problem for Hemivariational Inequalities: Existence and Multiplicity of Solutions, *J. Math. Anal. Appl.* **197** (1996), 75-89.

[7]Z. Naniewicz, On Some Nonconvex Variational Problems Related to Hemivariational Inequalities, *Nonlinear Anal. T.M.A.* **13** (1989), 87-100.

[8]Z. Naniewicz, On the Pseudo-monotonicity of Generalized Gradient of Nonconvex Functions, *Applic. Anal.* **47** (1992), 151-172.

[9]Z. Naniewicz and P.D. Panagiotopoulos, *Mathematical Theory of Hemivariational Inequalities*, Marcel Dekker, N.York (1995).

[10]P.D. Panagiotopoulos, *Inequality Problems in Mechanics and Applications. Convex and Nonconvex Energy Functions*, Birkhäuser Verlag, Boston,Basel, 1985 (Russian Transl. MIR Publ. Moscow, 1989).

[11]P.D. Panagiotopoulos, Coercive and Semicoercive Hemivariational Inequalities, *Nonlinear Anal. T.M.A.* **16** (1991), 209-231.

[12]P.D. Panagiotopoulos, *Hemivariational Inequalities. Applications to Mechanics and Engineering* , Springer Verlag, N.York, Berlin (1993).

[13]P.D. Panagiotopoulos, (1988), *Hemivariational Inequalities and Their Applications*, in J.J. Moreau, P.D. Panagiotopoulos and G.Strang (eds), *Topics in Nonsmooth Mechanics*, Birkhäuser Verlag, Boston, Basel.

[14]P.D. Panagiotopoulos, (1988), *Nonconvex Superpotentials and Hemivariational Inequalities. Quasidifferentiability in Mechanics*, in J.J.Moreau, P.D.Panagiotopoulos (eds), *Nonsmooth Mechanics and Applications* , CISM Courses and Lectures Nr. **302**, Springer Verlag, Wien, New York.

[15]P.D. Panagiotopoulos, Nonconvex Problems of Semipermeable Media and Related Topics, *ZAMM* **65** (1985), 29-36.

[16]P.H. Rabinowitz. Variational Methods for Nonlinear Eigenvalue Problems. in: Eigenvalues of Nonlinear Problems (CIME, Varena, 1974), *Ed. Cremonese, Rome* (1974) 141-195.

[17]P.H. Rabinowitz, Minimax Methods in Critical Point Theory with Applications to Differential Equations, *CBMS Reg. Conf. Ser. in Math.* **65** (1986), *Amer. Math. Soc.*, Providence, R.I.

[18]A. Szulkin, Minimax Principles for Lower Semicontinuous Functions and Applications to Nonlinear Boundary Value Problems, *Ann. Inst. H. Poincaré, Analyse Nonlinéaire* **3** (1986), 77-109.

7. Resonant Eigenvalue Problems for Hemivariational Inequalities

The present chapter deals with the mathematical study of the resonant case of eigenvalue problems for hemivariational inequalities. One obtains new Landesman - Lazer type results allowing unbounded nonlinearities and resonance at multiple eigenvalues. A buckling problem in Mechanics illustrates the practical motivation and the interest in these topics. We use two different methods: a nonsmooth version of the Saddle-Point Theorem due to K. C. Chang and the Leray-Schauder topological degree. The results of this chapter improve and extend all the known results related to the nonsmooth resonant eigenvalue problems.

7.1 Formulation of the Resonant Problem

The aim of the present chapter is the mathematical study of a new kind of nonlinear eigenvalue problems arising in the stability analysis of mechanical systems subjected to realistic nonmonotone boundary conditions [17]. Precisely, consider the nonsmooth eigenvalue problem: find $\lambda \in \mathbb{R}$ and $u \in V$ satisfying (cf. (5.62))

$$(P) \qquad a(u,v) - \lambda \int_{\Omega} u(x)v(x)dx + \int_{\Omega} j^{o}(x,u(x);v(x))dx \geq 0, \forall v \in V,$$

where $V \subset L^2(\Omega)$ is a real Hilbert space, $a(.,.)$ is a continuous bilinear form, $j(x,.)$ is a locally Lipschitz function. The notation $j^o(x,u(x);\, v(x))$ stands for the generalized directional derivative of $j(x,.)$ at $u(x)$ in the direction $v(x)$ (see Definition 1.1 in Chapter 1). This problem (P) is an eigenvalue problem for hemivariational inequalities. The resonant case in (P) holds, by definition, when λ is equal to an eigenvalue, say $\lambda = \lambda_k$, of the linearized eigenvalue problem

$$a(u,v) = \lambda \int_{\Omega} u(x)v(x)dx, \quad \forall v \in V.$$

This situation was not covered by the results in Chapter 6 (see also Goeleven, Motreanu and Panagiotopoulos [8]), where only the nonresonant case $\lambda \in (\lambda_k, \lambda_{k+1})$ has been studied. Here we deal with the existence of solutions of

(P) in the resonant case $\lambda = \lambda_k$ with different types of hypotheses for the nonlinearity $j(x, y)$. We obtain necessary conditions and sufficient conditions for the solvability of the eigenvalue problem (P) subject to the resonant constraint $\lambda = \lambda_k$, which reduce in the classical situation to the well-known Landesman-Lazer conditions [10], [12].

Beyond the theoretical interest, problem (P) at resonance, as well as at nonresonance, has a strong motivation coming from applications in Mechanics. This kind of inequality expressions in (P) have been introduced by P.D. Panagiotopoulos [15]-[17] to deal with problems in Mechanics whose variational forms are inequalities which express the principle of virtual work or power and involve the variations of nonconvex energy functions. Such formulations arise, for instance, in the study of adhesive joints in structural mechanics, the behavior of composites, the unilateral contact and nonmonotone friction problems in cracks, the adhesive grasping problem in robotics, etc (see [15]-[17] and the references cited in these books). Among the important mechanical phenomena leading to problem (P) at resonance, a typical example is offered by a model of beam buckling that is presented in the next section.

As shown in several works [8], [9], [13]-[17], problem (P) contains various mathematical problems, for instance the semilinear elliptic equations with discontinuities, whose variational study was a major contribution of Chang [5]. The minimax method developed by Chang is now one of the most efficient and appreciated tools to deal with hemivariational inequalities arising in Mechanics (see [13], [14], [17] and the references cited in these papers). The critical point approach of Chang [5] was included in Chapters 2 and 3 in more general theories. Concerning the field of nonsmooth eigenvalue problems, Theorem 5.4 of Chang [5] extends for partial differential equations with discontinuous nonlinearities at resonance the main result of Ahmad, Lazer and Paul [1] (see also Rabinowitz [18], [19]).

The purpose of the present chapter is to solve the general eigenvalue problem (P) of a hemivariational inequality at resonance with two precise goals: 1) to extend the part of the work of Chang [5] devoted to these topics and 2) to construct a Landesman-Lazer theory in the nonsmooth framework of problem (P). Using the theory of Chang and some ideas developed by Rabinowitz, we obtain new results in the field of hemivariational inequalities and as a consequence we show that the famous Landesman-Lazer conditions play in this context a role similar to the one known for partial differential equations. In addition, as it will be explained below by various comments and remarks in the text, we improve or complement even in the smooth case important known results. Our exposition follows closely the results of Goeleven, Motreanu and Panagiotopoulos [9].

Our main results are formulated in Theorems 7.1, 7.2, 7.3 and 7.4, where we obtain a new insight in problem (P) at resonance, concerning the role of the Landesman -Lazer conditions [8], [10] as well as of the other classical sufficient conditions of solvability like those due to Ahmad, Lazer and Paul [1], Chang [5], Gallouët and Kavian [7], [11], Landesman and Lazer [10], [12], Panagiotopoulos [17], Rabinowitz [18], [19]. Extensions of these known results are made in several directions. This is discussed in details by remarks in Sections 7.3 and 7.4. We emphasize here only two of the novelties of our approach. Firstly, we will deal with resonant problems with unbounded nonlinearities (see Theorem 7.1 and Remark 7.1, but also Theorem 7.2 and Remark 7.2), which is a subject not yet extensively studied. Exception is the recent work of Robinson and Landesman [20] where the authors have obtained various new results in the field of resonance, some of them being applicable to problems involving unbounded nonlinearities. Note also that it is mentioned explicitely in the recent paper of Robinson and Landesman [20], p. 1558, that all the previous results treated resonance with bounded nonlinearities. Compared now with the growth condition of Robinson and Landesman [20] used at resonance with the first eigenvalue λ_1, we see that our assumption (H_2) is more general, because it allows a growth for the potential $j(x, y)$ of at most $c|y|^\sigma + d(x)|y| + e(x)$ with $\sigma < 2$, $c > 0$, $e \in L^2(\Omega)$ and $d \in L^2(\Omega)$, while in [20] it is necessary to have a growth of at most $p(x)|y| + c_0$ with $p \in L^2(\Omega)$ and $c_0 \in \mathbb{R}$. However, one cannot give additional comparisons, since in [20] the linear part is defined by a nonselfadjoint operator. Secondly, it seems that for a first time one obtains conditions of Landesman-Lazer type in the nonsmooth resonant problems, at eigenvalues λ_k which are not simple. We point out that our assertion refers to the nonsmooth setting (see Theorems 7.1, 7.2 and Propositions 7.1, 7.2). Concerning this point we recall that Remark 5.3 in Chang [5] requires that $\dim(N(L)) = 1$, a fact which is not the case in our presentation.

In this chapter we use two different approaches: Chang's nonsmooth minimax results [5] (see Chapter 2) in Section 7.3 and Leray-Schauder degree in Section 7.4. These two methods complete each other and address to different situations. Since the setting is locally Lipschitz, the results of Chang [5] (see Chapter 2) based on the calculus with generalized gradients of Clarke [6] are extensively used. To fulfill our aims of applicability and generality, (consider for example the case of nonsmooth variational systems), we work with potentials j defined on $\Omega \times \mathbb{R}^N$, with $N \geq 1$, and thus the solutions of the eigenvalue problem (P) are vector-valued functions $u : \Omega \to \mathbb{R}^N$. In the part based on the use of Leray-Schauder degree for the derivation of Landesman-Lazer type results we work with $N = 1$. For these two cases there is in our results a clear distinction between the expressions of necessary and sufficient conditions for the solvability at resonance of problem (P).

7.2 Buckling of Sandwich Beams

We consider two elastic beams (linear elasticity) of length l measured along the axis Ox of the coordinate system yOx, and with the same cross-section. The beams, numbered here by $i = 1, 2$, are simply supported at their

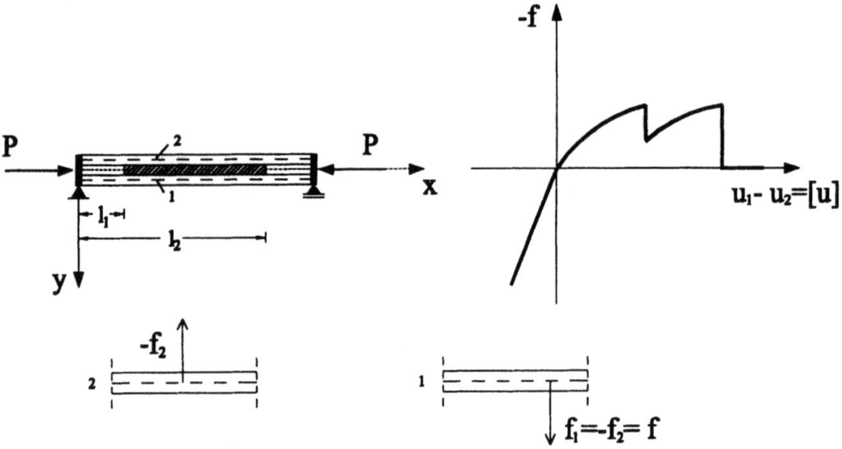

Fig. 7.1 Sandwich Beam.

ends $x = 0$ and $x = l$ and on the interval (l_1, l_2), $l_1 < l_2 < l$, are connected with an adhesive material of negligible thickness (Fig. 7.1). The displacements of the i-th beam are denoted by $x \to u_i(x)$, $i = 1, 2$, and the behaviour of the adhesive material is described by a nonmonotone possibly multivalued law between $-f(x)$ and $[u(x)]$, where $x \to f(x)$ denotes the reaction force per unit length vertical to the Ox axis, due to the adhesive material (cf. [15] p.87 and [17] p.110) and $[u] = u_1 - u_2$ is the relative deflection of the two beams. Recall that u_i is referred to the middle line of the beam i (the dotted lines in Fig. 7.1) and that each beam has constant thickness which remains the same after the deformation. The adhesive material can sustain a small tensile force before rupture (debonding). In Fig. 7.1 a rupture of zig-zag brittle type is depicted in the $(-f, u)$ diagramm. The beams are assumed to have the same moduli of elasticity E and let I be the moment of inertia of them. The sandwich beam is subjected to the compressive force P and we want to determine the buckling loading of it. This problem is a yet open problem in Engineering. From the large deflection theory of beams (cf. Section 5.5), we may write the following relations which describe the behaviour of the i-th beam:

$$a^2 u_i''''(x) + u_i''(x) = a^2 f_i(x) \qquad \text{on } (0, l); \tag{7.1}$$

$$u_i(0) = u_i(l) = 0, \qquad u_i''(0) = u_i''(l) = 0 \qquad i = 1, 2. \tag{7.2}$$

Here $a^2 := IE/P$. We assume that the $(-f, [u])$ graph results from a $L^\infty_{Loc}(\mathbb{R})$ function by filling - in the jumps, i.e. according to Proposition 1.7 there exists a locally Lipschitz function $j : \mathbb{R} \to \mathbb{R}$ such that

$$-f(x) \in \partial j([u(x)]), \quad \forall x \in (l_1, l_2). \tag{7.3}$$

As in Section 5.5 we set

$$V := H^2(\Omega) \cap H_0^1(\Omega), \qquad \Omega = (0, l).$$

It is a Hilbert space with the inner product $a(u,v) := \int_0^l u''(x)v''(x)dx$.

Let $L : V \to V^*$ be the linear compact operator defined by

$$\langle Lu_i, v_i \rangle := \int_0^l u_i'(x)v_i'(x)dx, \forall u_i, v_i \in V.$$

The superpotential law (7.3) implies that

$$j^0([u(x)]; y) \geq -f(x)y, \qquad \forall x \in (l_1, l_2), \forall y \in \mathbf{R}.$$

Multiplying (7.1) by $v_i(x) - u_i(x)$, integrating over $(0, l)$ and adding the resulting relations for $i = 1, 2$, implies by taking into account the boundary condition (7.2), the hemivariational inequality

$$\begin{aligned}
&u = \{u_1, u_2\} \in V \times V, \\
&a^2 \sum_{i=1}^{2} \int_0^l u_i''(x)[v_i''(x) - u_i''(x)]dx - \sum_{i=1}^{2} \int_0^l u_i'(x)[v_i'(x) - u_i'(x)]dx \\
&\qquad\qquad + a^2 \int_{l_1}^{l_2} j^0([u(x)]; [v(x)] - [u(x)])dx \geq 0, \\
&\qquad\qquad \forall v = \{v_1, v_2\} \in V \times V.
\end{aligned}$$

Thus buckling of the beam occurs if $\lambda := 1/a^2$ is an eigenvalue for the following hemivariational inequality

$$\sum_{i=1}^{2} a_i(u_i, v_i - u_i) - \lambda \sum_{i=1}^{2} \langle Lu_i, v_i - u_i \rangle + \int_{l_1}^{l_2} j^0([u(x)]; [v(x)] - [u(x)])dx \geq 0, \tag{7.4}$$

$$\forall v = \{v_1, v_2\} \in V \times V.$$

Let us consider the linearized problem

$$\sum_{i=1}^{2} a_i(u_i, v_i) - \lambda \sum_{i=1}^{2} \langle Lu_i, v_i \rangle = 0, \qquad \forall v = \{v_1, v_2\} \in V \times V. \tag{7.4a}$$

In Chapter 5 (as well as in [8]), we have shown that if $\lambda_k < \lambda_{k+1}$ are two distinct eigenvalues for problem (7.4a) then suitable assumptions on the graph of j guarantee the existence of at least k pairs of solutions $(-u(\lambda_o), +u(\lambda_o))$ for problem (7.4) when $\lambda_o \in (\lambda_k, \lambda_{k+1})$. Under symmetry assumptions for $j(x, \cdot)$, $x \in \Omega$, one could have infinitely many solutions. Our aim in this chapter is now to discuss the resonance case, that is when $\lambda_o = \lambda_k$ in (7.4).

Note that analogously to the buckling problem of a sandwich beam we can consider the buckling problem for sandwich Timoshenko plates and for sandwich cylindrical shells.

7.3 Existence of Solutions

The theoretical basis of this Section in studying of problem (P) is the following nonsmooth version of Saddle-Point Theorem of Rabinowitz [18], [19] which is due to Chang [5]. Although it is a particular case of Corollary 2.4 and Corollary 3.3 for the sake of clarity we quote this result from Chang [5].

Theorem (Chang [5]). Let X be a reflexive Banach space and let $f : X \to \mathbb{R}$ be a locally Lipschitz function satisfying the Palais-Smale condition. Suppose $X = X_1 \oplus X_2$, with a finite dimensional subspace X_1, and there exist constants $b_1 < b_2$ and a neighborhood N of 0 in X_1 such that

$$f(x) \geq b_2, \forall x \in X_2$$

and

$$f(x) \leq b_1, \forall x \in \partial N.$$

Then f has a critical point.

The abstract result in the Theorem above will be applied to treat the eigenvalue problem (P) at resonance. We describe now the setting of our approach.

Let V be a real Hilbert space densely and compactly embedded in $L^2(\Omega; \mathbb{R}^N)$ for a bounded domain Ω in $\mathbb{R}^m$, $m, N \geq 1$, with the norm $\|\cdot\|_V$. Let $a : V \times V \to \mathbb{R}$ be a continuous, symmetric, bilinear form which is coercive

$$a(u, u) \geq \alpha \|u\|^2, \ \forall u \in V,$$

with a constant $\alpha > 0$. Thus

$$\|\cdot\|_V \ := \ a(\cdot, \cdot)^{\frac{1}{2}}$$

is an equivalent norm on V. Denote by

$$0 < \lambda_1 < \lambda_2 \leq ....$$

the sequence of eigenvalues (where $\lambda_n \to +\infty$) of the linear problem

$$a(u, v) = \lambda(u, v)_{L^2}, \ \forall v \in V. \tag{7.5}$$

In this statement we use essentially the compactness of the embedding $V \subset L^2(\Omega; \mathbb{R}^N)$. The fact that λ_1 is simple follows from Krein-Rutman Theorem (see Brézis [3]). We consider a basis $\{\Phi_i\}_{i\geq 1}$ for V consisting of eigenfunctions, where Φ_i corresponds to λ_i, i.e., $u = \Phi_i$ and $\lambda = \lambda_i$ in (7.5), which is normalized in the following sense

$$a(\Phi_i, \Phi_j) = \delta_{ij},$$

where $\delta_{ii} = 1$ and $\delta_{ij} = 0$ if $i \neq j$. Let $j : \Omega \times \mathbb{R}^N \to \mathbb{R}$ denote a (Carathéodory) function satisfying

(a) $j(.,y) : \Omega \to \mathbb{R}$ is measurable, $\forall y \in \mathbb{R}^N$;

(b) $j(x,.) : \mathbb{R}^N \to \mathbb{R}$ is locally Lipschitz, $\forall x \in \Omega$;

(c) $j(.,0) \in L^1(\Omega)$.

Let $k \geq 1$ be an integer for which we have $\lambda_k < \lambda_{k+1}$. Setting

$$X := \mathrm{span}\{\Phi_1, ..., \Phi_k\},$$

$$Y := \overline{\mathrm{span}\{\Phi_i\}_{i \geq k+1}},$$

the direct sum decomposition holds

$$V = X \oplus Y.$$

Then any element $v \in V$ is written uniquely as

$$v = v^- + v^o + v^+,$$

with

$$v^+ \in Y,\ v^- + v^o \in X,$$

$$v^- \in \mathrm{span}\{\Phi_i : \lambda_i < \lambda_k\},$$

$$v^o \in Z_o := \mathrm{span}\{\Phi_i : \lambda_i = \lambda_k\}.$$

If there exist eigenvalues $\lambda_i < \lambda_k$ in (7.5), for $k \geq 2$ we denote

$$\lambda^*_{k-1} := \max\{\lambda_i : \lambda_i < \lambda_k\},$$

and, for $k = 1$, we take λ^*_0 any fixed number in $(0, \lambda_1)$. Firstly, we admit that $j : \Omega \times \mathbb{R}^N \to \mathbb{R}$ verifies the conditions below

(H_1) $|z| \leq d(x) + c|y|^{\sigma-1}$, $\forall y \in \mathbb{R}^N$, a.e. $x \in \Omega$, $\forall z \in \partial_y j(x,y)$, with constants $c > 0$ and $1 \leq \sigma < 2$ and $d \in L^2(\Omega)$;

(H_2) $\|v\|_V^{-\sigma} \int_\Omega j(x,v)dx \to -\infty$ as $\|v\|_V \to \infty$, for $v \in Z_0$ if (H_1) holds with $1 < \sigma < 2$, and $\int_\Omega j(x,v)dx \to -\infty$ as $\|v\|_V \to +\infty$, for $v \in Z_0$ if $\sigma = 1$.

Theorem 7.1 Under (H_1) and (H_2), the hemivariational inequality

$$a(u,v) + \int_\Omega j^o(x, u(x); v(x))dx \geq \lambda_k (u,v)_{L^2},\ \forall v \in V, \tag{7.6}$$

has a solution.

Proof. We introduce the functional $I : V \to \mathbb{R}$ by

$$I(v) = \frac{1}{2}(a(v,v) - \lambda_k \|v\|_V^2) + \int_\Omega j(x,v)dx, \forall v \in V. \tag{7.7}$$

Due to (H_1) the functional I is locally Lipschitz and its generalized gradient is given by

$$\partial I(v) \subset a(v,.) - \lambda_k(v,.)_{L^2} + \int_\Omega \partial_y j(x,v)dx, \forall v \in V, \tag{7.8}$$

(cf. Theorem 1.3 in Chapter 1). Under hypothesis (H_1) we find constants $c_1 > 0$ and $c_2 \in \mathbb{R}$ such that for all $v = \sum_{i\geq k+1} \alpha_i \Phi_i \in Y$ one can write

$$\begin{aligned} I(v) &= \frac{1}{2}\sum_{i\geq k+1}(1-\lambda_k\|\Phi_i\|_{L^2}^2)\alpha_i^2 + \int_\Omega j(x,v)dx \\ &\geq \frac{1}{2}(1-\lambda_k/\lambda_{k+1})\|v\|^2 - c_1\|v\|^\sigma - c_2. \end{aligned}$$

Since $\sigma < 2$ we conclude that there exists a constant $b_2 \in \mathbb{R}$ provided

$$I(v) \geq b_2, \forall v \in Y. \tag{7.9}$$

For $v = \sum_{i\leq k} \alpha_i \Phi_i = v^- + v^o \in X$ we see from (H_1) and Lebourg's mean value theorem (see Theorem 1.1 in Chapter 1) that

$$\begin{aligned} I(v) &= \frac{1}{2}\sum_{i=1}^{k}(1-\lambda_k\|\Phi_i\|_{L^2}^2)\alpha_i^2 + \int_\Omega j(x,v)dx \\ &= \frac{1}{2}\sum_{i=1}^{k}(1-\lambda_k/\lambda_i)\alpha_i^2 + \int_\Omega (j(x,v)-j(x,v^o))dx + \int_\Omega j(x,v^o)dx \\ &\leq \frac{1}{2}(1-\lambda_k/\lambda_{k-1}^*)\|v^-\|^2 + \int_\Omega (d+c|v|^{\sigma-1})|v^-|dx + \int_\Omega j(x,v^o)dx. \end{aligned}$$

It follows that there exist constants $a_1 > 0, a_2 > 0$ and $a_3 > 0$ such that

$$I(v) \leq \frac{1}{2}(1-\lambda_k/\lambda_{k-1}^*)\|v^-\|^2 + a_1\|v^-\|^\sigma + \int_\Omega j(x,v^o)dx + a_2\|v_o\|^\sigma + a_3.$$

Here we have applied the Young inequality under the integral sign. Let $\|v\|_V \to +\infty$ for $v \in X$. If $\|v^-\| \to +\infty$, with the aid of (H_2), it is clear that $I(v) \to -\infty$. If $\|v^-\|$ remains bounded, then necessarily $\|v^o\| \to +\infty$ and by means of (H_2) we also conclude that $I(v) \to -\infty$. Consequently, there exist some constants $b_1 \in \mathbb{R}$ and $\delta > 0$ such that

$$I(v) \leq b_1, \ \forall v \in X, \ \|v\| = \delta, \tag{7.10}$$

and

$$b_1 < b_2. \tag{7.11}$$

We check now the Palais-Smale condition in the sense of Chang [5] (see also Chapter 2) for the locally Lipschitz functional I on V given by (7.7).

Let $v_n = v_n^- + v_n^o + v_n^+ \in V$ be a sequence in V such that $I(v_n)$ is bounded and

$$a(v_n,.) - \lambda_k(v_n,.)_{L^2} + w_n \to 0 \text{ in } V^*, \tag{7.12}$$

for some $w_n \in L^{\sigma/(\sigma-1)}(\Omega;\mathbb{R}^N)$ satisfying

$$w_n(x) \in \partial_y j(x, v_n(x)) \text{ for a.e. } x \in \Omega. \tag{7.13}$$

Relations (7.12), (7.13) are written according to (7.8). Let us treat firstly the case in (H_1), (H_2) where $1 < \sigma < 2$.

By (7.12), (7.13), for a sufficiently large n, we obtain the inequalities

$$\begin{aligned} \|v_n^+\| &\geq a(v_n, v_n^+) - \lambda_k (v_n, v_n^+)_{L^2} + \int_\Omega w_n v_n^+ dx \\ &\geq a(v_n^+, v_n^+) - \lambda_k \|v_n^+\|_{L^2}^2 - \int_\Omega (d + c|v_n|^{\sigma-1})|v_n^+| dx \\ &\geq \|v_n^+\|^2 - \lambda_k \|v_n^+\|_{L^2}^2 - c_1\|v_n^+\|^\sigma - c_2\|v_n^-\|^\sigma - c_3\|v_n^o\|^\sigma - c_o, \end{aligned} \tag{7.14}$$

with constants $c_i \geq 0, i = 0, 1, 2, 3$. Analogously, we get for large n that

$$\begin{aligned} \|v_n^-\| &\geq -a(v_n, v_n^-) + \lambda_k (v_n, v_n^-)_{L^2} - \int_\Omega w_n v_n^- dx \\ &\geq -\|v_n^-\|^2 + \lambda_k \|v_n^-\|_{L^2}^2 - \int_\Omega (d + c|v_n|^{\sigma-1})|v_n^-| dx \\ &\geq -\|v_n^-\|^2 + \lambda_k \|v_n^-\|_{L^2}^2 - d_1\|v_n^+\|^\sigma - d_2\|v_n^-\|^\sigma - d_3\|v_n^o\|^\sigma - d_o, \end{aligned} \tag{7.15}$$

for constants $d_i \geq 0, i = 0, 1, 2, 3$.

Knowing that there is a constant $M > 0$ such that $-I(v_n) \leq M$ we see that

$$\begin{aligned} M &\geq \frac{1}{2}(-a(v_n, v_n) + \lambda_k \|v_n\|_{L^2}^2) - \int_\Omega j(x, v_n) dx = \\ &= \frac{1}{2}(-\|v_n\|^2 + \lambda_k \|v_n\|_{L^2}^2) - \int_\Omega (j(x, v_n) - j(x, v_n^o)) dx - \int_\Omega j(x, v_n^o) dx \\ &\geq \frac{1}{2}(-\|v_n\|^2 + \lambda_k \|v_n\|_{L^2}^2) - e_1\|v_n^+\|^\sigma - e_2\|v_n^-\|^\sigma - \\ &\quad -e_3\|v_n^o\|^\sigma - e_o - \int_\Omega j(x, v_n^o) dx \end{aligned} \tag{7.16}$$

with constants $e_i \geq 0, i = 0, 1, 2, 3$.

From the inequalities (7.14)-(7.16) we get by addition the new inequality

$$\begin{aligned} &\|v_n^+\| + \|v_n^-\| + M \geq \\ &\quad \frac{1}{2}(\|v_n^+\|^2 - \lambda_k \|v_n^+\|_{L^2}^2) - \frac{3}{2}(\|v_n^-\|^2 - \lambda_k \|v_n^-\|_{L^2}^2) - \\ &\quad -C_1\|v_n^+\|^\sigma - C_2\|v_n^-\|^\sigma - C_3\|v_n^o\|^\sigma - C_0 - \int_\Omega j(x, v_n^o) dx \geq \\ &\quad \frac{1}{2}(1 - \lambda_k/\lambda_{k+1})\|v_n^+\|^2 + \frac{3}{2}(\lambda_k/\lambda_{k-1}^* - 1)\|v_n^-\|^2 - C_1\|v_n^+\|^\sigma - \\ &\quad -C_2\|v_n^-\|^\sigma - C_3\|v_n^o\|^\sigma - C_0 - \int_\Omega j(x, v_n^o) dx \end{aligned}$$

for n sufficiently large, with constants $C_i, i = 0, 1, 2, 3$.

Since $\sigma < 2$ and $\lambda_{k-1}^* < \lambda_k < \lambda_{k+1}$, in view of (H_2) we deduce that $\{v_n\}$ is a bounded sequence in V.

The remaining case $\sigma = 1$ with the respective part of assumption (H_2), represents a particular case of Theorem 7.2 below. This can be seen easily by a direct comparison of hypotheses using Lebourg's mean value theorem. According to the proof of Theorem 7.2 the boundedness of the sequence $\{v_n\}$ follows.

Now we handle property (7.12). The boundedness of $\{v_n\}$ in V and the compactness of the embedding $V \subset L^2(\Omega; I\!R^N)$ imply that along a subsequence $\{v_n\}$ converges strongly in $L^2(\Omega; I\!R^N)$. Due to (H_1) the functional $J : L^2(\Omega; I\!R^N) \to I\!R$ given by

$$J(v) = \int_\Omega j(x, v)dx, \ \forall v \in L^2(\Omega; I\!R^N) \tag{7.17}$$

is locally Lipschitz. Since we have

$$w_n \in \partial J(v_n) \tag{7.18}$$

(see also (7.13)) it follows that $\{w_n\}$ has a bounded subsequence in $L^2(\Omega; I\!R^N)$, so strongly convergent in V^* along a subsequence. Therefore, by (7.12), $\{v_n\}$ possesses a convergent subsequence in V.

All the assumptions of Chang's Theorem stated at the beginning of Section are verified for the locally Lipschitz functional $I : V \to \mathbb{R}$ in (7.7) (see also Corollaries 2.4 and 3.3 which extend this result). Indeed, the relations (7.9) - (7.11) and the Palais-Smale condition are the necessary requirements to conclude that the locally Lipschitz functional I on V admits a critical point $u \in V$, that is, we have

$$0 \in \partial I(u). \tag{7.19}$$

Taking into account (7.8) the element $u \in V$ satisfies (7.6). □

Remark 7.1 Theorem 7.1 must be regarded in relation with Theorem 5.4 of Chang [5] which extends to the nonsmooth case the important resonance results of Ahmad, Landesman and Paul [1], Rabinowitz ([18], [19], Theorem 4.12) and others. Theorem 7.1 extends in several directions Theorem 5.4 of Chang [5], i.e. for a general continuous, symmetric, coercive, bilinear form a on a general Hilbert space $V \subset L^2(\Omega; I\!R^N)$, for a general potential $j(x, .)$ defined on $I\!R^N$ with $N \geq 1$ which does not have for $N > 1$ the form of a primitive, and finally for an unbounded generalized gradient of $j(x, .)$. A specific feature of our result is the relationship between the growth condition in (H_1) and the rate of decreasing to infinity in (H_2). Just in the smooth case Theorem 7.1 is new. Among other things it is so, due to the relaxation of the growth for the nonlinearity. As it is explicitly written in Robinson and Landesman [20], p. 1558, only problems at resonance with bounded nonlinearities, that is in our notations $\partial j(x, y)$, have been solved. Relatively to this open problem, the paper [20] allowed the sublinear unboundedness of $g(x, y) := \partial j(x, y)$ with some further stringent requirements (for j smooth). Our Theorem 7.1 allows the subquadratic unboundedness of $\partial j(x, y)$ with respect to $y \in I\!R^N$, even in the nonsmooth setting. Theorem 7.1 completes the main result of Goeleven, Motreanu and Panagiotopoulos [8] (see Chapter 5) where the nonresonant case was analysed under the hypothesis that the involved functional is even. Applications in mechanics and engineering leading to resonant hemivariational problems of type (7.6) can be found in Naniewicz and Panagiotopoulos [15] and Panagiotopoulos [16], [17]. An interesting question related to the main result in Goeleven, Motreanu and Panagiotopoulos [8], [9] is the multiplicity of solutions in the resonant case considered here.

Corollary 7.1 Assume that (H_1) is satisfied and λ_k is simple. If $j(x,.)$ is positively homogeneous of order $\beta > \sigma$ and

$$\int_\Omega j(x, \pm\Phi_k(x))dx < 0, \tag{7.20}$$

then the hemivariational inequality (7.6) admits a solution.

Proof. Notice that

$$Z_o = \{t\Phi_k : t \in \mathbb{R}\}.$$

For any $t > 0$, we have

$$\int_\Omega j(x, \pm t\Phi_k(x))dx/\|t\Phi_k\|^\sigma = t^{\beta-\sigma}\int_\Omega j(x, \pm\Phi_k(x))dx.$$

Therefore inequalities (7.20) entail assumption (H_2) and we conclude by application of Theorem 7.1. □

We now present a version of Theorem 7.1 which allows a different type of Lipschitz condition for $j(x,.)$ depending on $x \in \Omega$. Let $j : \Omega \times \mathbb{R}^N \to \mathbb{R}$ denote a function satisfying the requirements (a) and (c). We assume that

(H_3) There exists $l \in L^2(\Omega; \mathbb{R}^N)$ such that

$$|j(x,y) - j(x,y')| \le l(x)|y - y'|, \ \forall y, y' \in \mathbb{R}^N;$$

and

(H_4) $\int_\Omega j(x, v(x))dx \to -\infty$ as $\|v\| \to +\infty$, $v \in Z_o$.

Theorem 7.2 If (H_3) and (H_4) are fulfilled, then the hemivariational inequality (7.6) admits a solution.

Proof. Let $I : V \to \mathbb{R}$ be the functional introduced in (7.7). Using assumption (H_3) we obtain for each $v = \sum_{i\ge k+1}\alpha_i\Phi_i \in Y$ that

$$\begin{aligned} I(v) &\ge \frac{1}{2}\sum_{i\ge k+1}(1 - \lambda_k\|\Phi_i\|^2_{L^2})\alpha_i^2 + \int_\Omega j(x,0)dx - \int_\Omega |j(x,v) - j(x,0)|dx \\ &\ge \frac{1}{2}(1 - \lambda_k/\lambda_{k+1})\|v\|^2 - c_1\|v\| - c_2, \end{aligned}$$

where $c_1 > 0$ and $c_2 \in \mathbb{R}$ are constants. Thus there exists $b_2 \in \mathbb{R}$ such that (7.9) holds.

For $v = \sum_{i\le k}\alpha_i\Phi_i = v^- + v^o \in X$ the computation below is carried out:

$$
\begin{aligned}
I(v) &= \frac{1}{2}\sum_{i=1}^{k}(1-\lambda_k/\lambda_i)\alpha_i^2 + \int_\Omega (j(x,v)-j(x,v^o))dx + \int_\Omega j(x,v^o)dx \\
&\le \frac{1}{2}(1-\lambda_k/\lambda_{k-1}^*)\|v^-\|^2 + C_o\|v^-\| + \int_\Omega j(x,v^o)dx,
\end{aligned}
$$

with a constant $C_o > 0$ and λ_{k-1}^* having the meaning above. We use assumption (H_4) to assert, as in the proof of Theorem 4.1 that constants $b_1 \in \mathbb{R}$ and $\delta > 0$ can be found to satisfy (7.10), (7.11).

We check now the Palais-Smale condition in the sense of Chang [5] for the locally Lipschitz functional $I : V \to \mathbb{R}$ in (7.7) (see Chapters 2 and 3 for more details). Let $v_n = v_n^- + v_n^o + v_n^+ \in V$ be a sequence in V such that $I(v_n)$ is bounded and (7.12) holds for some $w_n \in L^2(\Omega; \mathbb{R}^N)$ satisfying (7.18) and (7.13) (see also (7.22)). For n sufficiently large, it is seen that

$$
\begin{aligned}
\|v_n^+\| &\ge a(v_n, v_n^+) - \lambda_k(v_n, v_n^+)_{L^2} + \int_\Omega w_n v_n^+ dx \\
&\ge \|v_n^+\|^2 - \lambda_k\|v_n^+\|_{L^2}^2 - C\|v_n^+\|
\end{aligned}
$$

with constant $C > 0$. Analogously, for a large n one obtains

$$
\begin{aligned}
\|v_n^-\| &\ge -a(v_n, v_n^-) + \lambda_k(v_n, v_n^-)_{L^2} - \int_\Omega w_n v_n^- dx \\
&\ge -\|v_n^-\|^2 + \lambda_k\|v_n^-\|_{L^2}^2 - D\|v_n^-\|
\end{aligned}
$$

with a constant $D > 0$. Since there is a constant $M > 0$ such that $-I(v_n) \le M$, it is allowed to write

$$
\begin{aligned}
M &\ge \frac{1}{2}(-a(v_n,v_n) + \lambda_k\|v_n\|_{L^2}^2) - \int_\Omega (j(x,v_n) - j(x,v_n^o))dx - \int_\Omega j(x,v_n^o)dx \\
&\ge \frac{1}{2}(-\|v_n\|^2 + \lambda_k\|v_n\|_{L^2}^2) - C_1(\|v_n^+\| + \|v_n^-\|) - \int_\Omega j(x,v_n^o)dx,
\end{aligned}
$$

with constant $C_1 > 0$. Combining the foregoing inequalities it follows

$$
\begin{aligned}
&\|v_n^+\| + \|v_n^-\| + M \ge \\
&\ge \frac{1}{2}(\|v_n^+\|^2 - \lambda_k\|v_n^+\|_{L^2}^2) - \frac{3}{2}(\|v_n^-\|^2 - \lambda_k\|v_n^-\|_{L^2}^2) \\
&-C_1\|v_n^+\| - C_2\|v_n^-\| - \int_\Omega j(x,v_n^o)dx \\
&\ge \frac{1}{2}(1-\lambda_k/\lambda_{k+1})\|v_n^+\|^2 + \frac{3}{2}(\lambda_k/\lambda_{k-1}^* - 1)\|v_n^-\|^2 \\
&-C_1\|v_n^+\| - C_2\|v_n^-\| - \int_\Omega j(x,v_n^o)dx, \qquad (7.21)
\end{aligned}
$$

for n sufficiently large, with constants $C_1 > 0, C_2 > 0$. This implies that $\{v_n\}$ is bounded. Indeed, suppose $\{v_n\}$ is unbounded. Then, clearly, $\|v_n^-\| \to +\infty$ or $\|v_n^+\| \to +\infty$ leads to a contradiction. If $\|v_n^-\|$ and $\|v_n^+\|$ remain bounded, then necessarily $\|v_n^o\| \to +\infty$. In this case (H_4) leads to a contradiction. Hence the boundedness of $\{v_n\}$ is proved. Now, the argument in the proof of Theorem 4.1

relying on relations (7.12), (7.13), (7.17), (7.18) shows the strong convergence of a subsequence $\{v_n\}$ in V. Since assumption (H_3) ensures inclusion (7.8) (see Theorem 1.3 in Chapter 1), we conclude as in the proof of Theorem 7.1 the solvability of (7.19), and thus (7.6) has a solution. □

Remark 7.2 Theorem 7.2 extends also, in the case of the resonant problem (7.6), Theorem 5.4 of Chang [5], and thus the related results of Ahmad, Lazer and Paul [1], Rabinowitz [18], [19] and others. This is true just for the one-dimensional case $N = 1$, in our notation with $\Omega \subset \mathbb{R}^N$ and $j : \Omega \times \mathbf{R} \to \mathbb{R}$ in the form of a primitive

$$j(x,t) = \int_0^t \Phi(x,s)ds, \ \forall (x,t) \in \Omega \times \mathbb{R},$$

where $\Phi : \Omega \times \mathbf{R} \to \mathbb{R}$ satisfies the measurability condition (2.4) in Chang [5]. To see this we compare assumptions (H_3), (H_4) with those of Theorem 5.4 in Chang [5]. Since in Chang [5] it is assumed that Φ is bounded on $\Omega \times \mathbb{R}$, our hypothesis (H_3) clearly holds, while for j in the form above, our hypothesis (H_4) coincides with the corresponding one of Theorem 5.4 in Chang [5], because $Z_0 = N(L)$ in Chang's notation. Notice that in the smooth setting this assumption was firstly introduced by Ahmad, Lazer and Paul [1]. We discuss the comparison of hypothesis (H_4) with a possible nonsmooth generalization of the hypothesis used in Rabinowitz [18], [19] to study the resonance: $j(x,y) \to -\infty$ as $|y| \to \infty$, $y \in \mathbb{R}^N$, uniformily with respect to $x \in \Omega$. Clearly, the statement above implies (H_4) as can be readily checked using an argument similar to that in Lemma 4.21 of Rabinowitz [19]. The converse is not generally true as simple examples show. For instance, take $j : \Omega \times \mathbb{R}^N \to \mathbb{R}$ given by $j(x,y) = -|y|$ for $x \neq x_0 \in \Omega$ and 0 otherwise. Then the foregoing statement fails at $x = x_0$, but (H_4) holds since Z_0 is finite dimensional. Thus (H_4) is more general than this possible generalization of the condition of Rabinowitz [18], [19] in the smooth case at resonance.

7.4 Landesman-Lazer conditions

Let us consider the problem (7.6) in the setting of $N = 1$. Thus the nonlinear part in (7.6) is determined by a function $j : \Omega \times \mathbf{R} \to \mathbf{R}$ satisfying $(a), (b), (c)$ of Section 7.3. Defining

$$j_-(x) := \inf\{z \in \partial_y j(x,y) : y \in \mathbb{R}\}, \ \forall x \in \Omega$$

and

$$j_+(x) := \sup\{z \in \partial_y j(x,y) : y \in \mathbb{R}\}, \ \forall x \in \Omega,$$

we assume that j_- and j_+ belong to $L^2(\Omega)$. Recall that Z_o denotes the eigenspace associated to the eigenvalue λ_k in (7.5) (see Section 7.3).

Given $q \in V$, the notations q_+ and q_- represent the positive and the negative parts of q, i.e.

$$q_+ = \max\{0, q\}$$

and

$$q_- = \max\{0, -q\}.$$

Accordingly, we consider the following sets

$$\Omega_+(q) := \{x \in \Omega : q(x) > 0\}$$

and

$$\Omega_-(q) := \{x \in \Omega : q(x) < 0\}.$$

Proposition 7.1 A necessary condition for the existence of a solution $u \in V$ of eigenvalue hemivariational inequality (7.6) is

$$\int_\Omega (j_- \Phi_+ - j_+ \Phi_-)dx \le 0 \le \int_\Omega (j_+ \Phi_+ - j_- \Phi_-)dx, \ \forall \Phi \in Z_o.$$

Proof. If we put in (7.6) $v = \pm\Phi \in Z_o$, $\Phi \ne 0$, we obtain

$$\int_\Omega j^o(x, u, \pm\Phi)dx \ge 0. \tag{7.22}$$

Then (7.22) and a well known property of the generalized gradient (see (ii) of Proposition 1.4) imply

$$\begin{aligned} 0 &\le \int_\Omega j^o(x, u; \Phi)dx \\ &\le \int_\Omega \max\{z\Phi(x) : z \in \partial_y j(x, u(x))\}dx \\ &= \int_{\Omega_+(\Phi)} j_+(x)\Phi(x)dx + \int_{\Omega_-(\Phi)} j_-(x)\Phi(x)dx \\ &= \int_\Omega (j_+(x)\Phi_+(x) - j_-(x)\Phi_-(x))dx. \end{aligned}$$

In a similar way, working with (7.22) entering $-\Phi$, we establish the left inequality in the statement. □

We point out that our hypothesis (H_2) (or (H_4)) is closely related to the generalized Landesman-Lazer conditions of Proposition 7.1.

Proposition 7.2 If (H_3) and (H_4) hold, then the conclusion of Proposition 7.1 is valid with strict inequalities.

Proof. Take $\Phi \in Z_o$. For all $t > 0$, Lebourg's Mean Value Theorem (see Theorem 1.1) yields

$$\begin{aligned} \int_\Omega j(x, t\Phi(x))dx &= \int_\Omega (j(x, t\Phi(x)) - j(x, 0))dx + \int_\Omega j(x, 0)dx \\ &\ge t\int_{\Omega_+(\Phi)} j_- \Phi dx + t\int_{\Omega_-(\Phi)} j_+ \Phi dx + \int_\Omega j(x, 0)dx \\ &= t\int_\Omega (j_- \Phi_+ - j_+ \Phi_-)dx + \int_\Omega j(x, 0)dx. \end{aligned}$$

Letting $t \to +\infty$, we see from (H_4) that

$$\int_\Omega (j_-\Phi_+ - j_+\Phi_-)dx < 0.$$

Proceeding in the same way with $-\Phi$ in place of Φ we derive from (H_4) that

$$\int_\Omega (j_+\Phi_+ - j_-\Phi_-)dx > 0.$$

□

The next partial converse statement of Proposition 7.1 extends in fact the main result in Gallouët and Kavian [7], so in particular the Landesman-Lazer Theorem [10], [12]. See also Chang [5] and Ahmad, Lazer and Paul [1].

Theorem 7.3 Assume that the growth condition in (H_1) holds with $\sigma = 2$ and let there exist $j_+^\infty \in L^2(\Omega)$, $j_-^\infty \in L^2(\Omega)$ and a continuous map $W : L^2(\Omega) \to L^2(\Omega)$ such that

$$\int_\Omega (j_+^\infty(x)\Phi_+(x) - j_-^\infty(x)\Phi_-(x))dx > 0,\ \forall \Phi \in Z_o\backslash\{0\}, \tag{7.23}$$

$$j_+^\infty(x) = \inf_{\{z_n\}} \liminf_{n\to\infty} z_n(x) \text{ (resp. } j_-^\infty(x) = \sup_{\{z_n\}} \limsup_{n\to\infty} z_n(x)), \tag{7.24}$$

for all sequences $z_n \in \partial_y j(x, \zeta_n) \in \mathbb{R}$ with $\zeta_n \to \infty$ (resp.$\zeta_n \to -\infty$), a.e. $x \in \Omega$, and

$$W(v)(x) \in \partial_y j(x, v(x)),\ \forall v \in V,\ \text{a.e. } x \in \Omega. \tag{7.25}$$

Then the hemivariational inequality (7.6) has a solution.

Proof. Fix a number $\varepsilon > 0$ such that the interval $[\lambda_k - \varepsilon, \lambda_k)$ does not contain an eigenvalue of the linear problem (7.5).

In order to develop an argument involving the Leray-Schauder degree (see e.g. Kavian [11]) we need an a priori estimate. Precisely, we must prove that there exists $r > 0$ such that for all $(u, t) \in V \times [0, 1]$ and $w \in L^2(\Omega)$ verifying

$$a(u, .) - \lambda_k u + tw + (1 - t)\varepsilon u = 0, \tag{7.26}$$

$$w(x) \in \partial_y j(x, u(x)) \text{ for a.e. } x \in \Omega, \tag{7.27}$$

then

$$\|u\|_{L^2} < r. \tag{7.28}$$

Arguing by contradiction we suppose that there are sequences $(u_n, t_n) \in V \times [0, 1]$ and $w_n \in L^2(\Omega)$ which satisfy

$$a(u_n, .) - \lambda_k u_n + t_n w_n + (1 - t_n)\varepsilon u_n = 0, \tag{7.29}$$

$$w_n(x) \in \partial_y j(x, u_n(x)) \text{ for a.e. } x \in \Omega, \tag{7.30}$$

$$\|u_n\|_{L^2} \to \infty \text{ as } n \to \infty. \tag{7.31}$$

If we set

$$v_n = (1/\|u_n\|_{L^2})u_n,$$

then (7.29) implies

$$a(v_n, .) - \lambda_k v_n + (t_n/\|u_n\|_{L^2})w_n + (1 - t_n)\varepsilon v_n = 0. \tag{7.32}$$

By (H_1) with $\sigma = 2$ and (7.30) there is $C > 0$ with

$$|w_n(x)| \leq C(1 + |u_n(x)|), \forall n \geq 1, \text{ a.e. } x \in \Omega.$$

Thus we find constants $C_1 > 0$ and $C_2 > 0$ provided

$$\|w_n\|_{L^2} \leq C_1 + C_2\|u_n\|_{L^2}.$$

This and (7.31) imply that

$$(1/\|u_n\|_{L^2})w_n \text{ is bounded in } L^2(\Omega). \tag{7.33}$$

From (7.32) and (7.33) it follows the boundedness of $a(v_n, .)$ in $L^2(\Omega)$, so in V^*. The inequality

$$\alpha\|v_n\|_V^2 \leq \|a(v_n, .)\|_{V^*}\|v_n\|_V$$

ensures that $\{v_n\}$ is bounded in V. The compactness of the embedding $V \subset L^2(\Omega)$ guarantees the existence of some $v^* \in L^2(\Omega)$ such that along a subsequence of $\{v_n\}$, denoted again by $\{v_n\}$, we have

$$v_n \to v^* \text{ in } L^2(\Omega), \tag{7.34}$$

$$v_n \to v^* \text{ a.e. in } \Omega. \tag{7.35}$$

Additionally, for an element $h \in L^2(\Omega)$, we know that

$$|v_n(x)| \leq h(x) \text{ for a.e. } x \in \Omega, \forall n \geq 1.$$

Then, in view of (7.34), (7.35) and Lebesgue Dominated Convergence Theorem, we derive from (7.32) the next convergence in $L^2(\Omega)$

$$a(v_n, .) = \lambda_k v_n - (t_n/\|u_n\|_{L^2})w_n - (1 - t_n)\varepsilon v_n \to \lambda_k v^* - (1 - t^*)\varepsilon v^* \tag{7.36}$$

for some $t^* \in [0, 1]$. In establishing (7.36) we made use of

$$j_-(x) \leq w_n(x) \leq j_+(x) \text{ for a.e. } x \in \Omega.$$

The continuity of the inclusion $L^2(\Omega) \subset V^*$ implies the convergence of $a(v_n, .)$ in V^*. By the coercivity of the bilinear form a we get the convergence (of a subsequence) of $\{v_n\}$ in V. Comparing with (7.34) we note that $v_n \to v^*$ in V, so by (7.36) we infer the equality

$$a(v^*, .) = \lambda_k v^* - (1 - t^*)\varepsilon v^*.$$

Since $\|v^*\|_{L^2} = 1, v^*$ is an eigenelement of (7.5) whose corresponding eigenvalue is $\lambda_k - (1-t^*)\varepsilon$. The choice of $\varepsilon > 0$ shows that necessarily we have $t^* = 1$ and thus $v^* \in Z_o$. Consequently, assumption (7.23) assures that

$$\int_\Omega (j_+^\infty v_+^* - j_-^\infty v_-^*)dx > 0. \tag{7.37}$$

On the other hand (7.32) enables us to write

$$a(v_n, v^*) - \lambda_k(v_n, v^*)_{L^2} + (t_n/\|u_n\|_{L^2})(w_n, v^*)_{L^2} + (1-t_n)\varepsilon(v_n, v^*)_{L^2} = 0.$$

Because $v^* \in Z_o$ this reduces to

$$-t_n \int_\Omega w_n v^* dx = \varepsilon(1-t_n)\|u_n\|_{L^2} \int_\Omega v_n v^* dx.$$

Then, by (7.34), it is clear that for n large enough

$$\int_\Omega w_n v^* dx \leq 0 \tag{7.38}$$

If $v^*(x) > 0$, from (7.35) it follows that $v_n(x) > 0$ when n is sufficiently large. Thus, by (7.31), we see that $u_n(x) \to +\infty$. Due to (7.24) we arrive at the conclusion that

$$j_+^\infty(x) \leq \liminf_{n\to\infty} w_n(x).$$

Analogously, if $v^*(x) < 0$ we obtain

$$j_-^\infty(x) \geq \limsup_{n\to\infty} w_n(x).$$

Therefore we conclude by Fatou's lemma that

$$\begin{aligned}
\liminf_{n\to\infty} \int_\Omega w_n v^* dx &= \liminf_{n\to\infty} \left(\int_{\Omega_+(v^*)} w_n v^* dx + \int_{\Omega_-(v^*)} w_n v^* dx\right) \\
&= \liminf_{n\to\infty} \left(\int_\Omega w_n v_+^* dx - \int_\Omega w_n v_-^* dx\right) \\
&\geq \int_\Omega (\liminf_{n\to\infty} w_n) v_+^* dx - \int_\Omega (\limsup_{n\to\infty} w_n) v_-^* dx \\
&\geq \int_\Omega (j_+^\infty v_+^* - j_-^\infty v_-^*) dx.
\end{aligned}$$

By (7.38) it is clear that

$$\int_\Omega (j_+^\infty v_+^* - j_-^\infty v_-^*)dx \leq 0.$$

This contradicts (7.37). It results the a priori estimate indicated in (7.26)-(7.28).

Take a number $r > 0$ as given in (7.26)-(7.28) and denote by $B(0,r)$ the open ball in $L^2(\Omega)$ centered at the origin and of radius r. Using $W \in C(L^2(\Omega), L^2(\Omega))$ as given in (7.25), we define the homotopy $H : L^2(\Omega) \times [0,1] \to L^2(\Omega)$ by

$$H(v,t) = A^{-1}(\lambda_k v - tW(v) - (1-t)\varepsilon v), \ \forall (v,t) \in L^2(\Omega) \times [0,1], \tag{7.39}$$

where A stands for the self-adjoint operator corresponding to the symmetric bilinear form a on V, that is

$$a(u,v) = \langle Au, v\rangle_V, \ \forall u, v \in V.$$

The compactness of the embedding $V \subset L^2(\Omega)$ implies that H in (7.39) is a compact homotopy. The a priori estimate in (7.26)-(7.28) ensures that the Leray-Schauder degree

$$\deg(id_{L^2(\Omega)} - H(.,t), B(0,r), 0),$$

for all $t \in [0,1]$, is well defined. The homotopy invariance of Leray-Schauder degree applied to H in (7.39) and the choice of $\varepsilon > 0$ lead to the equalities

$$\deg(id_{L^2(\Omega)} - H(.,1), B(0,r), 0) = \deg(id_{L^2(\Omega)} - H(.,0), B(0,r), 0) = \pm 1.$$

This yields an $u \in B(0,r)$ such that

$$u = A^{-1}(\lambda_k u - W(u)).$$

It turns out that $u \in V$ and

$$a(u,v) + \int_\Omega W(u)v\, dx = \lambda_k (u,v)_{L^2}, \ \forall v \in V. \tag{7.40}$$

It suffices now to invoke a standard property of the generalized gradient (see (ii) of Proposition 1.4) to derive from (7.25) and (7.40) that

$$\begin{aligned} \lambda_k(u,v)_{L^2} &\le a(u,v) + \int_\Omega \max\{z \cdot v(x) : z \in \partial_y j(x, u(x))\} dx \\ &= a(u,v) + \int_\Omega j^o(x, u(x); v(x)) dx, \ \forall v \in V. \end{aligned}$$

This completes the proof. □

Regarding the generalized Landesman-Lazer conditions of Proposition 7.1 a complementary result with respect to Theorem 7.3 is formulated below.

Theorem 7.4 Assume that the conditions before Proposition 7.1 and the growth condition in (H_1) with $\sigma = 2$ hold, and that there exist $\overline{j}_+^\infty \in L^2(\Omega)$, $\overline{j}_-^\infty \in L^2(\Omega)$ and a continuous map $W : L^2(\Omega) \to L^2(\Omega)$ fulfilling

$$\int_\Omega (\overline{j}_+^\infty(x)\Phi_+(x) - \overline{j}_-^\infty(x)\Phi_-(x)) dx < 0, \ \forall \Phi \in Z_0 \backslash \{0\}, \tag{7.41}$$

$$\overline{j}_+^\infty(x) = \sup_{\{z_n\}} \limsup_{n\to\infty} z_n \ (\text{resp. } \overline{j}_+^\infty(x) = \inf_{\{z_n\}} \liminf_{n\to\infty} z_n) \tag{7.42}$$

for all sequences $z_n \in \partial_y j(x, \zeta_n)$ with $\zeta_n \to +\infty$ (resp. $\zeta_n \to -\infty$), a.e. $x \in \Omega$, and (7.25). Then the conclusion of Theorem 7.3 holds true.

Proof. The argument consists in an appropriate modification of the proof of Theorem 7.3. We mention only the necessary changes. This time we fix a number $\varepsilon > 0$ such that the interval $(\lambda_k, \lambda_k + \varepsilon]$ be without eigenvalues of (7.5). Accordingly, for $(u,t) \in V \times [0,1]$ and $w \in L^2(\Omega)$, (7.26) is replaced by

$$a(u,.) - \lambda_k u + tw - (1-t)\varepsilon u = 0, \tag{7.43}$$

where w satisfies (7.27). Then the same reasoning as in the proof of Theorem 7.3, with (7.29), (7.32), (7.36) - (7.38) replaced by

$$a(u_n,.) - \lambda_k u_n + t_n w_n - (1-t_n)\varepsilon u_n = 0, \tag{7.44}$$

$$a(v_n,.) - \lambda_k v_n + (t_n/\|u_n\|_{L^2}) w_n - (1-t_n)\varepsilon v_n = 0, \tag{7.45}$$

$$a(v_n,.) \to \lambda_k v^* + (1-t^*)\varepsilon v^* \text{ in } L^2(\Omega), \tag{7.46}$$

$$\int_\Omega (\overline{j}_-^\infty v_+^* - \overline{j}_+^\infty v_-^*)dx < 0, \tag{7.47}$$

$$\int_\Omega w_n v^* dx \geq 0, \tag{7.48}$$

allows us to justify (7.28). Thus we are in the position to consider the Leray-Schauder degree

$$\deg(id_{L^2(\Omega)} - H(.,t), B(0,r), 0)$$

for the compact homotopy $H : L^2(\Omega) \times [0,1] \to L^2(\Omega)$ given by

$$H(v,t) = A^{-1}(\lambda_k v - tW(v) + (1-t)\varepsilon v), \ \forall (v,t) \in L^2(\Omega) \times [0,1]. \tag{7.49}$$

From now on the proof continues as in Theorem 7.3. □

We end this section with a verifiable sufficient condition for the existence of the continuous selection W arising in (7.25). Many other situations can also occur.

Proposition 7.3 Under assumption (H_1) with $\sigma = 2$, if the generalized gradient ∂J of the locally Lipschitz functional $J : L^2(\Omega) \to \mathbb{R}$ given by the formula (7.17) is continuous as a multivalued map from $L^2(\Omega)$ to $L^2(\Omega)$, then there exists $W \in C(L^2(\Omega), L^2(\Omega))$ satisfying (7.25). In particular, this is true if the integrand $j : \Omega \times \mathbb{R} \to \mathbb{R}$ verifies the Lipschitz condition: there is $k \in L^\infty(\Omega)$ such that

$$|z_1 - z_2| \leq k(x)|y_1 - y_2|$$

whenever $z_i \in \partial_y j(x, y_i), i = 1, 2$, a.e. $x \in \Omega$.

Proof. Since the values of the multivalued map ∂J are closed and convex sets, the continuity of ∂J suffices to ensure the existence of a continuous mapping W with $W(v) \in \partial J(v)$ for each $v \in L^2(\Omega)$ (see, e.g., Aubin and Cellina [2]). Then the basic property involving the generalized gradient of an integral functional (see Theorem 1.3) implies (7.25). Concerning the second part of the statement it is straightforward to check that the imposed Lipschitz condition for j ensures the fact that ∂J as a multivalued map is Lipschitz with the Lipschitz constant $\|k\|_{L\infty}$. □

Remark 7.3 The foregoing results can be easily extended to the case in which the $\lambda\langle u, v\rangle$ – eigenvalue problem includes a linear compact operator, i.e. it is a $\lambda\langle Lu, v\rangle$ – eigenvalue problem, as it is the case for the applications given in Section 7.2.

Remark 7.4 The necessary conditions and the sufficient conditions we have proved for the existence of solutions to problem (P), seem to be the first Landesman-Lazer type conditions in the nonsmooth case at resonance for eigenvalues λ_k of multiplicity > 1. In Chang [5] one obtains Landesman-Lazer type conditions in the case where $j(x,t)$ has the form of the primitive in Remark 7.2 with $\Phi(x,t) = g(t) - h(x)$, for g possessing limits at $\pm\infty$ and $h \in L^2(\Omega)$, provided the eigenvalue of the linear part is simple (see Chang [5], Remark 5.3). In many aspects our results at resonance are new also for the smooth case, due to the fact that we made a clear distinction between the role played by the lower and by the upper bounds of the nonlinearity and the limits at $\pm\infty$, whenever these exist. Compare for instance our results, where six functions $j_+, j_-, j_+^\infty, j_-^\infty, \overline{j}_+^{\infty}, \overline{j}_-^{\infty}$ are used, with the theorem of Gallouët and Kavian [7], [11]. We stress that we do not demand the existence of the limits at $\pm\infty$ for the nonlinearity. In this respect we mention the use in Robinson and Landesman [12] of limsup and liminf at $\pm\infty$.

To the authors' knowledge Ref. [9] is the first paper dealing with the application of the method of topological degree in nonsmooth eigenvalue problems for hemivariational inequalities. The essential technical contribution was the use of appropriate continuous selections which made possible the construction of the homotopies needed.

References

[1]S. Ahmad, A.C. Lazer and J.L. Paul, Elementary Critical Point Theory and Perturbations of Elliptic Boundary Value Problems at Resonance, *Indiana Univ.. Math. J.* **25**(1976), 933-944.

[2]J.P. Aubin and A. Cellina, *Differential Inclusions*, Springer-Verlag, Berlin, 1984.

[3]H. Brézis, *Analyse Fonctionnelle. Théorie et Applications*, Masson, Paris, 1992.

[4]B. Budianski, *Theory of Buckling and Postbuckling. Behavior of Elastic Structures, in: Advances in Applied Mechanics* (ed. by Chia-Shun Yih), Academic Press, London, 1994.

[5]K.C. Chang, Variational Methods for Non-Differentiable Functionals and their Applications to Partial Differential Equations, *J. Math. Anal. Appl.* **80** (1981), 102-129.

[6]F.H. Clarke, *Nonsmooth Analysis and Optimization*, Wiley & Sons, New York, 1984.

[7]M. Gallouët and O. Kavian, Resonance for Jumping Nonlinearities, *Commun. P.D.E.* **7**(1982), 325-342.

[8]D. Goeleven, D. Motreanu and P. D. Panagiotopoulos, Multiple Solutions for a Class of Eigenvalue Problems in Hemivariational Inequalities, *Nonlinear Analysis T.M.A.* **29** (1997), 9-26.

[9]D. Goeleven, D. Motreanu and P.D. Panagiotopoulos, Eigenvalue Problems for Variational-Hemivariational Inequalities at Resonance, *Nonlinear Analysis T.M.A.* **33** (1998), 161 180.

[10]P. Hess, On a Theorem of Landesman and Lazer, *Indiana Univ. Math. J.* **23** (1974), 827-829.

[11]O. Kavian, *Introduction à la Théorie des Points Critiques et Applications aux Problèmes Elliptiques*, Springer Verlag, Paris, 1993.

[12]E.M. Landesman and A.C. Lazer, Nonlinear Perturbations of Linear Elliptic Boundary Value Problems at Resonance, *J. Math. Mech.* **19** (1970), 609-623.

[13]D. Motreanu and P.D. Panagiotopoulos, Hysteresis: The eigenvalue Problem for Hemivariational Inequalities, in: Models of Hysteresis, *Pitman Research Notes in Mathematics* **286**, London, Harlow, 1993, pp. 102-117.

[14]D. Motreanu and P.D. Panagiotopoulos, Double Eigenvalue Problems for Hemivariational Inequalities, *Arch. Rational Mech. Anal.* **140** (1997), 225-252.

[15]Z. Naniewicz and P.D. Panagiotopoulos, *Mathematical Theory of Hemivariational Inequalities and Applications*, Marcel Dekker, Inc., New York, 1995.

[16]P.D. Panagiotopoulos, *Inequality Problems in Mechanics and Applications. Convex and Nonconvex Energy Functionals*, Birkhaüser, Basel, 1985.

[17] P.D. Panagiotopoulos, *Hemivariational Inequalities. Applications to Mechanics and Engineering*, Springer-Verlag, New York, 1993.

[18] P.H. Rabinowitz, Some Minimax Theorems and Applications to Nonlinear Partial Differential Equations, Nonlinear Analysis: A collection of papers in honor of E. Röthe, *Academic Press*, New-York, 1978, pp. 161-177.

[19] P.H. Rabinowitz, Minimax Methods in Critical Point Theory with Applications to Differential Equations, CBMS Reg. Conf. Ser. Math. **65**, *Amer. Math. Soc., Providence*, R.I., 1986.

[20] S.B. Robinson and E.M. Landesman, A General Approach to Solvability Conditions for Semilinear Elliptic Boundary Value Problems at Resonance, *Diff. Int. Equations.* **8** (1995), 1555-1569.

8. Double Eigenvalue Problems for Hemivariational Inequalities

In the present chapter we study a new type of eigenvalue problem, called "double eigenvalue problems", which arise in the theory of hemivariational inequalities [3],[4],[12]. We give existence results, as well as some qualitative properties for the solutions of these eigenvalue problems. Three different approaches to the double eigenvalue problems for general hemivariational inequalities are developed: minimization, minimax methods and (sub)critical point theory on a sphere.

These techniques are generalizations of some minimax methods of Rabinowitz [20] and of the nonsmooth critical point theory of Chang [5]. A basic role is played in our approach, by some results and ideas from Ekeland [9], Szulkin [23], from the critical point theory and the Riemannian geometry. To treat the nonvariational structure of the problem caused by the presence of (possibly) nonsymmetric operators in the hemivariational inequality, we use specific modifications of the Palais-Smale condition. We study the existence but also certain qualitative properties of the solutions, for instance the multiplicity and the location. The results are illustrated with various applications concerning problems in Mechanics.

8.1 Double Eigenvalue Problems in L^2-spaces

Let us formulate the following double eigenvalue problem for a hemivariational inequality: find $u_1, u_2 \in V$ and $\lambda_1, \lambda_2 \in \mathbb{R}$ such that

$$\begin{aligned} a_1(u_1, v_1) &+ \tilde{C}_1(u_1, v_1) + a_2(u_2, v_2) + \tilde{C}_2(u_2, v_2) \\ &+ \int_\Omega j_y^0(x, (u_1 - u_2)(x); (v_1 - v_2)(x)) d\Omega \\ &\geq \lambda_1 (B_1 u_1, v_1)_V + \lambda_2 (B_2 u_2, v_2)_V \quad \forall\, v_1, v_2 \in V\,. \end{aligned}$$

An interesting special form of the previous problem is the following

$$\begin{aligned} a_1(u_1, v_1) &+ (C_1(u_1), v_1)_V + a_2(u_2, v_2) + (C_2(u_2), v_2)_V \\ &+ \int_\Omega j_y^0(x, (u_1 - u_2)(x); (v_1 - v_2)(x)) d\Omega \\ &\geq \lambda_1 (B_1 u_1, v_1)_V + \lambda_2 (B_2 u_2, v_2)_V \quad \forall\, v_1, v_2 \in V\,. \end{aligned}$$

For the sake of brevity we denote the first problem by $(\tilde{P})$ (resp. the second by (P)). We always impose a compactness assumption for the mappings $\tilde{C}_1, \tilde{C}_2$ in $(\tilde{P})$ (or C_1, C_2 in (P)). V denotes a real Hilbert space with the scalar product $(\cdot,\cdot)_V$ and the norm $\|\cdot\|_V$. We suppose for the present that V is densely and compactly imbedded in $L^2(\Omega;\mathbf{R}^N)$, where Ω is an open bounded subset of $\mathbf{R}^m$ of measure $|\Omega|$ with $m, N \geq 1$. The boundary $\partial\Omega$ of Ω is assumed to be regular (e.g. a Lipschitz boundary). Thus there is a positive constant C_Ω satisfying

$$\|v\|_{L^2} \leq C_\Omega \|v\|_V \,, \quad \forall\, v \in V \,. \tag{8.1}$$

There are given two continuous symmetric bilinear forms $a_1, a_2 : V \times V \to \mathbf{R}$ and two self-adjoint bounded linear operators $B_1, B_2 : V \to V$. The mappings $\tilde{C}_1, \tilde{C}_2 : V \times V \to \mathbf{R}$ in $(\tilde{P})$ are assumed to be of the form

$$\tilde{C}_i(u;v) = f_i^0(u;v) = \limsup_{\substack{t\downarrow 0\\ w\to u}} \frac{1}{t}(f_i(w+tv) - f_i(w)), \tag{8.2}$$

for all $u, v \in V$, corresponding to the locally Lipschitz functions $f_1, f_2 : V \to \mathbf{R}$. The superscript 0 denotes always Clarke's generalized directional derivative. We consider a Carathéodory function $j : \Omega \times \mathbf{R}^N \to \mathbf{R}$ with the properties

(i) $j(\cdot, y)$ is measurable in Ω for the each $y \in \mathbf{R}^N$ and $j(\cdot, 0)$ is essentially bounded on Ω;

(ii) $j(x,\cdot)$ is locally Lipschitz in $\mathbf{R}^N$ for each $x \in \Omega$.

The notation j_y^0 in $(\tilde{P})$ and (P) denotes Clarke's generalized directional derivative of j with respect to the second variable $y \in \mathbf{R}^N$. The functions above j, f_1, f_2 satisfy the growth conditions

$$|z| \leq c(1+|y|), \quad \forall\, y \in \mathbf{R}^N,\ z \in \partial_y j(x,y) \text{ and a.e. } x \in \Omega, \tag{8.3}$$

$$f_i(v) \geq -c_i\|v\|_V^2 - c_i' \,, \quad \forall\, v \in V,\ i = 1,2, \tag{8.4}$$

where c, c_1, c_2, c_1', c_2' are positive constants. The symbol $|\cdot|$ represents the Euclidean norm of $\mathbf{R}^N$. We denote by $\partial_y j(x,y)$ in (8.3) the Clarke's generalized gradient of j with respect to the y-variable.

From (8.3), (i) and Lebourg's mean value theorem (see Theorem 1.1) we obtain that

$$|j(x,y)| \leq |j(x,0)| + |j(x,y) - j(x,0)| \leq M + c(1+|y|)|y| \,,$$

for a.e. $x \in \Omega$ and $y \in \mathbf{R}^N$, where M is a positive constant. Thus constants $b_1 \geq 0$ and $b_2 \geq 0$ can be determined such that

$$|j(x,y)| \leq b_1 + b_2|y|^2 \text{ for a.e } x \in \Omega \,, \quad \forall\, y \in \mathbf{R}^N \,. \tag{8.5}$$

The first existence result for the eigenvalue problem $(\tilde{P})$ is given in the following theorem.

Theorem 8.1 Assume (8.3),(8.4) and the following compactness assumption for the mappings $\tilde{C}_1, \tilde{C}_2$ in (8.2): if $\{v_n^i\} \subset V$ and $\{w_n^i\} \subset V^*$, with $i = 1, 2$, are sequences such that $\{v_n^i\}$ are bounded and

$$w_n^i \in \partial f_i(v_n^i), \quad \forall\, n \geq 1, \ i = 1, 2, \tag{8.6}$$

there exists a convergent subsequence of $\{w_n^i\}$ in V^*, $i = 1, 2$. Then for all $\lambda_1, \lambda_2 \in \mathbf{R}$ which verify

$$\inf_{\substack{v \in V \\ \|v\|_V = 1}} (a_i(v, v) - \lambda_i (B_i v, v)_V) - c_i - 4 b_2 C_\Omega^2 > 0, \ i = 1, 2, \tag{8.7}$$

there exist $u_1, u_2 \in V$ solving $(\tilde{P})$.

Proof. Fix $\lambda_1, \lambda_2 \in \mathbf{R}$ as in (8.7). We introduce the functionals $J : L^2(\Omega; \mathbf{R}^N) \to \mathbf{R}$ and $I : V \times V \to \mathbf{R}$ by

$$J(v) = \int_\Omega j(x, v(x)) d\Omega, \quad \forall\, v \in L^2(\Omega; \mathbf{R}^N)\ , \tag{8.8}$$

$$\begin{aligned} I(v_1, v_2) \ &= \ \frac{1}{2}[a_1(v_1, v_1) + a_2(v_2, v_2) - \lambda_1 (B_1 v_1, v_1)_V - \lambda_2 (B_2 v_2, v_2)_V] \\ &\quad + f_1(v_1) + f_2(v_2) + J(v_1 - v_2), \quad \forall\, v_1, v_2 \in V\ . \end{aligned} \tag{8.9}$$

The existence of J in (8.8) is assured by (8.5). Using (8.1),(8.4),(8.5) we obtain the relation

$$\begin{aligned} I(v_1, v_2) \ &\geq \ \frac{1}{2}[\inf_{\substack{v \in V \\ \|v\|_V = 1}} (a_1(v, v) - \lambda_1 (B_1 v, v)_V) \|v_1\|_V^2 \\ &\quad + \inf_{\substack{v \in V \\ \|v\|_V = 1}} (a_2(v, v) - \lambda_2 (B_2 v, v)_V) \|v_2\|_V^2 - c_1 \|v_1\|_V^2 - c_1' \\ &\quad - c_2 \|v_2\|_V^2 - c_2' - b_1 |\Omega| - 4 b_2 C_\Omega^2 (\|v_1\|_V^2 + \|v_2\|_V^2), \quad \forall\, v_1, v_2 \in V\ . \end{aligned} \tag{8.10}$$

Taking into account (8.7) it follows that the functional I is bounded below on $V \times V$. Due to the hypothesis (8.3) the functional J in (8.8) is Lipschitz continuous on bounded sets of $L^2(\Omega; \mathbf{R}^N)$ and its generalized gradient satisfies

$$\partial J(v) \subset \int_\Omega \partial_y j(x, v(x)) d\Omega, \quad \forall\, v \in L^2(\Omega; \mathbf{R}^N) \tag{8.11}$$

(cf. Theorem 1.3). The density of V in $L^2(\Omega; \mathbf{R}^N)$ allows to apply Corollary 1.2. Thus we obtain

$$\partial(J\,|_V)(v) \subset \partial J(v), \quad \forall\, v \in V\ . \tag{8.12}$$

Combining (8.9),(8.11),(8.12) and using the addition property of generalized gradients we obtain

$$\begin{aligned}\partial I(v_1,v_2)(w_1,w_2) \subset\; & a_1(v_1,w_1)+a_2(v_2,w_2)-\lambda_1(B_1v_1,w_1)_V \\ & -\lambda_2(B_2v_2,w_2)_V+\partial f_1(v_1)w_1+\partial f_2(v_2)w_2 \\ & +\int_\Omega \partial_y j(x,(v_1-v_2)(x))(w_1(x)-w_2(x))d\Omega, \\ & \forall\, v_1,v_2,w_1,w_2\in V\ .\end{aligned} \tag{8.13}$$

We check now that the locally Lipschitz functional I satisfies the Palais-Smale condition in the sense of Chang [5]. Accordingly, let a sequence $\{v_n^1,v_n^2\}$ in $V\times V$ fulfill

$$I(v_n^1,v_n^2)\le M\ , \tag{8.14}$$

with a constant $M>0$ and, for some $J_n\in\partial I(v_n^1,v_n^2)$,

$$J_n\to 0 \ \text{ in } \ V^*\times V^* \ \text{ as } \ n\to\infty\ . \tag{8.15}$$

From (8.7),(8.10) and (8.14) one finds easily that $\{v_n^1,v_n^2\}$ is bounded in $V\times V$. On the other hand, due to (8.13) and (8.15), one gets $z_n^i\in\partial f_i(v_n^i)$, $i=1,2$, and $w_n\in L^2(\Omega;\mathbf{R}^N)$ such that

$$w_n(x)\in\partial_y j(x,v_n^1(x)-v_n^2(x)) \ \text{ for a.e. } \ x\in\Omega \tag{8.16}$$

and

$$\begin{aligned} J_n \;=\; & (a_1(v_n^1,\cdot)-\lambda_1(B_1v_n^1,\cdot)_V+z_n^1,a_2(v_n^2,\cdot) \\ & -\lambda_2(B_2v_n^2,\cdot)_V+z_n^2)+G^*w_n\to 0 \ \text{ in } V^*\times V^* \text{ as } n\to\infty,\end{aligned} \tag{8.17}$$

where $G:V\times V\to V$ is the map

$$G(v_1,v_2)=v_1-v_2,\quad \forall\, v_1,v_2\in V\ .$$

The boundedness of $\{v_n^1,v_n^2\}$ in $V\times V$ and the compactness of the embedding $V\subset L^2(\Omega;\mathbf{R}^N)$ imply that for a subsequence one can suppose that $v_n^i\to v_i$, $i=1,2$, in $L^2(\Omega;\mathbf{R}^N)$ with some elements $v_1,v_2\in L^2(\Omega;\mathbf{R}^N)$. Since J is Lipschitz near v_1-v_2 and $w_n\in\partial J(v_n^1-v_n^2)$, $n\ge 1$, we infer that $\{w_n\}$ is bounded in $L^2(\Omega;\mathbf{R}^N)$, and thus convergent along a subsequence in V^* due to the compactness of the inclusion $L^2(\Omega;\mathbf{R}^N)\subset V^*$. The compactness assumption of Theorem 8.1, where w_n^i in (8.6) is replaced by z_n^i, shows that $\{z_n^i\}$ has a convergent subsequence in V^*, $i=1,2$. Then (8.17) ensures that the sequence

$$\{a_1(v_n^1,\cdot)-\lambda_1(B_1v_n^1,\cdot)_V,a_2(v_n^2,\cdot)-\lambda_2(B_2v_n^2,\cdot)_V\} \tag{8.18}$$

possesses a convergent subsequence in $V^*\times V^*$. Notice that (8.7) implies

$$\inf_{\substack{v\in V\\ \|v\|_V=1}} (a_i(v,v)-\lambda_i(B_iv,v)_V)>0,\ i=1,2\ . \tag{8.19}$$

It is clear that the inequality below holds

$$\inf_{\substack{v\in V\\ \|v\|_V=1}} (a_i(v,v)-\lambda_i(B_iv,v)_V)\|v_n^i-v_m^i\|_V^2\le$$

$$\leq \|a_i(v_n^i - v_m^i, \cdot) - \lambda_i(B_i(v_n^i - v_m^i, \cdot)_V\|_{V^*} \, \|v_n^i - v_m^i\|_V, \quad i = 1, 2 ,$$

for all integers $m, n \geq 1$. Then (8.19) permits us to conclude that $\{v_n^1, v_n^2\}$ is a Cauchy sequence, and thus convergent in $V \times V$. Hence the Palais-Smale condition for the functional I in (8.9) is true.

The assumptions of Palais Minimization Theorem as it is formulated by Chang for locally Lipschitz functions (see Chang [5]) are verified for the functional I given by (8.9). It follows that there exists $\{u_1, u_2\} \in V \times V$ with

$$I(u_1, u_2) = \inf_{V \times V} I . \tag{8.20}$$

In particular, from (8.20) we obtain that

$$0 \in \partial I(u_1, u_2) . \tag{8.21}$$

Taking into account (8.13),(8.21) and Proposition 1.4 (cf. Clarke [6] p.27), the conclusion of Theorem 8.1 follows. □

The final part of this section deals with a variant of eigenvalue problem $(\tilde{P})$, where the integral term is replaced by an integral over the boundary $\Gamma = \partial\Omega$ of Ω. In this setting V is simply a Hilbert space and $j : \Gamma \times \mathbb{R}^N \to \mathbb{R}$ is a Carathéodory function satisfying the preceding conditions (i),(ii) and (8.3), where in place of a.e. $x \in \Omega$ we ask a.e. $x \in \Gamma$. Thus (8.5) holds for a.e. $x \in \Gamma$. We keep the same meaning for the data $a_i, B_i, \tilde{c}_i$ for $i = 1, 2$. The specific difference lies in the fact that we assume the existence of a compact linear operator $L : V \to L^2(\Gamma; \mathbb{R}^N)$. This can be considered as a generalized trace operator. For simplicity we pose

$$\hat{v} := Lv, \quad \forall\, v \in V .$$

Now the following double eigenvalue problem is formulated: find $u_1, u_2 \in V$ and $\lambda_1, \lambda_2 \in \mathbb{R}$ such that

$$(\tilde{P'}) \qquad a_1(u_1, v_1) + \tilde{C}_1(u_1, v_1) + a_2(u_2, v_2) + \tilde{C}_2(u_2, v_2)$$
$$+ \int_\Gamma j_y^0(w, (\hat{u}_1 - \hat{u}_2)(w); (\hat{v}_1 - \hat{v}_2)(w))d\Gamma$$
$$\geq \lambda_1(B_1 u_1, v_1)_V + \lambda_2(B_2 u_2, v_2)_V , \qquad \forall\, v_1, v_2 \in V .$$

The existence result concerning problem $(\tilde{P'})$ is formulated below.

Theorem 8.2 Under the assuptions of Theorem 8.1 (with the only modification that it is "a.e.in Γ" in place of "a.e. in Ω)", for every $\lambda_1, \lambda_2 \in \mathbb{R}$ with

$$\inf_{\substack{v \in V \\ \|v\|_V = 1}} (a_i(v, v) - \lambda_i(B_i v, v)_V) - c_i - 4b_2\|L\|^2) > 0, \; i = 1, 2, \tag{8.22}$$

there exist $u_1, u_2 \in V$ which solve $(\tilde{P'})$.

Proof. The method of proof follows the same lines as in Theorem 8.1. For fixed $\lambda_1, \lambda_2 \in \mathbb{R}$ satisfying (8.22) we define $\tilde{J}: L^2(\Gamma; \mathbb{R}^N) \to \mathbb{R}$ by

$$\tilde{J}(v) = \int_\Gamma j(x, v(x))d\Gamma, \quad \forall\, v \in L^2(\Gamma; \mathbb{R}^N), \tag{8.23}$$

and $\tilde{I}: V \times V \to \mathbb{R}$ by

$$\begin{aligned} \tilde{I}(v_1, v_2) &= \frac{1}{2}[a_1(v_1, v_1) + a_2(v_2, v_2) - \lambda_1(B_1 v_1, v_1)_V - \lambda_2(B_2 v_2, v_2)_V] \\ &\quad + f_1(v_1) + f_2(v_2) + \tilde{J}(\hat{v}_1 - \hat{v}_2), \quad \forall\, v_1, v_2 \in V. \end{aligned} \tag{8.24}$$

$\tilde{J}$ and $\tilde{I}$ are well defined and locally Lipschitz. From (8.4),(8.5),(8.23) and (8.24) the estimate

$$\begin{aligned} \tilde{I}(v_1, v_2) &\geq \frac{1}{2}[\inf_{\substack{v \in V \\ \|v\|_V = 1}} (a_1(v, v) - \lambda_1(B_1 v, v)_V) - c_1 - 4b_2\|L\|^2]\|v_1\|_V^2 \\ &\quad + \frac{1}{2}[\inf_{\substack{v \in V \\ \|v\|_V = 1}} (a_2(v, v) - \lambda_2(B_2 v, v)_V) - c_2 - 4b_2\|L\|^2]\|v_2\|_V^2 \\ &\quad - c_1' - c_2' - b_1|\Gamma|, \qquad \forall\, v_1, v_2 \in V \end{aligned} \tag{8.25}$$

results. The inequalities (8.22) and (8.25) ensure that $\tilde{I}$ is bounded below on $V \times V$.

As in the corresponding part of the proof of Theorem 8.1 we can prove that the generalized gradient $\partial\tilde{I}$ is determined by

$$\begin{aligned} \partial\tilde{I}(v_1, v_2)(w_1, w_2) &\subset a_1(v_1, w_1) + a_2(v_2, w_2) - \lambda_1(B_1 v_1, w_1)_V \\ &\quad - \lambda_2(B_2 v_2, w_2)_V + \partial f_1(v_1)w_1 + \partial f_2(v_2)w_2 \\ &\quad + \int_\Gamma \partial_y j(x, \hat{v}_1(x) - \hat{v}_2(x))(\hat{w}_1(x) - \hat{w}_2(x))d\Gamma, \\ &\qquad \forall\, v_1, v_2, w_1, w_2 \in V. \end{aligned} \tag{8.26}$$

In order to check that the locally Lipschitz functional $\tilde{I}$ in (8.25) verifies the Palais-Smale condition let us take a sequence $\{v_n^1, v_n^2\}$ in $V \times V$ and a sequence $\tilde{J}_n \in \partial\tilde{I}(v_n^1, v_n^2)$ such that, for a constant $M > 0$,

$$\tilde{I}(v_n^1, v_n^2) \leq M \tag{8.27}$$

and

$$\tilde{J}_n \to 0 \ \text{ in } \ V^* \times V^*. \tag{8.28}$$

From (8.22),(8.25) and (8.27) we obtain that the sequence $\{v_n^1, v_n^2\}$ is bounded in $V \times V$. Relations (8.26),(8.28), the boundedness of $\{v_n^1, v_n^2\}$ and the compactness of $L, \partial f_1, \partial f_2$, and the reasoning in the proof of Theorem 8.1, enable us to find that $\{v_n^1, v_n^2\}$ possesses a convergent subsequence. This yields the validity of the Palais-Smale condition for $\tilde{I}$ in the sense of Chang [5]. Applying the Palais Minimization Theorem as in [5] we see that there exists $(u_1, u_2) \in V \times V$ satisfying (8.20) with $\tilde{I}$ in place of I. Then (u_1, u_2) solves $(\tilde{P}')$. □

8.2 The Case of L^p-spaces with $p > 2$

To allow more general growth conditions of polynomial type than the ones considered in (8.4),(8.5) we may work in a L^p-space with $p > 2$. Then we replace the minimization approach with a minimax method.

Assume that V is a Hilbert space which is densely and compactly imbedded in $L^p(\Omega; \mathbf{R}^N)$ for a bounded domain $\Omega \subset \mathbf{R}^m$, with $p \geq 2$, $N \geq 1$, $m \geq 1$. Instead of (8.3) we suppose that the function $j : \Omega \times \mathbf{R}^N \to \mathbf{R}$ fulfills the following conditions (cf. also [11]):

$$|z| \leq c(1 + |y|^{p-1}), \quad \forall\, y \in \mathbf{R}^N, \quad \forall\, z \in \partial_y j(x, y), \text{ for a.e } x \in \Omega, \tag{8.29}$$

with a positive constant c ;

$$j(x, y) \geq \mu^{-1} j_y^0(x, y; y) - k|y|^\sigma - k_0, \quad \forall\, y \in \mathbf{R}^N, \quad \text{for a.e. } x \in \Omega, \tag{8.30}$$

with constants $\mu > 2$, $1 \leq \sigma < 2$, $k \geq 0$, $k_o \geq 0$;

$$\liminf_{y \to 0} |y|^{-2} j(x, y) \geq 0 \text{ uniformly with respect to } x \in \Omega; \tag{8.31}$$

$$\liminf_{n \to \infty} n^{-\sigma} \int_\Omega j(x, n u_0(x)) d\Omega = -\infty \tag{8.32}$$

for some $u_0 \in V$.
Assume further that

$$\int_\Omega j(x, 0) d\Omega = 0. \tag{8.33}$$

The data a_1, a_2, B_1, B_2 have the same meaning as in Section 8.1. The mappings $\tilde{C}_1, \tilde{C}_2$ are determined by (8.2), where f_1, f_2 are locally Lipschitz functionals on V verifying $f_1(0) = f_2(0) = 0$,

$$f_i(v) \geq \mu^{-1} f_i^0(v; v) - k_i \|v\|_V^\sigma - \bar{k}_i, \quad \forall\, v \in V,\ i = 1, 2, \tag{8.34}$$

with constants $k_1, k_2, \bar{k}_1, \bar{k}_2 \geq 0$ and the numbers $1 \leq \sigma < 2$ and $\mu > 2$ of (8.2), and

$$\liminf_{v \to 0} f_i(v) \|v\|_V^{-2} \geq 0, \ i = 1, 2. \tag{8.35}$$

The main result of this section is the following:

Theorem 8.3 Assume that the conditions (8.29)-(8.33) for j and (8.2), (8.34), (8.35) and the compactness requirement (8.6) of Theorem 8.1 are satisfied. Then for all $\lambda_1, \lambda_2 \in \mathbf{R}$ that verify (8.19) there exists a solution $(u_1, u_2) \in V \times V \setminus \{(0, 0)\}$ of problem $(\tilde{P})$.

Proof. By (8.29) and Lebourg's mean value theorem (cf. Theorem 1.1) there exist constants $b_1, b_2 \geq 0$ for which one has

$$|j(x, y)| \leq b_1 + b_2 |y|^p, \quad \forall y \in \mathbf{R}^n \text{ and for a.e. } x \in \Omega. \tag{8.36}$$

Choose $\lambda_1, \lambda_2 \in \mathbb{R}$ to satisfy (8.19). We define the functionals $J : L^p(\Omega; \mathbb{R}^N) \to \mathbb{R}$ and $I : V \times V \to \mathbb{R}$ by (8.8),(8.9), respectively. The estimate (8.36) implies that J and I are well defined and locally Lipschitz. Moreover, by (8.29) and Theorem 1.3 the generalized gradient ∂J fulfills the inclusion

$$\partial J(u) \subset \int_\Omega \partial_y j(x, u(x)) d\Omega, \quad \forall\, u \in L^p(\Omega; \mathbb{R}^N)\,. \tag{8.37}$$

We check now that the locally Lipschitz functional I satisfies the Palais-Smale condition in the sense of Chang [5]. To this end let $\{v_n^1, v_n^2\} \subset V \times V$ denote a sequence for which (8.14),(8.15) hold, with $J_n \in \partial I(v_n^1, v_n^2)$. Using (8.13) and (8.37) we find that

$$z_n \in \partial_y j(\cdot, (v_n^1, v_n^2)(\cdot)) \subset L^{p/(p-1)}(\Omega; \mathbb{R}^N), \quad n \geq 1, \tag{8.38}$$

and $J_n^i \in \partial f_i(v_n^i)$, $n \geq 1$, and $i = 1, 2$, such that

$$\langle J_n; w_1, w_2 \rangle = a_1(v_n^1, w_1) + a_2(v_n^2, w_2) - \lambda_1 (B_1 v_n^1, w_1)_V -$$
$$-\lambda_2 (B_2 v_n^2, w_2)_V + \langle J_n^1, w_1 \rangle_V + \langle J_n^2, w_2 \rangle_V + \int_\Omega z_n (w_1 - w_2) d\Omega$$

for all $w_1, w_2 \in V$. Then, from (8.9), (8.13)-(8.15) and (8.34), we obtain

$$\begin{aligned}
&M + \frac{1}{\mu}(\| v_n^1 \|_V + \| v_n^2 \|_V) \geq \\
&\frac{1}{2}[a_1(v_n^1, v_n^1) + a_2(v_n^2, v_n^2) - \lambda_1 (B_1\, v_n^1, v_n^1)_V - \\
&- \lambda_2 (B_2\, v_n^2, v_n^2)_V] + f_1(v_n^1) + f_2(v_n^2) + \int_\Omega j(x, (v_n^1 - v_n^2)) d\Omega - \\
&- \frac{1}{\mu}[a_1(v_n^1, v_n^1) + a_2(v_n^2, v_n^2) - \lambda_1 (B_1\, v_n^1, v_n^1)_V - \\
&- \lambda_2 (B_2\, v_n^2, v_n^2)_V + \langle J_n^1, v_n^1 \rangle_V + < J_n^2, v_n^2 >_V + \\
&+ \int_\Omega z_n (v_n^1 - v_n^2) d\Omega] \geq (\frac{1}{2} - \frac{1}{\mu})[\inf_{\substack{v \in V \\ \|v\|_V = 1}} (a_1(v, v) - \\
&- \lambda_1 (B_1 v, v)_V) \| v_n^1 \|_V^2 + \inf_{\substack{v \in V \\ \|v\|_V = 1}} (a_2(v, v) - \lambda_2 (B_2 v, v)_V) \| v_n^2 \|_V^2] - \\
&- \tilde{k}(\| v_n^1 \|_V^\sigma + \| v_n^2 \|_V^\sigma) - \tilde{k}_0, \quad n \geq 1,
\end{aligned} \tag{8.39}$$

with new constants $\tilde{k}, \tilde{k}_0 \geq 0$. By the choice of λ_1, λ_2 in (8.19) and for $\sigma < 2$, $\mu > 2$, we see from (8.39) that the sequence $\{v_n^1, v_n^2\}$ is bounded in $V \times V$.

The boundedness of $\{v_n^1, v_n^2\}$ in $V \times V$ and the compactness of the imbedding $V \subset L^p(\Omega; \mathbb{R}^N)$ ensure the convergence in $L^p(\Omega; \mathbb{R}^N)$ of subsequences denoted again by v_n^1, v_n^2. Since J is locally Lipschitz on $L^p(\Omega; \mathbb{R}^N)$ and the imbedding $L^{p/(p-1)}(\Omega; \mathbb{R}^N) \subset V^*$ is compact, we see that the sequence $\{z_n\}$ of (8.38) contains a convergent subsequence in V^*. The compactness assumption in (8.6) for $\tilde{C}_1, \tilde{C}_2$ yields the convergence of $\{J_n^i\}$, $i = 1, 2$, along subsequences. Thus (8.15) implies the convergence in $V^* \times V^*$ of the sequence (8.18) (see (8.17) for a similar situation). Then as in the proof of Theorem 8.1 applying the argument based on (8.19) we can show that $\{v_n^1, v_n^2\}$ has a convergent subsequence in $V \times V$. Therefore the functional I satisfies the Palais-Smale condition in the sense of Chang [5].

We claim now that there exist constants $\rho > 0$ and $\delta > 0$ with the property

$$I(v_1, v_2) \geq \rho \quad \text{whenever} \quad \|v_1\|_V^2 + \|v_2\|_V^2 = \delta \,. \tag{8.40}$$

Towards this, fix an $\varepsilon > 0$. By (8.31) one finds some $\delta > 0$ such that

$$j(x, y) \geq -\varepsilon |y|^2, \quad \forall\, x \in \Omega \text{ and } |y| \leq \delta \,. \tag{8.41}$$

Then (8.36) and (8.41) imply that

$$-j(x, y) \leq \varepsilon |y|^2 + (b_1 \delta^{-p} + b_2)|y|^p, \quad \forall\, y \in \mathbb{R}^n \text{ and for a.e. } x \in \Omega \,. \tag{8.42}$$

Without loss of generality we may suppose in (8.36) that $p > 2$. Then, by (8.42) and the continuity of the inclusion $V \subset L^p(\Omega)$, one obtains for a $\delta > 0$ appropriately small that

$$-J(v_1, v_2) \leq \varepsilon a(\|v_1\|_V^2 + \|v_2\|_V^2) \text{ if } \|v_1\|_V^2 + \|v_2\|_V^2 \leq \delta \,, \tag{8.43}$$

where $a > 0$ is a constant independent of ε. Then (8.9),(8.35) and (8.43) show that for small $\alpha_1 > 0$ and $\alpha_2 > 0$

$$\begin{aligned} I(v_1, v_2) \;\geq\; & \frac{1}{2}[\inf_{\substack{v \in V \\ \|v\|_V = 1}} (a_1(v, v) - \lambda_1 (B_1 v, v)_V) - \alpha_1 - \varepsilon a] \|v_1\|_V^2 \\ & + \frac{1}{2}[\inf_{\substack{v \in V \\ \|v\|_V = 1}} (a_2(v, v) - \lambda_2 (B_2 v, v)_V) - \alpha_2 - \varepsilon a] \|v_2\|_V^2 \end{aligned}$$

if $\|v_1\|_V^2 + \|v_2\|_V^2 = \delta$, for $\delta > 0$ appropriately small. Choosing $\alpha_1 > 0, \alpha_2 > 0$ and $\varepsilon > 0$ sufficiently small, in view of assumption (8.19), the coefficients of $\|v_1\|_V^2, \|v_2\|_V^2$ in the inequality above are positive . It follows that the claim in (8.40) is valid.

We show now that the element $u_0 \in V$ indicated in (8.32) satisfies the inequality

$$I(nu_0, 0) \leq 0 \tag{8.44}$$

for infinitely many integers $n \geq 1$.

Firstly, we note that

$$\partial_s (|sy|^{-\mu} j(x, sy)) = \mu |y|^{-\mu} s^{-1-\mu} (\mu^{-1} \partial_y j(x, sy) sy - j(x, sy)) \tag{8.45}$$

for $y \in \mathbb{R}^N \setminus \{0\}$, $s > 0$ and $x \in \Omega$, where the symbol ∂_s denotes the generalized gradient with respect to s. Lebourg's mean value theorem, hypothesis (8.30) and relation (8.45) imply

$$
\begin{aligned}
&(l+1)^{-\mu}|y|^{-\mu}j(x,(l+1)y) - l^{-\mu}|y|^{-\mu}j(x,ly) \\
&\quad \le \max_{l\le s\le l+1} \partial_s(|sy|^{-\mu}j(x,sy)) \le \mu|y|^{-\mu}\bar{s}^{-1-\mu}(k|\bar{s}y|^{\sigma}+k_0) \\
&\quad \le \mu|y|^{-\mu}(kl^{\sigma-\mu-1}|y|^{\sigma}+k_0 l^{-1-\mu}),
\end{aligned}
$$

for a.e. $x \in \Omega$, $y \in \mathbb{R}^N \setminus \{0\}$ and any integer $l \ge 1$, with some $l \le \bar{s} \le l+1$. Taking the sum of the previous inequalities for $l = 1, ..., n-1$, it follows that

$$
j(x,ny) \le n^{\mu}[j(x,y) + \mu(k|y|^{\sigma}\sum_{i\ge 1} i^{\sigma-\mu-1} + k_0\sum_{i\ge 1} i^{-\mu-1})] \,. \tag{8.46}
$$

Notice that the series in (8.46) are convergent. Analogously, one has

$$
\partial_s(\|sv\|_V^{-\mu}f_i(sv)) = \mu s^{-\mu-1}\|v\|_V^{-\mu}(\mu^{-1}\partial f_i(sv)sv - f_i(sv)) \tag{8.47}
$$

for $v \in V \setminus \{0\}$, $s > 0$, $i = 1,2$, where the functions $f_1, f_2 : V \to \mathbb{R}$ are the ones considered in (8.2). Then one obtains from (8.34) and (8.47) that

$$
\begin{aligned}
&(l+1)^{-\mu}\|v\|_V^{-\mu}f_i((l+1)v) - l^{-\mu}\|v\|_V^{-\mu}f_i(lv) \\
&\quad \le \max_{l\le s\le l+1} \partial_s(\|sy\|_V^{-\mu}f_i(sv)) \le \mu\|v\|_V^{-\mu}\bar{s}^{-1-\mu}(k_i\|\bar{s}v\|_V^{\sigma}+\bar{k}_i) \\
&\quad \le \mu\|v\|_V^{-\mu}(k_i l^{\sigma-\mu-1}\|v\|_V^{\sigma}+\bar{k}_i l^{-1-\mu})
\end{aligned}
$$

for $v \in V \setminus \{0\}$, $l \ge 1$, $i = 1,2$, with some $l \le \bar{s} \le l+1$. The sum of the first $n-1$ inequalities yields

$$
f_i(nv) \le n^{\mu}[f_i(v) + \mu(k_i\|v\|_V^{\sigma}\sum_{l\ge 1} l^{\sigma-\mu-1} + \bar{k}_i\sum_{l\ge 1} l^{-\mu-1})], \tag{8.48}
$$

for $v \in V$, $n \ge 1$, $i = 1,2$.

From (8.9),(8.46) and (8.48) it turns out that

$$
\begin{aligned}
I(nv,0) \;\le\; &\frac{1}{2}n^2[\alpha_1(v,v) - \lambda_1(B_1v,v)_V] \\
&+n^{\mu}(J(v,0) + \tilde{k}\|v\|_{L^{\sigma}}^{\sigma} + \tilde{k}_0), \quad \forall\, v \in V,\ n \ge 1,
\end{aligned} \tag{8.49}
$$

where $\tilde{k}, \tilde{k}_0$ are new positive constants. By (8.32) it is known that some $u_0 \in V$ and $n_0 \ge 1$ can be found such that

$$
J(n_0u_0, 0) \le -\tilde{k}n_0^{\sigma}\|u_0\|_{L^{\sigma}}^{\sigma} - \tilde{k}_0 \,. \tag{8.50}
$$

From (8.50) and (8.49) we deduce that (8.44) is valid because $\mu > 2$.

The Palais-Smale condition and assertions (8.40),(8.44) permit us to apply to the functional I in (8.9) the Mountain Pass Theorem of Ambrosetti and Rabinowitz ([1],[20]) in the Chang's variant [5] for the locally Lipschitz case (see also Motreanu and Panagiotopoulos [13]). Thus there exists a critical point $(u_1,u_2) \in V \times V$ in the sense of Chang [5], i.e., relation (8.21) holds. From(8.13),(8.37) and Proposition 1.4 it results that $(u_1,u_2) \in V \times V$ is a solution of problem $(\tilde{P})$ with eigenvalues λ_1, λ_2 satisfying (8.19). In addition, it is known from Chang [5] that the critical value $I(u_1, u_2)$ admits the following minimax characterization

$$I(u_1, u_2) = \inf_{g \in \Gamma} \max_{t \in [0,1]} I(g(t)), \tag{8.51}$$

where

$$\Gamma = \{g \in C([0,1],\ V \times V) :\ g(0) = (0,0),\ g(1) = (nu_0, 0)\}\ , \tag{8.52}$$

with a prescribed, sufficiently large, number n and $u_0 \in V$ appearing in (8.44). By (8.40),(8.51),(8.52) we note that $I(u_1, u_2) \geq \rho > 0$. Since, by (8.33), $I(0,0) = 0$, we see that $(u_1, u_2) \in V \times V$ represents a nontrivial solution of problem $(\tilde{P})$. The proof is thus complete. □

We turn now to problem $(\tilde{P}')$ where the integral is taken over the boundary $\Gamma = \partial\Omega$ of the domain $\Omega \subset \mathbf{R}^N$. As in Theorem 8.2 we introduce a compact linear operator $L:\ V \to L^p(\Gamma; \mathbf{R}^N)$ denoted $Lv = \hat{v},\ v \in V$, with $p > 2$, to compensate the lack of a compactness imbedding assumption for the space V. Let $j:\ \Gamma \times \mathbf{R}^N \to \mathbf{R}$ denote a Charathéodory function that satisfies (i),(ii) of Section 8.1 for x on Γ. Assume also that j verifies (8.29)-(8.31) for a.e. $x \in \Gamma$,

$$\liminf_{n \to \infty} n^{-\sigma} \int_\Gamma j(x, n\hat{u}_0(x))d\Gamma = -\infty \tag{8.32'}$$

with a prescribed $u_0 \in V$, and

$$\int_\Gamma j(x, 0)d\Gamma = 0\ . \tag{8.33'}$$

Theorem 8.4 Let the assumptions we have made for the function j : $\Gamma \times \mathbf{R}^N \to \mathbf{R}$ hold, as well as the same hypotheses for the mappings a_1, a_2, B_1, B_2, $\tilde{C}_1$, $\tilde{C}_2$ as in Theorem 8.3. Let $L:\ V \to L^p(\Gamma; \mathbf{R}^N)$, where $\Gamma = \partial\Omega,\ p > 2$ and $N \geq 1$ be a compact linear operator. Then for each $\lambda_1, \lambda_2 \in \mathbf{R}$ satisfying (8.19) there exist eigenfunctions $(u_1, u_2) \in V \times V \setminus \{(0,0)\}$ of problem $(\tilde{P}')$.

Proof. We follow the proof of Theorem 8.3 and thus we indicate only the necessary modifications. We introduce the functionals $\tilde{J}:\ L^p(\Gamma; \mathbf{R}^N) \to \mathbf{R}$ and $\tilde{I}:\ V \times V \to \mathbf{R}$ by the expressions given for $p = 2$ in (8.23),(8.24), respectively. The generalized gradient of an integral functional fulfills the relation

$$\partial\tilde{J}(v) \subset \int_\Gamma \partial_y j(x, \hat{v}(x))d\Gamma,\ \ \forall\, v \in L^p(\Gamma; \mathbf{R}^N). \tag{8.53}$$

Making use of the hypotheses for j and of (8.53) we derive as in the proof of Theorem 8.3 the boundedness of any sequence $\{v_n^1, v_n^2\}$ in $V \times V$ verifying (8.27),(8.28). Then the compactness of L allows to admit that a subsequence of $\{\hat{v}_n^1, \hat{v}_n^2\}$ converges in $L^p(\Gamma; \mathbf{R}^N)^2$. Employing essentially the same argument as in the proof of Theorem 8.3 and using the compactness of the adjoint operator L^* we infer the convergence of a subsequence of $\{v_n^1, v_n^2\}$ in $V \times V$. Hence the Palais-Smale condition for $\tilde{I}$ holds.

Notice that (8.42) (assuming without loss of generality that $p > 2$) gives rise to the estimate (8.43). Then one obtains as in the proof of Theorem 8.3 that assertion (8.40) holds for $\tilde{I}$ in place of I.

From (8.46) and (8.48) it results that

$$\begin{aligned}\tilde{I}(nv,0) \leq & \frac{1}{2}n^2(a_1(v,v)-\lambda_1(B_1v,v)_V) \\ & +n^\mu(\tilde{J}(\hat{v})+\tilde{k}\|\hat{v}\|^\sigma_{L^\sigma}+\tilde{k}_0), \quad \forall\, v \in V\,,\end{aligned} \tag{8.54}$$

where $\tilde{k}, \tilde{k}_0$ are positive constants. By (8.32$'$) we may take $u_0 \in V$ and $n_0 \geq 1$ provided

$$\tilde{J}(n_0\hat{u}_0) \leq -\tilde{k}\|\hat{u}_0\|^\sigma_{L^\sigma} - \tilde{k}_0\,.$$

Then (8.54) leads to (8.44) with I replaced by $\tilde{I}$. From now on the proof is the same as in the final part of Theorem 8.3. □

8.3 The Eigenvalue Problem (*P*)

This section deals with the variant (P) of problem $(\tilde{P})$. As in Section 8.2, V is a real Hilbert space, with the scalar product $(\cdot,\cdot)_V$ and the associated norm $\|\cdot\|_V$, which is densely and compactly imbedded in $L^p(\Omega;\mathbb{R}^N)$ for some $p \geq 2$, an integer $N \geq 1$ and a bounded domain Ω in $\mathbb{R}^m$, $m \geq 1$ as in the previous sections. Let us fix a constant $c_0 > 0$ such that

$$\|v\|_{L^p} \leq c_0\|v\|_V\,, \quad \forall\, v \in V\,. \tag{8.55}$$

Let $a_1, a_2 : V \times V \to \mathbb{R}$ be two continuous symmetric bilinear forms on V and let $B_1, B_2 : V \to V$ be two bounded self-adjoint linear operators that are coercive in the following sense

$$(B_iv,v)_V \geq b_i\|v\|^2_V, \quad \forall\, v \in V,\ i=1,2, \tag{8.56}$$

for constants $b_1 > 0$ and $b_2 > 0$. For fixed positive numbers a, b, r we consider the submanifold $S^{a,b}_r$ of $V \times V$ described as follows

$$S^{a,b}_r = \{(v_1,v_2) \in V \times V :\ a(B_1v_1,v_1)_V + b(B_2v_2,v_2)_V = r^2\}\,. \tag{8.57}$$

The manifold $S^{a,b}_r$ is of spherelike type. By (8.57) it is seen that the tangent space $T_{(u_1,u_2)}S^{a,b}_r$ is equal to

$$T_{(u_1,u_2)}S^{a,b}_r = \{(v_1,v_2) \in V \times V :\ a(B_1u_1,v_1)_V + b(B_2u_2,v_2)_V = 0\}\,, \tag{8.58}$$

for every $(u_1,u_2) \in S^{a,b}_r$. The geodesic distance on $S^{a,b}_r$ is denoted by $d(\cdot,\cdot)$, that is, for any points $x, y \in S^{a,b}_r$, $d(x,y)$ is equal to the infimum of the lengths of all paths on $S^{a,b}_r$ joining x and y. Let $C_1, C_2 : V \to V$ be two maps that are compact in the sense that the closures of $C_1(pr_1S^{a,b}_r)$ and of $C_2(pr_2S^{a,b}_r)$ are compact in V. Let $j : \Omega \times \mathbb{R}^N \to \mathbb{R}$ satisfy (i),(ii) in Section 8.1 and (8.29) in Section 8.2.

Consider the next strong form of problem (P): given $a > 0, b > 0, r > 0$ find $u_1, u_2 \in V$ and $\lambda_1, \lambda_2 \in \mathbb{R}$ such that

$$(P_{r,a,b})\begin{cases} a_1(u_1, v_1) + (C_1(u_1), v_1)_V + a_2(u_2, v_2) + (C_2(u_2), v_2)_V \\ \quad + \int_\Omega j_y^0(x, (u_1 - u_2)(x); (v_1 - v_2)(x)d\Omega \\ \geq \lambda_1(B_1u_1, v_1)_V + \lambda_2(B_2u_2, v_2)_V, \quad \forall\, v_1, v_2 \in V, \\ \\ a(B_1u_1, u_1)_V + b(B_2u_2, u_2)_V = r^2 . \end{cases}$$

Two additional assumptions are needed in our present approach.

(H_1) There exist a Fréchet differentiable function $g : V \times V \to \mathbb{R}$ and a function $h : V \times V \to \mathbb{R}$ which is lower semicontinuous (l.s.c.) on $S_r^{a,b}$ such that

$$(C_1(u_1), v_1)_V + (C_2(u_2), v_2)_V \geq g'(u_1, u_2)(v_1, v_2)$$
$$+h(u_1 + v_1, u_2 + v_2) - h(u_1, u_2), \tag{8.59}$$

$$g(u_1, u_2) + h(u_1, u_2) \geq c_1 , \tag{8.60}$$
$$h(\exp_{(u_1,u_2)}(tv_1, tv_2)) \leq (1 - t)h(u_1, u_2) + th(u_1 + v_1, u_2 + v_2) , \tag{8.61}$$

for $(u_1, u_2) \in S_r^{a,b}$, $(v_1, v_2) \in T_{(u_1,u_2)}S_r^{a,b}$, $0 < t < 1$, where c_1 is a real constant.

(H_2) For every sequence $\{u_n^1, u_n^2\} \subset S_r^{a,b}$ with $u_n^i \to u_i$ weakly in V, $a_i(u_n^i, u_n^i) + (C(u_n^i), u_n^i)_V \to \alpha_i$, $i = 1, 2$, and for all $w \in L^{p/(p-1)}(\Omega; \mathbb{R}^N)$ with

$$w(x) \in \partial_y j(x, u_1(x) - u_2(x)) \text{ for a.e. } x \in \Omega \tag{8.62}$$

such that

$$a_1(u_n^1, \cdot) - \lambda_0 a(B_1u_n^1, \cdot)_V, a_2(u_n^2, \cdot) - \lambda_0 b(B_2u_n^2, \cdot)_V \tag{8.63}$$

are convergent sequences in V^*, where

$$\lambda_0 = r^{-2}(\alpha_1 + \alpha_2 + \int_\Omega < w(x), u_1(x) - u_2(x) > d\Omega),$$

there exists a convergent subsequence of $\{u_n^1, u_n^2\}$ in $V \times V$ (thus in $S_r^{a,b}$).

The notation $\exp_{(u_1,u_2)}$ in (8.61) means the exponential mapping $\exp_{(u_1,u_2)}$: $T_{(u_1,u_2)}S_r^{a,b}$
$\to S_r^{a,b}$ of $S_r^{a,b}$ regarded as a Riemannian manifold with the Riemannian structure induced by $V \times V$, namely

$$\exp_{(u_1,u_2)}(v_1, v_2) = \gamma_{(v_1,v_2)}(1), \quad \forall\, (v_1, v_2) \in T_{(u_1,u_2)}S_r^{a,b},$$

where $\gamma_{(v_1,v_2)}$ denotes the unique geodesic of $S_r^{a,b}$ determined by the initial conditions

$$\gamma_{(v_1,v_2)}(0) = (u_1, u_2) \text{ and } \gamma'_{(v_1,v_2)}(0) = (v_1, v_2)\ .$$

For the theory of geodesics we refer to [2][10]. The main result concerning the eigenvalue problem $(P_{r,a,b})$ is the following:

Theorem 8.5 Under the hypotheses (H_1),(H_2), the eigenvalue problem $(P_{r,a,b})$ admits for every positive numbers a, b, r, a solution $(u_1, u_2) \in S_r^{a,b}$, $\lambda_1, \lambda_2 \in \mathbf{R}$ with $\lambda_1 = \lambda a$, $\lambda_2 = \lambda b$, where

$$\lambda = r^{-2}[a_1(u_1,u_2)+(C_1(u_1),u_1)_V+a_2(u_2,u_2)+(C_2(u_2),u_2)_V+\int_\Omega w(u_1-u_2)d\,\Omega]$$

for certain $w \in L^{p/(p-1)}(\Omega;\mathbf{R}^N)$ satisfying (8.62).

Proof. Since we supposed that (8.29) is true, the functional $J:\ L^p(\Omega;\mathbf{R}^N) \to \mathbf{R}$ of (8.8) is locally Lipschitz. Set $F:\ V \times V \to \mathbf{R}$, where

$$F(v_1, v_2) = \frac{1}{2}[a_1(v_1, v_1) + a_2(v_2, v_2)] + g(v_1, v_2),\quad \forall\ (v_1, v_2) \in V \times V, \tag{8.64}$$

and $I:\ V \times V \to \mathbf{R}$ where

$$I(v_1, v_2) = F(v_1, v_2) + J(v_1 - v_2) + h(v_1, v_2),\quad \forall\ (v_1, v_2) \in V \times\ V. \tag{8.65}$$

From (8.36),(8.55), I satisfies the estimate

$$\begin{aligned} I(v_1, v_2)\ \geq\ & -\frac{1}{2}(\|a_1\|\|v_1\|_V^2 + \|a_2\|\|v_2\|_V^2) + g(v_1, v_2) \\ & + h(v_1, v_2) - 2^p b_2 c_0^p(\|v_1\|_V^p + \|v_2\|_V^p) - b_1|\Omega|,\quad \forall v_1, v_2 \in V\ . \end{aligned} \tag{8.66}$$

Therefore I is bounded from below on $S_r^{a,b}$ because of (8.60),(8.66).

The functional I in (8.65) is l.s.c. on $S_r^{a,b}$. Consequently, one can apply to I Ekeland's variational principle on the complete metric space $S_r^{a,b}$ endowed with the geodesic distance d (see [9]). Then there exists a sequence $\{u_n^1, u_n^2\} \subset S_r^{a,b}$ such that

$$I(u_n^1, u_n^2) \leq \inf_{S_r^{a,b}} I + \frac{1}{n} \tag{8.67}$$

and

$$I(w_1, w_2) \geq I(u_n^1, u_n^2) - (1/n)d((w_1, w_2), (u_n^1, u_n^2)),\quad \forall\ (w_1, w_2) \in S_r^{a,b}. \tag{8.68}$$

With

$$(w_1, w_2) = \exp_{(u_n^1,u_n^2)}(tv_1, tv_2),\quad (v_1, v_2) \in T_{(u_n^1,u_n^2)}S_r^{a,b}, \tag{8.69}$$

with $t > 0$, the inequality (8.68) becomes

$$F(\exp_{(u_n^1,u_n^2)}(tv_1, tv_2)) - F(u_n^1, u_n^2) + J(w_1 - w_2) - J(u_n^1 - u_n^2)+$$

$$+h(\exp_{(u_n^1,u_n^2)}(tv_1, tv_2)) - h((u_n^1, u_n^2)) \geq -\frac{1}{n}d(\exp_{(u_n^1,u_n^2)}(tv_1, tv_2), (u_n^1, u_n^2)),$$

where (8.69) is used. Taking into account (8.61) we get

$$\begin{aligned} & F(\exp_{(u_n^1,u_n^2)}(tv_1,tv_2)) - F(u_n^1,u_n^2) + J(w_1 - w_2) \\ & -J(u_n^1 - u_n^2) + t(h(u_n^1 + v_1, u_n^2 + v_2) - h(u_n^1,u_n^2)) \\ & \geq -\frac{1}{n} d(\exp_{(u_n^1,u_n^2)}(tv_1,tv_2),(u_n^1,u_n^2)), \quad \forall\, (v_1,v_2) \in T_{(u_n^1,u_n^2)} S_r^{a,b}. \end{aligned} \tag{8.70}$$

We recall the following properties of the exponential mapping:

$$\frac{d}{dt}(\exp_{(u_n^1,u_n^2)}(tv_1,tv_2)) \mid_{t=0} = (v_1,v_2), \tag{8.71}$$

$$d(\exp_{(u_n^1,u_n^2)}(tv_1,tv_2),(u_n^1,u_n^2)) = t(\|v_1\|_V^2 + \|v_2\|_V^2)^{1/2}, \tag{8.72}$$

for all $(v_1, v_2) \in T_{(u_1,u_2)} S_r^{a,b}$ and t small in (8.72). Setting (8.72) in (8.70), dividing by $t > 0$ and letting $t \to 0$ in (8.70) we obtain

$$\begin{aligned} a_1(u_n^1,v_1) &+ a_2(u_n^2,v_2) + g'(u_n^1,u_n^2)(v_1,v_2) + h(u_n^1 + v_1, u_n^2 + v_2) \\ &- h(u_n^1,u_n^2) + J^0(u_n^1 - u_n^2; v_1 - v_2) \geq -\frac{1}{n}(\|v_1\|_V^2 + \|v_2\|_V^2)^{1/2}, \end{aligned}$$

for all $(v_1, v_2) \in T_{(u_n^1,u_n^2)} S_r^{a,b}$. In writing the above inequality we have used (8.69),(8.71), the differentiability of g and the definition of J^0. Combining with (8.59) we derive

$$\begin{aligned} & a_1(u_n^1,v_1) + a_2(u_n^2,v_2) + (C_1(u_n^1),v_1)_V + (C_2(u_n^2),v_2)_V \\ & +J^0(u_n^1 - u_n^2; v_1 - v_2) \geq -\frac{1}{n}(\|v_1\|_V^2 + \|v_2\|_V^2)^{1/2}, \\ & \qquad \forall\, (v_1,v_2) \in T_{(u_n^1,u_n^2)} S_r^{a,b} \end{aligned} \tag{8.73}$$

The left-hand side of (8.73), is as a function of $(v_1, v_2) \in T_{(u_n^1,u_n^2)} S_r^{a,b}$ continuous, convex and vanishes at 0. Thus we can apply Lemma 1.3 in Szulkin [23] (Lemma 5.8) on the tangent space $T_{(u_n^1,u_n^2)} S_r^{a,b}$ and we obtain an element $(z_n^1, z_n^2) \in T_{(u_n^1,u_n^2)} S_r^{a,b}$ of norm ≤ 1 satisfying

$$\begin{aligned} a_1(u_n^1,v_1) &+ a_2(u_n^2,v_2) + (C_1(u_n^1),v_1)_V + (C_2(u_n^2),v_2)_V \\ &+ J^0(u_n^1 - u_n^2; v_1 - v_2) \geq \frac{1}{n}((z_n^1,v_1)_V + (z_n^2,v_2)_V), \end{aligned} \tag{8.74}$$

for $(v_1, v_2) \in T_{(u_n^1,u_n^2)} S_r^{a,b}$. Due to Corollary 1.2, the density of V in $L^p(\Omega; \mathbb{R}^N)$ implies that

$$\partial(J \mid_V)(u_1 - u_2) \subset \partial J(u_1 - u_2), \quad \forall\, (u_1,u_2) \in V \times V\,.$$

From (8.74) we find that

$$w_n \in \partial J(u_n^1 - u_n^2), \tag{8.75}$$

and thus the equality below holds

$$
\begin{aligned}
a_1(u_n^1, v_1) \ &+ \ a_2(u_n^2, v_2) + (C_1(u_n^1), v_1)_V + (C_2(u_n^2), v_2)_V \\
&+ \ \langle w_n, v_1 - v_2 \rangle_V = \frac{1}{n}[(z_n^1, v_1)_V + (z_n^2, v_2)_V] ,
\end{aligned}
$$

for $(v_1, v_2) \in T_{(u_n^1, u_n^2)} S_r^{a,b}$. The tangent space admits the following explicit description (see (8.58))

$$
T_{(u_n^1, u_n^2)} S_r^{a,b} = \{(v_1, v_2) \in V \times V : \ a(B_1 u_n^1, v_1)_V + b(B_2 u_n^2, v_2)_V = 0\} . \quad (8.76)
$$

We see from (8.76) that there exists $\lambda_n \in \mathbb{R}$ such that in $V^* \times V^*$ is

$$
\begin{aligned}
(a_1(u_n^1, \cdot) \ &+ \ (C_1(u_n^1) - \frac{1}{n} z_n^1, \cdot)_V, a_2(u_n^2, \cdot) + (C_2(u_n^2) - \frac{1}{n} z_n^2, \cdot)_V) \\
&+ \ G^* w_n = \lambda_n((a B_1 u_n^1, \cdot)_V, (b B_2 u_n^2, \cdot)_V), \ \ \forall\, n \geq 1, \qquad (8.77)
\end{aligned}
$$

where $G : V \times V \to V$ denotes the map $G(v_1, v_2) = v_1 - v_2$, $(v_1, v_2) \in V \times V$. The boundedness of $\{u_n^1, u_n^2\} \subset S_r^{a,b}$ and the compactness of the mappings C_1, C_2 on $S_r^{a,b}$ and of the imbedding $V \subset L^p(\Omega; \mathbb{R}^N)$ ensure that a subsequence can be found such that

$$
u_n^i \to u_i \text{ weakly in } V, \ i = 1, 2, \qquad (8.78)
$$

$$
u_n^i \to u_i \text{ strongly in } L^p(\Omega; \mathbb{R}^N), \ i = 1, 2, \qquad (8.79)
$$

$$
C_i(u_n^i) \text{ converges strongly in } V, \ i = 1, 2, \qquad (8.80)
$$

$$
a_i(u_n^i, u_n^i) + (C_i(u_n^i), u_n^i)_V \to \alpha_i, \ i = 1, 2, \qquad (8.81)
$$

with $(u_1, u_2) \in V \times V$ and $\alpha_1, \alpha_2 \in \mathbb{R}$. Since J is locally Lipschitz on $L^p(\Omega; \mathbb{R}^N)$, properties (8.75),(8.79) imply that $\{w_n\}$ is a bounded sequence in $L^{p/(p-1)}(\Omega; \mathbb{R}^N)$. Thus it has a convergent subsequence in V^*, such that

$$
w_n \to w \text{ strongly in } V^* \text{ and weakly in } L^{p/(p-1)}(\Omega; \mathbb{R}^N). \qquad (8.82)
$$

Proposition 1.4(iii) and (8.75),(8.79) yield

$$
w \in \partial J(u_1 - u_2). \qquad (8.83)
$$

Theorem 1.3 and (8.83) show that (8.62) holds.

On the other hand one obtains directly from (8.77) that

$$
\begin{aligned}
a_1(u_n^1, u_n^1) \ &+ \ a_2(u_n^2, u_n^2) + (C_1(u_n^1) - \frac{1}{n} z_n^1, u_n^1)_V \\
&+ \ (C_2(u_n^2) - \frac{1}{n} z_n^2, u_n^2)_V + \langle w_n, u_n^1 - u_n^2 \rangle_V = \lambda_n r^2, \ n \geq 1.
\end{aligned}
$$

This equality, (8.62) and (8.77)-(8.82) imply that

$$
\lambda_n \to \lambda_0 \text{ as } n \to \infty \qquad (8.84)
$$

with λ_0 given in (H_2). In addition, (8.77),(8.80) and (8.82) imply that

$$
a_1(u_n^1, \cdot) - \lambda_n a(B_1 u_n^1, \cdot)_V, a_2(u_n^2, \cdot) - \lambda_n b(B_2 u_n^2, \cdot)_V
$$

converges in $V^* \times V^*$. Combining this fact with (8.62) we arrive at the convergence of the sequences in (8.63). Therefore we may apply hypothesis (H_2). Consequently, we can extract subsequences such that

$$u_n^i \to u_i \text{ in } V,\ i = 1, 2, \tag{8.85}$$

for certain point $(u_1, u_2) \in S_r^{a,b}$. By (8.67) one obtains

$$\lim_{n\to\infty} I(u_n^1, u_n^2) = \inf_{S_r^{a,b}} I. \tag{8.86}$$

Since h is l.s.c. on $S_r^{a,b}$, relations (8.65),(8.85),(8.86) ensure that

$$h(u_1, u_2) \leq \liminf_{n\to\infty} h(u_n^1, u_n^2) = \inf_{S_r^{a,b}} I - F(u_1, u_2) - J(u_1 - u_2),$$

and thus

$$I(u_1, u_2) = \inf_{S_r^{a,b}} I. \tag{8.87}$$

Due to (8.87) it is permitted to write for all $t > 0$ and $(v_1, v_2) \in T_{(u_1,u_2)} S_r^{a,b}$ that

$$\begin{aligned} 0 &\leq I(\exp_{(u_1,u_2)}(tv_1, tv_2)) - I(u_1, u_2) \\ &\leq F(\exp_{(u_1,u_2)}(tv_1, tv_2)) - F(u_1, u_2) \\ &\quad + J(G(\exp_{(u_1,u_2)}(tv_1, tv_2))) - J(u_1, u_2) \\ &\quad + t(h(u_1 + v_1, u_2 + v_2) - h(u_1, u_2)), \end{aligned}$$

where G was given in (8.77) and the second inequality was derived from (8.61). Dividing by t and passing to the limit for $t \to 0$ it results from (8.71) that

$$\begin{aligned} &a_1(u_1, v_1) + a_2(u_2, v_2) + J^0(u_1 - u_2; v_1 - v_2) \\ &+ g'(u_1, u_2)(v_1, v_2) + h(u_1 + v_1, u_2 + v_2) - h(u_1, u_2) \geq 0, \\ &\qquad \forall\ (v_1, v_2) \in T_{(u_1,u_2)} S_r^{a,b}. \end{aligned} \tag{8.88}$$

From (8.59) and (8.88) we obtain

$$a_1(u_1, v_1) + (C_1(u_1), v_1)_V + a_2(u_2, v_2) + (C_2(u_2), v_2)_V + J^0(u_1 - u_2; v_1 - v_2) \geq 0,$$

for all $(v_1, v_2) \in T_{(u_1,u_2)} S_r^{a,b}$. This is equivalent to

$$-(a_1(u_1, \cdot) + (C_1(u_1), \cdot)_V, a_2(u_2, \cdot) + (C_2(u_2), \cdot)_V) \in \partial(J \mid_{S_r^{a,b}})(u_1 - u_2). \tag{8.89}$$

Thus an element $w \in V^*$ is obtained satisfying (8.83) and such that

$$a_1(u_1, v_1) + a_2(u_2, v_2) + (C_1(u_1), v_1)_V + (C_2(u_2), v_2)_V + < w, v_1 - v_2 >_V = 0,$$

$$\forall\ (v_1, v_2) \in T_{(u_1,u_2)} S_r^{a,b}.$$

The characterization of the tangent space $T_{(u_1,u_2)} S_r^{a,b}$ in (8.58) leads to the existence of $\lambda \in \mathbb{R}$ such that

$$a_1(u_1, v_1) + (C_1(u_1), v_1)_V + a_2(u_2, v_2) + (C_2(u_2), v_2)_V \qquad (8.90)$$
$$+\langle w, v_1 - v_2\rangle_V = \lambda(a(B_1u_1, v_1)_V + b(B_2u_2, v_2)_V), \quad \forall\, v_1, v_2 \in V.$$

Thus, by means of (8.37),(8.83),(8.90) and Proposition 1.4 we conclude that

$$\begin{aligned}&\lambda(a(B_1u_1, v_1)_V + b(B_2u_2, v_2)_V)\\ &\quad \le a_1(u_1, v_1) + (C_1(u_1), v_1)_V + a_2(u_2, v_2) + (C_2(u_2), v_2)_V\\ &\quad + \int_\Omega j_y^0(x, (u_1 - u_2)(x); (v_1 - v_2)(x))d\Omega\end{aligned}$$

for $v_1, v_2 \in V$. It suffices to take $\lambda_1 = \lambda a$ and $\lambda_2 = \lambda b$ in order to show that $(u_1, u_2) \in V \times V$ satisfies $(P_{r,a,b})$. The formula for λ given in the statement of Theorem 8.5 follows directly from (8.90). This completes the proof of Theorem 8.5. □

Finally, we study briefly a variant of the eigenvalue problem $(P_{r,a,b})$. Let V denote a Hilbert space for which there exist a compact linear operator $L : V \to L^p(\Gamma; \mathbf{R}^N)$, with $\Gamma = \partial\Omega$. We state the problem: find $u_1, u_2 \in V$ and $\lambda_1, \lambda_2 \in \mathbf{R}$ such that

$$(P'_{r,a,b})\begin{cases} a_1(u_1, v_1) + (C_1(u_1), v_1)_V + a_2(u_2, v_2) + (C_2(u_2), v_2)_V\\ \quad + \int_\Gamma j_y^0(x, (\hat{u}_1 - \hat{u}_2)(x); (\hat{v}_1 - \hat{v}_2)(x))d\Gamma\\ \ge \lambda_1(B_1u_1, v_1)_V + \lambda_2(B_2u_2, v_2)_V, \quad \forall\, v_1, v_2 \in V,\\ \\ a(B_1u_1, u_1)_V + b(B_2u_2, u_2)_V = r^2\ . \end{cases}$$

In the statement of $(P'_{r,a,b})$ we use the notation

$$\hat{u} := Lu, \quad \forall\, u \in V.$$

All the other notations are the same as in Theorem 8.5 with the only difference that now j is a mapping $j : \Gamma \times \mathbf{R}^N \to \mathbf{R}$ that satisfies (i),(ii) in Section 8.1 for $x \in \Gamma$. We need a new version of hypothesis (H_2).

(H_2') For every sequence $\{u_n^1, u_n^2\} \subset S_r^{a,b}$ with $u_n^i \to u_i$ weakly in V, $a_i(u_n^i, u_n^i) + (C(u_n^i), u_n^i)_V \to \alpha_i$, $i = 1, 2$, and for all $w \in L^{p/(p-1)}(\Gamma; \mathbf{R}^N)$ satisfying

$$w(x) \in \partial_y j(x, \hat{u}_1(x) - \hat{u}_2(x)) \text{ for a.e. } x \in \Gamma \qquad (8.62')$$

such that the sequences in (8.63) converge in V^*, where

$$\lambda_0 = r^{-2}(\alpha_1 + \alpha_2 + \int_\Gamma \langle w(x), \hat{u}_1(x) - \hat{u}_2(x)\rangle d\Gamma),$$

there exists a convergent subsequence of $\{u_n^1, u_n^2\}$ in $V \times V$ (thus in $S_r^{a,b}$).

The following Theorem holds regarding $(P'_{r,a,b})$:

Theorem 8.6 Under the same hypotheses as in Theorem 8.5 and, with the only difference that (H_2) is replaced by (H_2'), the eigenvalue problem $(P'_{r,a,b})$ admits, for every positive numbers a, b, r, a solution $(u_1, u_2) \subset S_r^{a,b}$, $\lambda_1, \lambda_2 \in \mathbf{R}$ with $\lambda_1 = \lambda a$, $\lambda_2 = \lambda b$, where

$$\begin{aligned}\lambda &= r^{-2}(a_1(u_1,u_1) + (C_1(u_1),u_1)_V + a_2(u_2,u_2) \\ &\quad + (C_2(u_2),u_2)_V + \int_\Gamma \langle w, \hat{u}_1 - \hat{u}_2\rangle d\Gamma),\end{aligned}$$

for certain $w \in L^{p/(p-1)}(\Gamma; \mathbf{R}^N)$ satisfying (8.62′).

Proof. The argument is the same as in Theorem 8.5. The functional J is replaced now by $\tilde{J} : L^p(\Gamma; \mathbf{R}^N) \to \mathbf{R}$ given by (8.23). Its generalized gradient satisfies then (8.53). Working on the manifold $S_r^{a,b}$ (cf. (8.57)) as in Theorem 8.5 with $\tilde{J}$ in the place of J we arrive at the equality (8.87), where $I : V \times V \to \mathbf{R}$ is the functional obtained by replacing in (8.65) J by $\tilde{J}$. Then the same procedure as in Theorem 8.5 leads to (8.90) with $L^* w$ in place of w. This equality and the final part of Theorem 8.5 complete the proof. □

Remark 8.1 The foregoing theorems hold with minimal obvious changes if $v_1 \in V_1$ and $v_2 \in V_2$, where V_1 and V_2 are different real Hilbert spaces defined on $\Omega_1 \subset \mathbf{R}^N$ and $\Omega_2 \subset \mathbf{R}^N$ respectively, such that $L_1 : V_1 \to L^2(\Gamma; \mathbf{R}^N)$ and $L_2 : V_2 \to L^2(\Gamma; \mathbf{R}^N)$ are compact and linear operators. Here Γ is a common subset of the boundaries of Ω_1 and Ω_2. This is the case, for instance, of two linear elastic bodies Ω_1 and Ω_2 in contact along Γ, where $V_1 = [W^{1,2}(\Omega_1)]^3$, $V_2 = [W^{1,2}(\Omega_2)]^3$.

8.4 Multiple Solutions for a Double Eigenvalue Hemivariational Inequality

In this section we prove a multiplicity result for the new type of eigenvalue problems, introduced and studied in the previous sections. The solutions are searched on a spherelike type manifold $S_r^{a,b}$, where $a, b, r > 0$ are given constants. It is obtained a result similar to the one obtained in Chapter 6.

For proving our multiplicity results we shall use some techniques of Lusternik-Schnirelmann type (including Krasnoselski's genus) in their appropriate variants and also some of the new analysis techniques developed in [12] by Motreanu and Panagiotopoulos.

We use the same notation as in the previous sections: Let V be a real Hilbert space, with the scalar product and the associated norm denoted by $(\cdot,\cdot)_V$ and $\|\cdot\|_V$, respectively. We shall suppose that V is densely and compactly imbedded in $L^p(\Omega; \mathbf{R}^N)$ for some $p \geq 2$, an integer $N \geq 1$ and a bounded domain $\Omega \subset \mathbf{R}^m, m \geq 1$. We denote by $\langle\cdot,\cdot\rangle_V$, $\langle\cdot,\cdot\rangle_{V\times V}$ and $\langle\cdot,\cdot\rangle$, the duality pairings on V, $V \times V$ and $\mathbf{R}^N$, respectively. Let us denote by $C_p(\Omega)$ the constant of the imbedding $V \subset L^p(\Omega; \mathbf{R}^N)$ which means that

$$\|v\|_{L^p} \leq C_p(\Omega)\|v\|_V, \quad \text{for all } v \in V. \tag{8.91}$$

Let $a_1, a_2 : V \times V \to \mathbb{R}$ be two continuous symmetric bilinear forms on V and let $B_1, B_2 : V \to V$ be two bounded self-adjoint linear operators which are coercive in the sense that

$$(B_i v, v)_V \geq b_i \|v\|_V^2, \quad \text{for all } v \in V,\ i = 1, 2, \tag{8.92}$$

for some constants $b_1, b_2 > 0$. For fixed positive numbers a, b, r we consider, as in (8.57), the submanifold $S_r^{a,b}$ of $V \times V$ i.e.

$$S_r^{a,b} = \{(v_1, v_2) \in V \times V \ : \ a(B_1 v_1, v_1)_V + b(B_2 v_2, v_2)_V = r^2\}.$$

Let $j : \Omega \times \mathbb{R}^N \to \mathbb{R}$ satisfy the following assumptions

(*i*) $j(\cdot, y)$ is measurable in Ω for each $y \in \mathbb{R}^N$ and $j(\cdot, 0)$ is essentially bounded on Ω;

(*ii*) $j(x, \cdot)$ is locally Lipschitz in $\mathbb{R}^N$ for a.e. $x \in \Omega$.

(H_1) There exist $\theta \in L^{\frac{p}{(p-1)}}(\Omega)$ and $\rho \in \mathbb{R}$ such that

$$|z| \leq \theta(x) + \rho |y|^{p-1}, \tag{8.93}$$

for all $(x, y) \in \Omega \times \mathbb{R}^N$ and each $z \in \partial_y j(x, y)$, for a.e. $x \in \Omega$.

Let us consider (this is a slight modification of the problem of the previous sections) a real function $C : S_r^{a,b} \times V \times V \to \mathbb{R}$ on which we impose no continuity assumption. We consider now the following double eigenvalue problem : Find $u_1, u_2 \in V$ and $\lambda_1, \lambda_2 \in \mathbb{R}$ such that

$$(P_{r,a,b}) \begin{cases} a_1(u_1, v_1) + a_2(u_2, v_2) + C((u_1, u_2), v_1, v_2) \\ + \int_\Omega j_y^0(x, (u_1 - u_2)(x); (v_1 - v_2)(x))dx \\ \geq \lambda_1 (B_1 u_1, v_1)_V + \lambda_2 (B_2 u_2, v_2)_V, \quad \forall\, v_1, v_2 \in V, \\ \\ a(B_1 u_1, u_1)_V + b(B_2 u_2, u_2)_V = r^2. \end{cases}$$

Three additional assumptions are needed. The first one is the following

(H_2) There exist two locally Lipschitz maps $f_i : V \to \mathbb{R}$, bounded on $\pi_i(S_r^{a,b})$, ($i = 1, 2, \pi_i$ denotes the projection operator) respectively, and such that the following inequality holds

$$C((u_1, u_2), v_1, v_2) \geq f_1^0(u_1; v_1) + f_2^0(u_2; v_2), \tag{8.94}$$

$$\forall\ (u_1, u_2) \in S_r^{a,b} \text{ and } \forall\ (v_1, v_2) \in T_{(u_1,u_2)} S_r^{a,b},$$

where by $T_{(u_1,u_2)} S_r^{a,b}$ we denote the tangent space

$$\{(v_1, v_2) \in V \times V \ : \ a(B_1 u_1, v_1)_V + b(B_2 u_2, v_2)_V = 0\}.$$

In addition we suppose that the sets

$$\{z \in V^* \ : \ z \in \partial f_i(u_i), u_i \in \pi_i(S_r^{a,b})\} \tag{8.95}$$

are relatively compact in $V^*, i = 1, 2$.

Let us define the map $(A_1, A_2) : V \times V \to V^* \times V^*$ by the relation

$$\langle (A_1, A_2)(u_1, u_2), (v_1, v_2) \rangle_{V\times V} = a_1(u_1, v_1) + a_2(u_2, v_2) \tag{8.96}$$

and the duality map $\Lambda : V \times V \to V^* \times V^*$ given by

$$\langle \Lambda(u_1, u_2), (v_1, v_2) \rangle_{V\times V} = a(B_1 u_1, v_1)_V + b(B_2 u_2, v_2)_V, \tag{8.97}$$

$$\forall\ u_1, u_2, v_1, v_2 \in V.$$

The following assumption must hold

(H_3) For every sequence $\{u_n^1, u_n^2\} \subset S_r^{a,b}$ with $u_n^i \to u_i$, weakly in V, for any $z_n^i \in \partial f_i(u_n^i)$, with

$$a_i(u_n^i, u_n^i) + \langle z_n^i, u_n^i \rangle_V \to \alpha_i \in \mathbb{R}, \tag{8.98}$$

$i = 1, 2$ and for all $w \in L^{\frac{p}{p-1}}(\Omega; \mathbb{R}^N)$ which satisfy the relation

$$w(x) \in \partial_y j(x, (u_1 - u_2)(x)) \text{ for a.e. } x \in \Omega, \tag{8.99}$$

if for λ_0 which is given by

$$\lambda_0 = r^{-2}(\alpha_1 + \alpha_2 + \int_\Omega \langle w(x), (u_1 - u_2)(x) \rangle dx), \tag{8.100}$$

the sequence

$$[(A_1, A_2) - \lambda_0 \cdot \Lambda]\,(u_n^1, u_n^2)$$

converges in $V^* \times V^*$, then there exists a convergent subsequence of $\{u_n^1, u_n^2\}$ in $V \times V$ (thus in $S_r^{a,b}$).

The last is a symmetry assumption

(H_4) j is even with respect to the second variable $y \in \mathbb{R}^N$, i.e.,

$$j(x, -y) = j(x, y), \text{ for a.e. } x \in \Omega \text{ and every } y \in \mathbb{R}^N,$$

and f_i is even on $\pi_i(S_r^{a,b})$ i.e.,

$$f_i(-u_i) = f_i(u_i), \text{ for all } (u_1, u_2) \in S_r^{a,b}, i = 1, 2.$$

We are now in position to formulate a multiplicity result.

Theorem 8.7 Let the hypotheses $(H_1), (H_2), (H_3)$ and (H_4) be satisfied. Then the double eigenvalue problem $(P_{r,a,b})$ has infinitely many pairs of solutions $\{\pm(u_n^1, u_n^2), (\lambda_n^1, \lambda_n^2)\} \subset S_r^{a,b} \times \mathbb{R}^2$ with $\lambda_n^1 = a \cdot \lambda_n$ and $\lambda_n^2 = b \cdot \lambda_n$, where

$$\lambda_n = r^{-2}(a_1(u_n^1, u_n^1) + a_2(u_n^2, u_n^2) + \langle z_n^1, \pm u_n^1 \rangle_V$$

$$+\langle z_n^2, \pm u_n^2\rangle_V + \int_\Omega \langle w_n(x), \pm(u_n^1 - u_n^2)(x)\rangle dx), \tag{8.101}$$

for some $z_n^i \in V^*$ and $w_n \in L^{\frac{p}{p-1}}(\Omega; \mathbf{R}^N)$ satisfying

$$z_n^i \in \partial f_i(\pm u_n^i),\ i = 1, 2$$

and

$$w_n(x) \in \partial_y j(x, \pm(u_n^1 - u_n^2)(x)),\ \textit{for a.e. } x \in \Omega,$$

for every $n \geq 1$.

Proof. Let us consider the locally Lipschitz functional $I : V \times V \to \mathbf{R}$ given by

$$\begin{aligned} I(u_1, u_2) &= \frac{1}{2} \cdot [a_1(u_1, u_1) + a_2(u_2, u_2)] \\ &\quad + f_1(u_1) + f_2(u_2) + J(u_1 - u_2), \end{aligned} \tag{8.102}$$

where $J : L^p(\Omega; \mathbf{R}^N) \to \mathbf{R}$ is defined by

$$J(u) = \int_\Omega j(x, u(x))dx,\ \forall u \in L^p(\Omega; \mathbf{R}^N).$$

Due to (H_1), J is a locally Lipschitz functional. From (H_4) it follows that I is even on $S_r^{a,b}$, i.e.,

$$I(-u_1, -u_2) = I(u_1, u_2),\ \text{for every}\ \ (u_1, u_2) \in S_r^{a,b}. \tag{8.103}$$

We claim that I is bounded from below on $S_r^{a,b}$. Notice first that, for a.e. $(x, y) \in \Omega \times \mathbf{R}^N$ we have

$$\begin{aligned} |j(x, y)| &\leq |j(x, 0)| + |j(x, y) - j(x, 0)| \\ &\leq |j(x, 0)| + \sup\{|z|\ :\ z \in \partial_y j(x, Y), Y \in [0, y]\} \cdot |y| \\ &\leq |j(x, 0)| + \theta(x)|y| + \rho|y|^p. \end{aligned}$$

Therefore

$$|J(u)| \leq \|j(\cdot, 0)\|_{L^1} + \|\theta\|_{L^{\frac{p}{p-1}}} \cdot \|u\|_{L^p} + \rho\|u\|_{L^p}^p.$$

Hence, $I_{|S_r^{a,b}}$ satisfies the estimate

$$\begin{aligned} (I_{|S_r^{a,b}})(v_1, v_2) &\geq -\frac{1}{2}(\|a_1\| \cdot \|v_1\|_V^2 + |a_2\| \cdot \|v_2\|_V^2) \\ &\quad + f_1(v_1) + f_2(v_2) - \|j(\cdot, 0)\|_{L^\infty} \cdot |\Omega| \\ &\quad - \|\theta\|_{L^{\frac{p}{p-1}}} \cdot C_p(\Omega) \cdot r \cdot (\frac{1}{\sqrt{ab_1}} + \frac{1}{\sqrt{bb_2}}) \\ &\quad - \rho \cdot C_p^p(\Omega) \cdot r^p(\frac{1}{\sqrt{ab_1}} + \frac{1}{\sqrt{bb_2}})^p. \end{aligned} \tag{8.104}$$

Taking into account that f_i are bounded on $\pi_i(S_r^{a,b})(i=1,2)$, as stated in the hypothesis (H_2), we conclude that I is bounded from below on $S_r^{a,b}$. To continue the proof it is adequate to remark at this moment that the expression of the generalized gradient $\partial(I_{|S_r^{a,b}})$ at the point $(u_1,u_2)\in S_r^{a,b}$ is given by the formula

$$\partial(I_{|S_r^{a,b}})(u_1,u_2)=\{\xi-r^{-2}\langle\xi,(u_1,u_2)\rangle_{V\times V}\cdot\Lambda(u_1,u_2)\ :\ \xi\in\partial I(u_1,u_2)\},$$

where $\Lambda: V\times V\to V^*\times V^*$ is the duality map given in (8.97). Here, the duality $\langle\cdot,\cdot\rangle_{V\times V}$ is taken for the norm on $V\times V$ defined by

$$\|(u_1,u_2)\|_{V\times V}:=\sqrt{a(B_1u_1,u_1)_V+b(B_2u_2,u_2)_V}, \tag{8.105}$$

for all $u_1,u_2\in V$. Our next step is to prove that I fulfills the Palais-Smale condition (in the sense of Chang [5]) on $S_r^{a,b}$. Accordingly, let us consider a sequence $\{u_n^1,u_n^2\}\subset S_r^{a,b}$ such that

$$\sup_n|(I_{|S_r^{a,b}})(u_n^1,u_n^2)|<+\infty$$

and such that there exists some sequence $J_n\subset V^*\times V^*$ fulfilling the conditions

$$J_n\in\partial I(u_n^1,u_n^2),\quad n\geq 1 \tag{8.106}$$

and

$$J_n-r^{-2}\langle J_n,(u_n^1,u_n^2)\rangle_{V\times V}\cdot\Lambda(u_n^1,u_n^2)\to 0, \tag{8.107}$$

strongly in $V^*\times V^*$. We have to prove that $\{u_n^1,u_n^2\}$ contains a convergent subsequence in $V\times V$. Under hypothesis (H_1), the functional J is Lipschitz continuous on bounded sets in $L^p(\Omega;\mathbf{R}^N)$ and its generalized gradient satisfies (cf. Theorem 1.3)

$$\partial J(v)\subset\int_\Omega\partial_y j(x,v(x))dx,\ \forall\ v\in L^p(\Omega;\mathbf{R}^N).$$

The density of V into $L^p(\Omega;\mathbf{R}^N)$ allows us to apply Corollary 1.2. Thus, we obtain

$$\partial(J_{|V})(v)\subset\partial J(v),\ \forall\ v\in V.$$

From $J_n\in\partial I(u_n^1,u_n^2)$ we derive that there exist $z_n^i\in\partial f_i(u_n^i), i=1,2$ and $w_n\in\partial(J_{|V})(u_n^1-u_n^2)$ such that

$$J_n=(a_1(u_n^1,\cdot)+z_n^1,a_2(u_n^2,\cdot)+z_n^2)+G^*w_n,$$

where $G: V\times V\to V$ is the map given by

$$G(v_1,v_2)=v_1-v_2.$$

By the above considerations we have that

$$w_n(x)\in\partial_y j(x,(u_n^1-u_n^2)(x))\text{ for a.e. }x\in\Omega.$$

Taking into account the relation (8.107), we obtain that

$$\Big(a_1(u_n^1,\cdot)+z_n^1, a_2(u_n^2,\cdot)+z_n^2\Big)+G^*w_n -$$
$$-r^{-2}\langle[(a_1(u_n^1,\cdot)+z_n^1, a_2(u_n^2,\cdot)+z_n^2)+G^*w_n],(u_n^1,u_n^2)\rangle_{V\times V}\cdot\Lambda(u_n^1,u_n^2)\to 0,$$
strongly in $V^*\times V^*$.

Due to the fact that the sequence $\{u_n^1,u_n^2\}$ is contained in $S_r^{a,b}$ and due to the coercivity property of B_1 and B_2 it follows easily that each sequence $\{u_n^1\}$ and $\{u_n^2\}$ is bounded in V. Thus, we conclude that there exists a subsequence again denoted by $\{u_n^1\}$ such that

$$u_n^i \rightharpoonup u_i, \text{ weakly in } V, \text{ for some } u_i\in V, (i=1,2).$$

The compactness assumptions in (H_2) imply that, again passing to a subsequence, if necessary,

$$z_n^i\to z_i, \text{ strongly in } V^*, \text{ for some } z_i\in V^*\ (i=1,2).$$

Also we have

$$w_n\in\partial(J_{|V})(u_n^1-u_n^2)\subset\partial J(u_n^1-u_n^2). \tag{8.108}$$

The compactness of the imbedding $V\subset L^p(\Omega;\mathbb{R}^N)$, yields the convergence

$$u_n^i\to u_i, \text{ strongly in } L^p(\Omega;\mathbb{R}^N),\ (i=1,2). \tag{8.109}$$

Since J is locally Lipschitz on $L^p(\Omega;\mathbb{R}^N)$, the above property ensures that $\{w_n\}$ is bounded in $L^{\frac{p}{p-1}}(\Omega;\mathbb{R}^N)$. Due to the reflexivity of $L^{\frac{p}{p-1}}(\Omega;\mathbb{R}^N)$ and the compactness of the imbedding $L^{\frac{p}{p-1}}(\Omega;\mathbb{R}^N)\subset V^*$, there exists $w\in L^{\frac{p}{p-1}}(\Omega;\mathbb{R}^N)$ such that

$$w_n\to w \text{ strongly in } V^* \text{ and weakly in } L^{\frac{p}{p-1}}(\Omega;\mathbb{R}^N).$$

Proposition 1.4 and the relations (8.108) and (8.109) imply that

$$w\in\partial J(u_1-u_2). \tag{8.110}$$

At this moment we may assume that

$$a_i(u_n^i,u_n^i) \text{ converges in } \mathbb{R},$$

and

$$\langle\Big[(z_n^1,z_n^2)+G^*w_n\Big],(u_n^1,u_n^2)\rangle_{V\times V},$$

possesses a convergent subsequence in $\mathbb{R}$. From (8.107), taking into account the convergences stated above we derive that

$$\Big(a_1(u_n^1,\cdot),a_2(u_n^2,\cdot)\Big)-\lambda_0\cdot\Lambda(u_n^1,u_n^2),$$

converges strongly in $V^*\times V^*$, where λ_0 is given by (8.100). Now we apply hypothesis (H_3) and it follows that $\{u_n^1,u_n^2\}$ has a convergent subsequence in

$V \times V$, and thus in $S_r^{a,b}$. Thus the Palais-Smale condition for the functional I on $S_r^{a,b}$ is satisfied.

Let us denote by Υ the family of closed subsets of $S_r^{a,b}$ which are symmetric with respect to the origin 0 of $V \times V$. Let us denote by $\gamma(S)$ the Krasnoselski's genus of the set $S \in \Upsilon$, that is, the smallest integer $k \in \mathbb{N} \cup \{+\infty\}$ for which there exists an odd continuous mapping from S into $\mathbb{R}^k \setminus \{0\}$. For every $n \geq 1$, set

$$\Gamma_n = \{S \subset S_r^{a,b} \; : \; S \in \Upsilon, \gamma(S) \geq n\}.$$

Let us define the corresponding minimax values of I over Γ_n

$$\beta_n = \inf_{S \subset \Gamma_n} \sup_{(u_1,u_2) \in S} I(u_1, u_2).$$

Each class Γ_n contains compact sets, for instance $S_r^{a,b} \cap F_{n+1}$, where F_{n+1} is a $(n+1)$ – dimensional linear subspace of $V \times V$. Since I is bounded from below it is clear that each β_n is a real number. Since the submanifold $S_r^{a,b}$ becomes the sphere S_r in the Hilbert space $V \times V$ endowed with the new norm defined in (8.105), we may apply now Theorem 3.2 of Chang [5]. This implies that $\beta_n \in \mathbb{R}$ are critical values of I on $S_r^{a,b}$. Hence there exists a critical point (u_n^1, u_n^2) (in fact $\pm(u_n^1, u_n^2)$, since I is even on $S_r^{a,b}$), i.e.,

$$0_{V \times V} \in \left(\partial I_{|S_r^{a,b}}\right)(\pm(u_n^1, u_n^2)) \tag{8.111}$$

with

$$I(\pm(u_n^1, u_n^2)) = \beta_n, \tag{8.112}$$

$n \geq 1$. From (8.111) it follows that there exists $\xi_n \in \partial I(\pm(u_n^1, u_n^2))$ such that

$$\xi_n - r^{-2}\langle \xi_n, \pm(u_n^1, u_n^2)\rangle_{V\times V} \cdot \Lambda(u_n^1, u_n^2) = 0. \tag{8.113}$$

Clarke's calculus and the inclusions stated before imply

$$\langle \xi_n, (v_1, v_2)\rangle_{V\times V} \in \partial I(\pm(u_n^1, u_n^2))(v_1, v_2) \subset a_1(\pm u_n^1, v_1) + a_2(\pm u_n^2, v_2)$$
$$+\partial f_1(u_n^1)v_1 + \partial f_2(u_n^2)v_2 + \int_\Omega \partial_y j(x, \pm(u_n^1 - u_n^2)(x))(v_1 - v_2)(x)dx.$$

Thus, there exists some $z_n^i \in \partial f_i(\pm u_n^i)$ $(i = 1, 2)$ and $w_n \in L^{\frac{p}{p-1}}(\Omega; \mathbb{R}^N)$ with

$$w_n(x) \in \partial_y j(x, \pm(u_n^1 - u_n^2)(x)) \;\text{ for a.e. } x \in \Omega,$$

such that

$$\begin{aligned}\langle \xi_n, (v_1, v_2)\rangle_{V\times V} &= a_1(\pm u_n^1, v_1) + a_2(\pm u_n^2, v_2) \\ &\quad +\langle z_n^1, v_1\rangle_V + \langle z_n^2, v_2\rangle_V + \int_\Omega \langle w_n(x), (v_1 - v_2)(x)\rangle dx.\end{aligned}$$

From (8.113) it results that

$$a_1(\pm u_n^1, v_1) + a_2(\pm u_n^2, v_2) + \langle z_n^1, v_1\rangle_V + \langle z_n^2, v_2\rangle_V$$

$$+ \int_\Omega \langle w_n(x), (v_1 - v_2)(x)\rangle dx - r^{-2}\{a_1(u_n^1, u_n^1) + a_2(u_n^2, u_n^2)$$

$$+\langle z_n^1, \pm u_n^1\rangle_V + \langle z_n^2, \pm u_n^2\rangle_V + \int_\Omega \langle w_n(x), \pm(u_n^1 - u_n^2)(x)\rangle dx\}\cdot$$

$$\cdot\Big(a(B_1(\pm u_n^1), v_1)_V + b(B_2(\pm u_n^2), v_2)_V\Big) = 0, \quad \forall\, v_1, v_2 \in V.$$

Using the notations introduced in the formulation of Theorem 8.7 we may write

$$\begin{aligned} &a_1(\pm u_n^1, v_1) + a_2(\pm u_n^2, v_2) + \langle z_n^1, v_1\rangle_V \\ &+\langle z_n^2, v_2\rangle_V + \int_\Omega \langle w_n(x), (v_1 - v_2)(x)\rangle dx \\ &= \lambda_n^1(B_1(\pm u_n^1, v_1))_V + \lambda_n^2(B_2(\pm u_n^2, v_2))_V, \ \forall\, v_1, v_2 \in V. \end{aligned}$$

Taking into account the definitions of z_n^1, z_n^2 and w_n and (H_2) we obtain that

$$\begin{cases} a_1(\pm u_n^1, v_1) + a_2(\pm u_n^2, v_2) + C(\pm(u_1, u_2), v_1, v_2) \\ + \int_\Omega j_y^0(x, \pm(u_n^1 - u_n^2)(x); (v_1 - v_2)(x))dx \\ \geq \lambda_n^1(B_1(\pm u_n^1), v_1)_V + \lambda_n^2(B_2(\pm u_n^2), v_2)_V, \ \forall\, v_1, v_2 \in V, \\ \\ a(B_1(\pm u_n^1), \pm u_n^1)_V + b(B_2(\pm u_n^2), \pm u_n^2)_V = r^2. \end{cases}$$

The above relations complete the proof of Theorem 8.7. □

We are now briefly concerned with a variant of the eigenvalue problem $(P_{r,a,b})$. Let V denote a Hilbert space and let be given a compact linear operator $L : V \to L^p(\Gamma; \mathbb{R}^N)$, with $\Gamma = \partial\Omega$. We consider the following problem: find $u_1, u_2 \in V$ and $\lambda_1, \lambda_2 \in \mathbb{R}$ such that

$$(P'_{r,a,b})\begin{cases} a_1(u_1, v_1) + a_2(u_2, v_2) + C((u_1, u_2), v_1, v_2) \\ + \int_\Gamma j_y^0(x, (\hat{u}_1 - \hat{u}_2)(x); (v_1 - v_2)(\omega))d\Gamma \\ \geq \lambda_1(B_1u_1, v_1)_V + \lambda_2(B_2u_2, v_2)_V, \ \forall\, v_1, v_2 \in V, \\ \\ a(B_1u_1, u_1)_V + b(B_2u_2, u_2)_V = r^2. \end{cases}$$

In the formulation of the above problem we have used as in Section 8.3 the notation

$$\hat{u} := Lu, \ \forall\, u \in V.$$

All the other notations have the same meaning as before, except j, which is now a mapping $j : \Gamma \times \mathbb{R}^N \to \mathbb{R}$ that satisfies (i) and (ii) for $x \in \Gamma$. The hypotheses (H_1), (H_4) are also assumed to hold with the obvious change that now, $x \in \Gamma$. We give now the hypothesis (H_3'), which replaces (H_3):

(H_3') For every sequence $\{(u_n^1, u_n^2)\} \subset S_r^{a,b}$ with $u_n^i \to u_i$ weakly in V, for any $z_n^i \in \partial f_i(u_n^i)$, with

$$a_i(u_n^i, u_n^i) + \langle z_n^i, u_n^i \rangle_V \to \alpha_i \in \mathbf{R}, \tag{8.114}$$

$i = 1, 2$ and for all $w \in L^{\frac{p}{p-1}}(\Gamma\ ;\ \mathbf{R}^N)$ which satisfy the relation

$$w(x) \in \partial_y j(x, (\hat{u}_1 - \hat{u}_2)(x)) \text{ for a.e. } x \in \Gamma, \tag{8.115}$$

if for λ_0 given by

$$\lambda_0 = r^{-2}(\alpha_1 + \alpha_2 + \int_\Gamma \langle w(x), (\hat{u}_1 - \hat{u}_2)(x) \rangle d\Gamma), \tag{8.116}$$

the sequence

$$[(A_1, A_2) - \lambda_0 \cdot \Lambda]\,(u_n^1, u_n^2)$$

converges in $V^* \times V^*$, then there exists a convergent subsequence of $\{u_n^1, u_n^2\}$ in $V \times V$ (thus in $S_r^{a,b}$).

Our multiplicity result concerning $(P'_{r,a,b})$ is the following

Theorem 8.8 Suppose that $(H_1), (H_2), (H_3'), (H_4)$ hold. Then the problem $(P'_{r,a,b})$ admits, for every positive numbers a, b, r, infinitely many pairs of solutions $\{\pm(u_n^1, u_n^2), (\lambda_n^1, \lambda_n^2)\} \subset S_r^{a,b} \times \mathbf{R}^2$ with $\lambda_n^1 = a \cdot \lambda_n$ and $\lambda_n^2 = b \cdot \lambda_n$, where

$$\begin{aligned} \lambda_n &= r^{-2}(a_1(u_n^1, u_n^1) + a_2(u_n^2, u_n^2) + \langle z_n^1, \pm u_n^1 \rangle_V \\ &\quad + \langle z_n^2, \pm u_n^2 \rangle_V + \int_\Gamma \langle w_n(x), \pm(\hat{u}_n^1 - \hat{u}_n^2)(x) \rangle d\Gamma), \end{aligned} \tag{8.117}$$

for some $z_n^i \in V^*$ and $w_n \in L^{\frac{p}{p-1}}(\Gamma\ ;\ \mathbf{R}^N)$ satisfying

$$z_n^i \in \partial f_i(\pm u_n^i),\ i = 1, 2$$

and

$$w_n(x) \in \partial_y j(x, \pm(\hat{u}_n^1 - \hat{u}_n^2)(x)),\ \textit{for a.e. } x \in \Gamma,$$

for every $n \geq 1$.

Proof. The arguments are the same as in the proof of Theorem 8.7. Only the functional J will be replaced with the corresponding functional $\hat{J} : L^p(\Gamma, \mathbf{R}^N) \to \mathbf{R}$ given by

$$\hat{J}(u) = \int_\Gamma j(x, u(x))dx,\ \forall u \in L^p(\Gamma; \mathbf{R}^N).$$

□

Remark 8.2 The proofs of Theorems 8.7 and 8.8 also hold, if $C((u_1, u_2), v_1, v_2)$ is replaced in the formulation of the problem $(P_{r,a,b})$ by the sum $(C_1(u_1), v_1)_V + (C_2(u_2), v_2)_V$, where $C_i : V \to V, i = 1, 2$, are two operators satisfying the condition (H_2)-in which we make the same replacement. The same remark holds for problem $(P'_{r,a,b})$.

8.5 Applications

In several problems in Mechanics and Engineering the cost or the weight of the structure is expressed as a linear function of the norm of the unknown function. Thus the constraint $\|u\|_V = r$ or $a\|u_1\|^2 + b\|u_2\|^2 = r^2$ imposed means that we have a system with prescribed cost or weight, or in some cases energy consumption. The stability analysis of such a system involving nonconvex nonsmooth potential functions (called also nonconvex superpotential) leads to the treatment of a double eigenvalue problem for hemivariational inequalities on a spherelike manifold. First we give some mathematical examples and then we proceed to an application from Mechanics.

8.5.1 Mathematical Examples

i) Let us consider the case of Problem $(P_{r,a,b})$ in which $C \equiv 0,\ B_1 = B_2 = I$ (the identity operator on V, $a = b = 1$. Moreover a_1, a_2 are coercive, in the sense that

$$a_i(v, v) \geq \bar{a}_i\|v\|_V^2,\ \forall v \in V, i = 1, 2,$$

for some constants $\bar{a}_1, \bar{a}_2 > 0$ and $j : \mathbb{R} \to \mathbb{R}$ is the primitive

$$j(t) = \int_0^t \varphi(\tau)d\tau,\ t \in \mathbb{R},$$

with $\varphi : \mathbb{R} \to \mathbb{R}$ even, locally bounded, measurable and satisfying the subcritical growth condition : for some $1 \leq p < \frac{2m}{m-2}$, if $m \geq 3$, $(1 \leq p < +\infty$, if $m = 1, 2)$, we have

$$|\varphi(t)| \leq c_1 + c_2|t|^{p-1},\ \forall t \in \mathbb{R}.$$

It is known that (cf. Chapter 1)

$$\partial j(t) \subset [\underline{\varphi}(t), \overline{\varphi}(t)],\ \forall t \in \mathbb{R},$$

where

$$\underline{\varphi}(t) = \lim_{\delta \to 0} \operatorname{essinf}\{\varphi(s)\ ;\ |t - s| < \delta\}$$

and

$$\overline{\varphi}(t) = \lim_{\delta \to 0} \operatorname{esssup}\{\varphi(s)\ ;\ |t - s| < \delta\}.$$

Suppose further that the sign condition of Chang [5]

$$\underline{\varphi}(t) > 0 \text{ if } t < 0 \text{ and } \overline{\varphi}(t) < 0 \text{ if } t > 0$$

holds. Then the assumptions $(H_1), (H_2), (H_3)$ and (H_4) are obviously satisfied and Theorem 8.7 applies on every sphere $\|v_1\|_V^2 + \|v_2\|_V^2 = r^2$ of $V \times V$.

ii) As an important application of Theorem 8.7, we consider [3][12] the coupled semilinear Poisson equations on a bounded domain Ω in $\mathbb{R}^N$ with a smooth boundary $\partial\Omega$ in the double eigenvalue problem

$$\Delta u_1 + \lambda_1 u_1 \in [\underline{\varphi}(u_1(x) - u_2(x)), \overline{\varphi}(u_1(x) - u_2(x))] \text{ for a.e. } x \in \Omega$$

$$\Delta u_2 + \lambda_2 u_2 \in [-\overline{\varphi}(u_1(x) - u_2(x)), -\underline{\varphi}(u_1(x) - u_2(x))] \text{ for a.e. } x \in \Omega$$

$$u_1 = u_2 = 0 \text{ on } \partial\Omega.$$

Here $\lambda_1, \lambda_2 \in \mathbb{R}$ are the eigenvalues, u_1, u_2 are the corresponding eigenfunctions and $\underline{\varphi}, \overline{\varphi}$ are determined in Example i) for the function $\varphi : \mathbb{R} \to \mathbb{R}$. We choose $V = H_0^1(\Omega)$,

$$a_1(u,v) = a_2(u,v) = \int_\Omega \nabla u \cdot \nabla v dx, \ \forall u, v \in H_0^1(\Omega),$$

$$(B_1 u, v)_{H_0^1} = (B_2 u, v)_{H_0^1} = \int_\Omega uv dx, \ \forall u, v \in H_0^1(\Omega),$$

$j : \mathbb{R} \to \mathbb{R}$ equal to the primitive of φ as in Example i) and, for simplicity, $C \equiv 0$. Notice that each eigensolution of the hemivariational inequality appearing in the formulation of the problem (P) or $(\tilde{P})$ represents a weak solution of the Dirichlet system above. Under the growth condition for φ as in the example i), and the assumptions on j given in the Section 8.2, Theorem 8.3 implies the existence of nontrivial double eigenfunctions.

8.5.2 Adhesively Connected Plates. Buckling for Given Cost or Weight

In the framework of the theory of elastic von Kármán plates, (cf. Section 4.5) we consider two or more such plates connected with an adhesive material, which is assumed to have negligible thickness (Fig. 8.1). The behaviour of the adhesive material may be described by a relation of the form

$$-f \in \partial j(u_1 - u_2), \tag{8.118}$$

(cf. [16], p. 109), where f denotes the gluing force and u_i, $i = 1,2$, denotes the deflection of the i-th plate. The graph of $\{f, u_1 - u_2\}$ may be a zig-zag graph with finite vertical branches (i.e. f is a multivalued function of the relative deflection $u_1 - u_2$) in the most general case. Concerning the derivation and study of the corresponding hemivariational inequalities we refer to [17],[19]. We assume that we have two plates Ω_1 and Ω_2, $\Omega_i \subset \mathbb{R}^2, i = 1,2$, which are adhesively connected on $\Omega' \subset \Omega_i, i = 1,2$. The plates have the boundaries Γ_1 and Γ_2 respectively and $\overline{\Omega} \cap \Gamma_i = \emptyset, i = 1,2$. The boundaries are assumed to be Lipschitzian and are not subjected to any loading on Ω_1 and Ω_2 vertical to the middle plate plane or parallel to it. We assume that $\Omega_1 \equiv \Omega_2$ as subsets of $\mathbb{R}^2$ and we denote both Ω_1 and Ω_2 by Ω. The plates are only subjected along their boundaries Γ_1 and Γ_2 to continuously distributed compressive forces, i.e.

$$\sigma_{\alpha\beta i} n_{\alpha i} = \lambda_i g_{\alpha i} \quad \alpha, \beta = 1,2, \ i = 1,2,$$

where $\sigma = \{\sigma_{\alpha\beta}\}$ denotes the stress tensor for the in plane action of the plate, $n = \{n_\alpha\}$ is the outer unit normal vector to Γ_1 or to $\Gamma_2, g_i = \{g_{1_i}, g_{2_i}\}$ is a given force distribution, which is self equilibrated, i.e. for each plate (cf. Section 4.5)

$$\int_{\Gamma_i} g_{\alpha i} ds = 0, \quad \int_{\Gamma_i} (x_1 g_{2i} - g_{1i} x_2) ds = 0, \quad i = 1, 2.$$

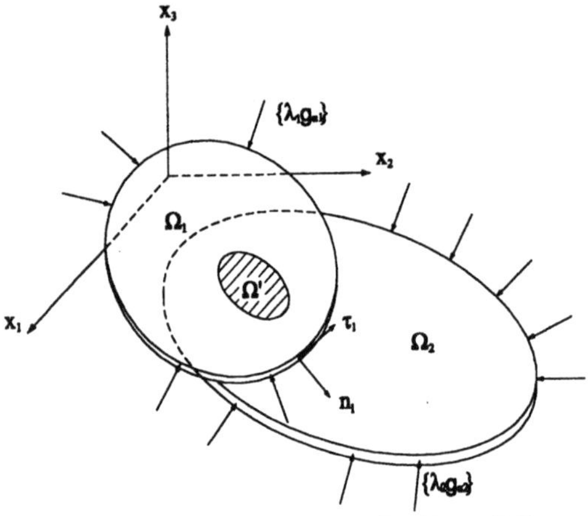

Fig. 8.1 On the Double Load Buckling of Plates.

Note that the eccentricity of the compressive forces with respect to the glued area is ignored. Here λ_i, $i = 1, 2$, is a real number which measures the magnitude of the compressive forces having the direction $g_i, i = 1, 2$, along the boundaries of the plates. These compressive forces may cause buckling of the composite plate with partial debonding of the adhesive material. As in [15], p. 455 and in [18] p. 234, where the analogous buckling problem for variational inequalities is formulated, the notion of "reduced variational solution" is introduced (cf. Section 4.5) and we obtain the following eigenvalue problem: Find $u_1, u_2 \in V$ and $\lambda_1, \lambda_2 \in \mathbf{R}$ such that

$$a_1(u_1, v_1) + a_2(u_2, v_2) + (C_1(u_1), v_1)_V + (C_2(u_2), v_2)_V$$
$$+ \int_{\Omega'} j_y^0(x, (u_1 - u_2)(x); (v_1 - v_2)(x)) dx \geq \lambda_1 (B_1 u_1, v_1)_V + \lambda_2 (B_2 u_2, v_2)_V,$$

for all $v_1, v_2 \in V$. Here V is the real Sobolev space $H^2(\Omega)$ with inner product $(\cdot,\cdot)_V$, $a_i(u_i, v_i)$ is the bending energy of the plate $i, (C_i(u_i), v_i)$, with $C_i(\cdot)$ a nonlinear compact operator, is the bending energy of the plate i due to the stretching of the same plate, $j^0(x, u_1 - u_2; v_1 - v_2)$ denotes the directional derivative in the sense of Clarke at the state $(u_1 - u_2)(x)$ and in the direction $(v_1 - v_2)(x)$ at x, and $(B_i u_i, v_i)$ is given by the relation (4.126), i.e.

$$(B_i u_i, v_i) = - \int_{\Omega_i} h_i \sigma^0_{\alpha\beta i} u_{i,\alpha} v_{i,\beta} dx \ \forall \ u_i, v_i \in V,$$

for $i = 1, 2$. Here h_i is the thickness of the plate i and $\sigma^0_{\alpha\beta_i}$ the stress field in the plane of the plate i caused by the forces $g_{\alpha_i}(\alpha, \beta = 1, 2,\ i = 1, 2)$. Moreover we note that on Γ_i, concerning the plate bending, boundary conditions which guarantee the coercivity of the bilinear forms $a_i(\cdot,\cdot), i = 1, 2$, are assumed to hold. For instance, the built-in boundary conditions $u_i = \frac{\partial u_i}{\partial n} = 0, i = 1, 2$, or the simple support boundary conditions $u_i = 0, M_i(u_i) = 0, i = 1, 2$, where M_i denotes the bending element of the i–th plate. Further we shall not need for the operators B_i the property of coercivity (this property is a consequence of the assumption that each subdomain of the plate is subjected to compressive forces, (cf. [15] p. 457)). Further we express the total cost or weight of the structure by the form $\sum_{i=1}^{2} a_i (B_i u_i, v_i) = r^2$, where a_i are given positive constants. Theorems 8.1 and 8.3 may be applied to the present problem, if the corresponding hypotheses are fulfilled. The remaining theorems hold for B coercive. In the von Kármán plates model B is compact. But coercivity and compactness are compatible only in a finite dimensional space. Thus for any discretized model of the von Kármán plates for which the coercivity assumption for B is fullfilled, one can verify the validity of the remaining Theorems (e.g. Theorem 8.5 etc.) if the corresponding additional assumptions are fulfilled. The passage to the continuous plate problem is a still open problem.

8.5.3 Sandwich Beams with Double Buckling Load

In the framework of a classical bending theory of beams we consider a sandwich beam consisting of two (or more) simply supported beams connected with an adhesive material. Each beam is subjected to a complessive force P_i. All the beams have the length l. Then the behaviour of the adhesive material may be described by a relation of the form (8.118) i.e.

$$-f \in \partial j(u_1 - u_2),$$

where $f = f(x)$ denotes the bonding force at the point $x \in (0, l)$, $y \to j(y)$, is a nonconvex nonsmooth locally Lipschitz function expressing the bonding energy, $(u_1 - u_2)(x)$ denotes the relative deflection at $x \in (0, l)$ of the two beams respectively. The relation between f and $u_1 - u_2$ is a possibly multivalued nonmonotone relation, which takes into account the possibility of partial debonding of the adhesive material, i.e. the graph $\{f, u_1 - u_2\}$ may be a zig-zag graph with filled-in vertical branches of finite length. In this model the adhesive material, is assumed to have, for the sake of simplicity, negligible thickness. Concerning the derivation and study of the corresponding hemivariational inequalities we refer to Section 7.2, where we set $a_i^2 = \frac{1}{\lambda_i} = \frac{P_i}{E_i I_i}$. Then the corresponding double eigenvalue problem reads:
Find $u_1, u_2 \in H^2(0, l)$ and $\lambda_1, \lambda_2 \in \mathbf{R}$ such that

$$a_1(u_1, v_1) + a_2(u_2, v_2) + \int_0^l j^0((u_1 - u_2)(x); (v_1 - v_2)(x))dx$$

$$\geq \lambda_1(\tilde{L}_1 u_1, v_1)_V + \lambda_2(\tilde{L}_2 u_2, v_2)_V \quad \forall\, v_1, v_2 \in H^2(0,l) \; .$$

Here $H^2(0,l)$ is the well-known Sobolev space and $a_i(u_i, v_i)$ is the bending energy of the i - beam, which is coercive. Moreover, $(\tilde{L}_i u_i, v_i)_V = \int_0^l u_i' v_i' dx$. Thus $\tilde{L}_1, \tilde{L}_2$ are linear compact operators. Theorems 8.1 and 8.3 hold for the present problem provided j verifies the required assumptions.

8.5.4 Sandwich Timoshenko Plates, Cylindrical Shells and Their Buckling

We refer now to the Sections 4.6 and 4.7. Assume that n thin plates are adhesively connected and they are subjected to the compressive loads $P_1, P_2, ..., P_n$ (cf. eq. (4.148)). We introduce the nonconvex superpotential law

$$-q_{i,i+1} \in \partial j_{i,i+1}([\zeta]_{i,i+1}) \qquad [\zeta]_{i,i+1} = \zeta_i - \zeta_{i+1}$$

which describes the interface behaviour between the i and $i+1$ plates caused by the gluing material. The thickness of this gluing sheet is assumed to be negligible. In the above formula $j_{i,i+1}$ results analogously to (4.95)(4.97). The plates are assumed to by simply supported. Then as in Section 8.5.3 and using the notation of Section 4.6 we are led to the following problem: Find $\lambda_i \in \mathbb{R}$ $i = 1, ..., n$ and $\zeta_i \in Z_i$ such that

$$\sum_{i=1}^{n} a_i(\zeta_i, z_i - \zeta_i) + \sum_{\mu=1}^{n-1} \int_{\Omega_\mu} j_\mu^0([\zeta]_\mu, [z-\zeta]_\mu) d\Omega \geq \sum_{i=1}^{n} \lambda_i (L_i \zeta_i, z_i - \zeta_i) \quad \forall\, z_i \in Z_i \; .$$

Here μ enumerates the interfaces and L_i is defined by (4.151).

Analogously in the case of cylindrical shells we are led to the problem: Find $\lambda_i \in \mathbb{R}$ $i = 1, ..., n$ and $w_i \in V$ (cf. (4.159)) such that

$$\sum_{i=1}^{n} a_i(w_i, v_i - w_i) + \sum_{i=1}^{n} \frac{E_i h_i}{a^2} \int_0^l w_i (v_i - w_i) dx + \sum_{\mu=1}^{n-1} \int_{l_1}^{l_2} j_\mu^0([w]_\mu, [v]_\mu - [w]_\mu) dx \geq$$

$$\geq \sum_{i=1}^{n} \lambda_i (L_i w_i, v_i - w_i) \qquad \forall\, v_i \in V \; .$$

Here all the shells have the same length and they are simply supported. For both plate and shell problem Theorems 8.1 and 8.3 may be easily verified. For the corresponding discretized problem also Theorems 8.5 and 8.7 are valid.

8.6 A Perturbation Result for a Double Eigenvalue Hemivariational Inequality

8.6.1 The Mathematical Theory

In this section we prove a perturbation result for the double eigenvalue problem for hemivariational inequalites studied in Section 8.4. The perturbation is made in the nonsmooth and nonconvex term of a double eigenvalue problem on a spherlike type manifold. For our aim we use some techniques related to the Lusternik-Schnirelman theory (including Krasnoselski's genus). The same functional frame work and the same notations are used as in Section 8.4. Moreover we assume that the hypotheses $(H_1)(H_2)(H_3)$ and (H_4) of the same section still hold.

Let us consider the following double eigenvalue problem: Find $u_1, u_2 \in V$ and $\lambda_1, \lambda_2 \in \mathbb{R}$ such that

$$(P^1_{r,a,b}) \begin{cases} a_1(u_1,v_1) + a_2(u_2,v_2) + C((u_1,u_2),v_1,v_2) \\ + \int\limits_\Omega j^0_y(x,(u_1-u_2)(x);(v_1-v_2)(x))dx \\ \geq \lambda_1(B_1u_1,v_1)_V + \lambda_2(B_2u_2,v_2)_V, \ \forall\, v_1,v_2 \in V, \\ \\ a(B_1u_1,u_1)_V + b(B_2u_2,u_2)_V = r^2. \end{cases}$$

The hypotheses $(H_1),(H_2),(H_3)$ and (H_4), imply that (cf. Section 8.4) that the double eigenvalue problem $(P^1_{r,a,b})$ admits infinitely many pairs of solutions $\{\pm(u^1_n,u^2_n),(\lambda^1_n,\lambda^2_n)\} \subset S^{a,b}_r \times \mathbb{R}^2$. Moreover, it is found the expression of the eigenvalues λ^1_n and λ^2_n. The aim of this section is to study the following problems: what happens if we perturb $(P^1_{r,a,b})$. For the proof of our main result [12] we need some notions of Algebraic Topology which may be found in [22]. We recall now only some basic definitions.

Let X be a metric space and $A \subset X$. We said that a map $r: X \to A$ is a retraction, if it is continuous, and fulfills $r_{|A} = Id$. A retraction r is called to be a strong deformation retraction, if there exists a homotopy $F: X \times [0,1] \to X$ of $i \circ r$ and Id_X such that $F(x,t) = F(x,0)$, for each $(x,t) \in A \times [0,1]$. Here i stands for the inclusion map of A in X. We call X to be weakly locally contractible, if every point has a contractible neighbourhood in X. Let $\xi : X \to \mathbb{R}$ be a locally Lipschitz functional. Set, for every $a \in \mathbb{R}$

$$[\xi \leq a] := \{u \in X; \xi(u) \leq a\}.$$

Let us fix $a, b \in \mathbb{R}$ with $a \leq b$. The pair $([\xi \leq b],[\xi < a])$ is called trivial if, for every neighbourhoods $[a', a'']$ of a and $[b', b'']$ of b, there exist some closed sets A and B such that $[\xi \leq a'] \subset A \subset [\xi \leq a'']$, $[\xi \leq b'] \subset B \subset [\xi \leq b'']$ and such that A is a strong deformation retract of B.

The next notion is essentialy due to M. Degiovanni and S. Lancelotti (see [8]).

A real number c is said to be an essential value of ξ, if, for every $\epsilon > 0$, there exist $a, b \in (c-\epsilon, c+\epsilon)$, with $a < b$ and such that the pair $([\xi \leq b],[\xi \leq a])$ is not trivial.

Let ϕ be an arbitrary element in V^* and $g: \Omega \times \mathbb{R}^N \to \mathbb{R}$ a Caratheodory function which is locally Lipschitz with respect to the second variable and such that $g(\cdot,0) \in L^1(\Omega)$. Let us consider the following non-symmetric perturbed double eigenvalue problem: find $(u_1,u_2) \in V \times V$ and $(\lambda_1,\lambda_2) \in \mathbb{R}^2$ such that

$$(P^2_{r,a,b})\begin{cases} a_1(u_1,v_1)+a_2(u_2,v_2)+C((u_1,u_2),v_1,v_2) \\ +\int_\Omega\{j^0_y(x,(u_1-u_2)(x);(v_1-v_2)(x)) \\ +g^0_y(x,(u_1-u_2)(x);(v_1-v_2)(x))\}dx \\ +<\phi,v_1>_V+<\phi,v_2>_V\geq \\ \geq \lambda_1(B_1u_1,v_1)_V+\lambda_2(B_2u_2,v_2)_V,\ \forall\, v_1,v_2\in V, \\ \\ a(B_1u_1,u_1)_V+b(B_2u_2,u_2)_V=r^2. \end{cases}$$

Fix $\delta>0$. We impose to g the growth condition

(H_5) There exist $\theta_1\in L^{\frac{p}{(p-1)}}(\Omega)$ and $\delta>0$ such that

$$|z|\leq\theta_1(x)+\delta|y|^{p-1}, \tag{8.119}$$

for a.e. $(x,y)\in\Omega\times\mathbf{R}^N$ and each $z\in\partial_y g(x,y)$.

Let us denote by J and G the (locally Lipschitz, due to (H_1) and (H_5)) functionals from $L^p(\Omega;\mathbf{R}^N)$ into $\mathbf{R}$, defined by

$$J(u)=\int_\Omega j(x,u(x))dx \text{ and } G(u)=\int_\Omega g(x,u(x))dx.$$

We associate to the problems $(P^1_{r,a,b})$ and $(P^2_{r,a,b})$ the energy functions $I_1,I_2:V\times V\to\mathbf{R}$, defined by

$$\begin{aligned} I_1(u_1,u_2) &= \frac{1}{2}\cdot[a_1(u_1,u_1)+a_2(u_2,u_2)] \\ &\quad +f_1(u_1)+f_2(u_2)+J(u_1-u_2), \end{aligned} \tag{8.120}$$

and

$$I_2(u_1,u_2)=I_1(u_1,u_2)+G(u_1-u_2)+\langle\phi,u_1\rangle_V+\langle\phi,u_2\rangle_V, \tag{8.121}$$

for all $u_1,u_2\in V$.

We denote by Υ the family of closed and symmetric, with respect to the origin 0 of $V\times V$, subsets of $S^{a,b}_r$. Let us denote, as usually, by $\gamma(S)$ the Krasnoselski's genus of the set $S\in\Upsilon$, that is, the smallest integer $k\in\mathbf{N}\cup\{+\infty\}$ for which there exists an odd continuous mapping from S into $\mathbf{R}^k\setminus\{0\}$. For every $n\geq 1$, set

$$\Gamma_n=\{S\subset S^{a,b}_r\ :\ S\in\Upsilon,\gamma(S)\geq n\}.$$

Recall that the corresponding minimax values of I_1 over Γ_n

$$\beta_n=\inf_{S\subset\Gamma_n}\sup_{(u_1,u_2)\in S}\{I_1(u_1,u_2)\}, \tag{8.122}$$

are critical values of I_1 on $S^{a,b}_r$ (see [3, Theorem 1] and Section 8.4).

The first result of this section concerns the functional I_1.

Lemma 8.1 Let $s := \sup_{(u_1,u_2)\in S_r^{a,b}} \{I_1(u_1,u_2)\}$. Then the supremum is not achieved and $\lim_{n\to\infty} \beta_n = s$. Moreover, there exists a sequence $\{b_n\}$ of essential values of the restriction of I_1 to $S_r^{a,b}$, strictly increasing to s.

Proof. This result was proved by Radulescu in [21] (see Lemma 1) by using the ideas of M. Degiovanni and S. Lancelotti (see [8], Theorem 2.12). The only difference is that now, we work not on a sphere but on the Riemannian manifold $S_r^{a,b}$. It is sufficient to point out that this is a weakly locally contractible space as the usual sphere in V is. Moreover, I_1 satisfies the Palais-Smale condition on $S_r^{a,b}$ as was proved in Section 8.5. With these observations, the proof of the Lemma 8.1 is the same as the one in [21]. □

Further we need two aditional assumptions

(H_6) The following inequalities hold

$$\|\theta_1\|_{L^{\frac{p}{p-1}}} \leq \delta,\ \|g(\cdot,0)\|_{L^1} \leq \delta \text{ and } \|\phi\|_{V^*} \leq \delta, \tag{8.123}$$

with δ as in (8.119).

The second assumption is actually a variant of the compactness hypothesis (H_3).

(H_7) For every sequence $\{u_n^1, u_n^2\} \subset S_r^{a,b}$ with $u_n^i \to u_i$ weakly in V, for any $z_n^i \in \partial f_i(u_n^i)$, with

$$a_i(u_n^i, u_n^i) + \langle z_n^i, u_n^i\rangle_V + < \phi, u_n^i >_V \to \alpha_i \in \mathbf{R}, \tag{8.124}$$

$i = 1, 2$ and for all $w, z \in L^{\frac{p}{p-1}}(\Omega; \mathbf{R}^N)$ which satisfy the relations

$$w(x) \subset \partial_y j(x, (u_1 - u_2)(x)),$$

$$z(x) \in \partial_y g(x, (u_1 - u_2)(x)), \text{ for a.e. } x \in \Omega, \tag{8.125}$$

such that

$$[(A_1, A_2) - \lambda_0 \cdot \Lambda]\,(u_n^1, u_n^2)$$

converges in $V^* \times V^*$, where,

$$\lambda_0 = r^{-2}(\alpha_1 + \alpha_2 + \int_\Omega \langle w(x) + z(x), (u_1 - u_2)(x)\rangle dx), \tag{8.126}$$

there exists a convergent subsequence of $\{u_n^1, u_n^2\}$ in $V \times V$.

The next Lemma shows that if $\delta > 0$ is sufficiently small in the hypotheses (H_5) and (H_6), then I_2 is a perturbation very close to I_1 on $S_r^{a,b}$.

Lemma 8.2 For every $\epsilon > 0$, there exists $\delta_0 > 0$ such that, for all $\delta \leq \delta_0$ we have

$$\sup_{(u_1,u_2)\in S_r^{a,b}} |I_1(u_1,u_2) - I_2(u_1,u_2)| < \epsilon.$$

Proof. By using mainly the Lebourg's mean value theorem for locally Lipschitz functionals and the hypothesis (H_5) we find

$$|G(u)| \leq \|g(\cdot,0)\|_{L^1} + \|\theta_1\|_{L^{\frac{p}{p-1}}} \cdot \|u\|_{L^p} + \delta\|u\|_{L^p}^p.$$

Taking into account the hypothesis (H_6) and the fact that $(u_1,u_2) \in S_r^{a,b}$ we derive that

$$\begin{aligned}|I_1(u_1,u_2) - I_2(u_1,u_2)| &= |G(u_1-u_2) + <\phi, u_1>_V + <\phi, u_2>_V| \\ &\leq \|g(\cdot,0)\|_{L^1} + \|\theta_1\|_{L^{\frac{p}{p-1}}} \cdot C_p(\Omega)\cdot r \cdot (\frac{1}{\sqrt{ab_1}} + \frac{1}{\sqrt{bb_2}}) \\ &+\delta \cdot C_p^p(\Omega)\cdot r^p (\frac{1}{\sqrt{ab_1}} + \frac{1}{\sqrt{bb_2}})^p + \delta\cdot r\cdot(\frac{1}{\sqrt{ab_1}} + \frac{1}{\sqrt{bb_2}}) < \epsilon,\end{aligned}$$

for $\delta > 0$ small enough. □

Lemma 8.3 The functional I_2 satisfies the Palais-Smale condition on $S_r^{a,b}$.

Proof. It is analogous to the proof of the corresponding step of the Theorem 8.7. (cf. also in this respect [4]). We recall the expression of the generalized gradient $\partial(I_{2|S_r^{a,b}})$ at the point $(u_1,u_2) \in S_r^{a,b}$ given in the proof of Theorem 8.7, i.e. for $I = I_2$

$$\partial(I_{2|S_r^{a,b}})(u_1,u_2) = \{\xi - r^{-2}\langle\xi,(u_1,u_2)\rangle_{V\times V}\cdot\Lambda(u_1,u_2) \;:\; \xi\in\partial I_2(u_1,u_2)\},$$

where $\Lambda : V\times V \to V^*\times V^*$ is the appropriate duality map given in (8.97). The duality $\langle\cdot,\cdot\rangle_{V\times V}$ is taken for the norm defined in (8.105)

$$\|(u_1,u_2)\|_{V\times V} := \sqrt{a(B_1u_1,u_1)_V + b(B_2u_2,u_2)_V},\ \forall\, u_1,u_2\in V.$$

Let us consider a sequence $\{u_n^1,u_n^2\}\subset S_r^{a,b}$ such that

$$\sup_n |(I_{2|S_r^{a,b}})(u_n^1,u_n^2)| < +\infty$$

and such that there exists some sequence $J_n \subset V^*\times V^*$ fulfilling the conditions

$$J_n \in \partial I_2(u_n^1,u_n^2),\quad n\geq 1$$

and

$$J_n - r^{-2}\left\langle J_n,(u_n^1,u_n^2)\right\rangle_{V\times V}\cdot\Lambda(u_n^1,u_n^2)\to 0, \tag{8.127}$$

strongly in $V^*\times V^*$. It suffices to prove now that $\{u_n^1,u_n^2\}$ contains a convergent subsequence in $V\times V$. Under hypothesis (H_1) the functionals J and G are Lipschitz continuous on bounded sets in $L^p(\Omega;\mathbb{R}^N)$ and their generalized gradients satisfy (cf. Theorem 1.3)

$$\partial J(v) \subset \int_{\Omega} \partial_y j(x, v(x))dx$$

and

$$\partial G(v) \subset \int_{\Omega} \partial_y g(x, v(x))dx, \ \forall\ v \in L^p(\Omega; \mathbb{R}^N).$$

The density of V into $L^p(\Omega; \mathbb{R}^N)$ allows us to apply Corollary 1.2. Thus, we obtain

$$\partial(J_{|V})(v) \subset \partial J(v),$$

and

$$\partial(G_{|V})(v) \subset \partial G(v), \ \forall\ v \in V.$$

From $J_n \in \partial I_2(u_n^1, u_n^2)$ we derive that there exists $z_n^i \in \partial f_i(u_n^i)(i = 1, 2),\ w_n \in \partial(J_{|V})(u_n^1 - u_n^2)$ and $z_n \in \partial(G_{|V})(u_n^1 - u_n^2)$ such that

$$J_n = (a_1(u_n^1, \cdot) + z_n^1 + \phi, a_2(u_n^2, \cdot) + z_n^2 + \phi) + K^*(w_n) + K^*(z_n),$$

where $K : V \times V \to V$ is the map given by

$$K(v_1, v_2) = v_1 - v_2.$$

Thus we obtain that

$$w_n(x) \in \partial_y j(x, (u_n^1 - u_n^2)(x))$$

and

$$z_n(x) \in \partial_y g(x, (u_n^1 - u_n^2)(x)), \text{ for a.e. } x \in \Omega.$$

Using (8.127) we may write

$$\begin{aligned} &\Big(a_1(u_n^1, \cdot) + z_n^1 + \phi, a_2(u_n^2, \cdot) + z_n^2 + \phi\Big) + K^*(w_n) + K^*(z_n) \\ &- r^{-2}\langle[(a_1(u_n^1, \cdot) + z_n^1 + \phi, a_2(u_n^2, \cdot) + z_n^2 + \phi) + K^*(w_n) + K^*(z_n)], \\ &\qquad (u_n^1, u_n^2)\rangle_{V\times V} \cdot \Lambda(u_n^1, u_n^2) \to 0, \ \text{ strongly in } V^* \times V^*. \end{aligned}$$

The sequence $\{u_n^1, u_n^2\}$ is contained in $S_r^{a,h}$ and due to the coercivity property of B_1 and B_2 it results that each sequence $\{u_n^1\}$ and $\{u_n^2\}$ is bounded in V. Thus, we may conclude that there exists a subsequence such that

$$u_n^i \rightharpoonup u_i, \text{ weakly in } V, \text{ for some } u_i \in V (i = 1, 2).$$

The compactness assumption in the hypothesis (H_2) implies that, again up to a subsequence,

$$z_n^i \to z_i, \ \text{ strongly in } V^*, \text{ for some } z_i \in V^* \ (i = 1, 2).$$

We have also

$$\begin{aligned} w_n &\in \partial(J_{|V})(u_n^1 - u_n^2) \subset \partial J(u_n^1 - u_n^2), \\ z_n &\in \partial(G_{|V})(u_n^1 - u_n^2) \subset \partial G(u_n^1 - u_n^2). \end{aligned} \tag{8.128}$$

The compactness of the imbedding $V \subset L^p(\Omega;\mathbf{R}^N)$ leads to the convergence

$$u_n^i \to u_i, \quad \text{strongly in } L^p(\Omega;\mathbf{R}^N), \ (i=1,2). \tag{8.129}$$

Since J and G are locally Lipschitz on $L^p(\Omega;\mathbf{R}^N)$, the above property ensures that $\{w_n\}$ and $\{z_n\}$ are bounded in $L^{\frac{p}{p-1}}(\Omega;\mathbf{R}^N)$. Due to the reflexivity of $L^{\frac{p}{p-1}}(\Omega;\mathbf{R}^N)$ and the compactness of the imbedding $L^{\frac{p}{p-1}}(\Omega;\mathbf{R}^N) \subset V^*$, there exist $w, z \in L^{\frac{p}{p-1}}(\Omega;\mathbf{R}^N)$ such that, up to a subsequence,

$$w_n \to w \ \text{ strongly in } V^* \ \text{ and weakly in } L^{\frac{p}{p-1}}(\Omega;\mathbf{R}^N)$$

and

$$z_n \to z \ \text{ strongly in } V^* \ \text{ and weakly in } L^{\frac{p}{p-1}}(\Omega;\mathbf{R}^N).$$

From Proposition 1.4 and from (8.128) and (8.129) we obtain

$$\begin{aligned} w &\in \partial J(u_1 - u_2), \\ z &\in \partial G(u_1 - u_2). \end{aligned} \tag{8.130}$$

Now we may suppose that

$$a_i(u_n^i, u_n^i) \ \text{ converges in } \mathbf{R}, \ i = 1, 2,$$

and

$$\left\langle \left[(z_n^1 + \phi, z_n^2 + \phi) + K^*(w_n) + K^*(z_n) \right], (u_n^1, u_n^2) \right\rangle_{V\times V}$$

possesses a convergent subsequence in $\mathbf{R}$. From (8.127) and the foregoing convergences we obtain that

$$\left(a_1(u_n^1, \cdot), a_2(u_n^2, \cdot) \right) - \lambda_0 \cdot \Lambda(u_n^1, u_n^2),$$

converges strongly in $V^* \times V^*$, where λ_0 is the one given in (H_7). Hypothesis (H_7) implies that $\{u_n^1, u_n^2\}$ has a convergent subsequence in $V \times V$, and accordingly in $S_r^{a,b}$. Thus the Palais-Smale condition for the functional I_2 on $S_r^{a,b}$ is satisfied. □

Lemma 8.4 If $u = (u_1, u_2)$ is a critical point of $I_{2|S_r^{a,b}}$ then there exists a pair $(\lambda_1, \lambda_2) \in \mathbf{R}^2$ such that $((u_1, u_2), (\lambda_1, \lambda_2))$ is a solution of the problem $(P^2_{r,a,b})$.

Proof. u is a critical point for $I_{2|S_r^{a,b}}$. Thus

$$0_{V\times V} \in \left(\partial I_{2|S_r^{a,b}} \right)(u_1, u_2). \tag{8.131}$$

Taking into account the expression of the generalized gradient of the restriction of I_2 at $S_r^{a,b}$ (cf. Chapter 1), we may conclude the existence of an element $\xi \in \partial I_2(u_1, u_2)$ such that

$$\xi - r^{-2} \langle \xi, (u_1, u_2) \rangle_{V\times V} \cdot \Lambda(u_1, u_2) = 0. \tag{8.132}$$

Further we obtain (see the inclusions in the proof of Lemma 8.3)

$$\partial I_2(u_1,u_2)(v_1,v_2) \subset a_1(u_1,v_1)+a_2(u_2,v_2)+\partial f_1(u_1)v_1$$
$$+\partial f_2(u_2)v_2+\int_\Omega \partial_y j(x,(u_1-u_2)(x))(v_1-v_2)(x)dx$$
$$+\int_\Omega \partial_y g(x,(u_1-u_2)(x))(v_1-v_2)(x)dx+\langle\phi,v_1\rangle_V+\langle\phi,v_2\rangle_V,$$

for all $v_1,v_2\in V$. Accordingly, there exists some $z_i\in\partial f_i(u_i)$ $(i=1,2)$ and $w,z\in L^{\frac{p}{p-1}}(\Omega;\mathbb{R}^N)$ with

$$w(x)\in\partial_y j(x,(u_1-u_2)(x)) \text{ for a.e. } x\in\Omega,$$

and

$$z(x)\in\partial_y g(x,(u_1-u_2)(x)) \text{ for a.e. } x\in\Omega,$$

such that

$$\begin{aligned}\langle\xi,(v_1,v_2)\rangle_{V\times V} &= a_1(u_1,v_1)+a_2(u_2,v_2)+\langle z_1,v_1\rangle_V+\langle z_2,v_2\rangle_V\\ &\quad+\int_\Omega\langle w(x),(v_1-v_2)(x)\rangle dx+\int_\Omega\langle z(x),(v_1-v_2)(x)\rangle dx\\ &\quad+\langle\phi,v_1\rangle_V+\langle\phi,v_2\rangle_V.\end{aligned}$$

Relation (8.132) implies that

$$\begin{aligned}&a_1(u_1,v_1)+a_2(u_2,v_2)+\langle z_1,v_1\rangle_V+\langle z_2,v_2\rangle_V\\ &+\int_\Omega\langle w(x),(v_1-v_2)(x)\rangle dx+\int_\Omega\langle z(x),(v_1-v_2)(x)\rangle dx\\ &+\langle\phi,v_1\rangle_V+\langle\phi,v_2\rangle_V-r^{-2}[a_1(u_1,u_1)+a_2(u_2,u_2)+\langle z_1,u_1\rangle_V+\langle z_2,u_2\rangle_V\\ &+\int_\Omega\langle w(x),(u_1-u_2)(x)\rangle dx+\int_\Omega\langle z(x),(u_1-u_2)(x)\rangle dx\\ &+\langle\phi,u_1\rangle_V+\langle\phi,u_2\rangle_V]\cdot(a(B_1u_1,v_1)_V+b(B_2u_2,v_2)_V)=0,\end{aligned}$$

for all $v_1,v_2\in V$. Now we may set

$$\begin{aligned}\lambda &= r^{-2}[a_1(u_1,u_1)+a_2(u_2,u_2)+\langle z_1,u_1\rangle_V+\langle z_2,u_2\rangle_V\\ &\quad+\int_\Omega\langle(w+z)(x),(u_1-u_2)(x)\rangle dx+\langle\phi,u_1\rangle_V+\langle\phi,u_2\rangle_V],\end{aligned}$$

and we obtain

$$\int_\Omega \langle (w+z)(x), (v_1 - v_2)(x)\rangle dx$$
$$\leq \int_\Omega \max\{\langle \mu_1, (v_1 - v_2)(x)\rangle; \mu_1 \in \partial_y j(x, (u_1 - u_2)(x))\}$$
$$+ \int_\Omega \max\{\langle \mu_2, (v_1 - v_2)(x)\rangle; \mu_2 \in \partial_y g(x, (u_1 - u_2)(x))\}$$
$$= \int_\Omega j_y^0(x, (u_1 - u_2)(x); (v_1 - v_2)(x))dx + \int_\Omega g_y^0(x, (u_1 - u_2)(x); (v_1 - v_2)(x))dx.$$

The last equality holds because of Proposition 1.4(ii). The choice of $z_i (i = 1, 2)$, z and w, and the notation $\lambda_1 = \lambda a$ and $\lambda_2 = \lambda b$, together with (H_2) lead us to the desired result, and the proof is complete. □

Now we can prove the main perturbation result.

Theorem 8.9 Let the hypotheses $(H_1) - (H_7)$ be fulfilled. Then, for every $n \geq 1$, there exists $\delta_n > 0$ such that, for each $\delta \leq \delta_n$, the problem $(P^2_{r,a,b})$ admits at least n distinct solutions.

Proof. Fix $n \geq 1$. Due to Lemma 8.4 it is enough to prove the existence of a $\delta_n > 0$ such that, for every $\delta \leq \delta_n$, the functional $I_{2_{|S_r^{a,b}}}$ has at least n distinct critical values. We may use Lemma 8.1, which implies that it is possible to consider a sequence $\{b_n\}$ of essential values of $I_{1_{|S_r^{a,b}}}$, strictly increasing to s. Choose an arbitrary $\epsilon_0 < \frac{1}{2} \min_{1 \leq i \leq n-1} (b_{i+1} - b_i)$. Let us now apply Theorem 2.6 from [20] to the functionals $I_{1_{|S_r^{a,b}}}$ and $I_{2_{|S_r^{a,b}}}$. Thus, for every $1 \leq i \leq n-1$, there exists $\eta_i > 0$ such that the relation

$$\sup_{(u_1,u_2)\in S_r^{a,b}} |I_1(u_1, u_2) - I_2(u_1, u_2)| < \eta_i$$

implies the existence of an essential value c_i of $I_{2_{|S_r^{a,b}}}$ in $(b_i - \epsilon_0, b_i + \epsilon_0)$. By taking $\epsilon = \min\{\epsilon_0, \eta_1, \cdots, \eta_{n-1}\}$ in Lemma 8.2, we obtain that a $\delta_n > 0$ exists such that

$$\sup_{(u_1,u_2)\in S_r^{a,b}} |I_1(u_1, u_2) - I_2(u_1, u_2)| < \epsilon,$$

if $\delta \leq \delta_n$ in (H_5) and (H_6). Accordingly the functional $I_{2_{|S_r^{a,b}}}$ admits at least n distinct essential values $c_1, c_2, \cdots, c_n$ in $(-\infty, b_n + \epsilon)$. To finish the proof it suffices to show that $c_1 \cdots, c_n$ are critical values of $I_{2_{|S_r^{a,b}}}$. If c_i does not have this property, then we show first that there exists $\epsilon > 0$ such that $I_{2_{|S_r^{a,b}}}$ has no critical value in $(c_i - \epsilon, c_i + \epsilon)$. Suppose the contrary. Then, there exists a sequence $\{d_n\}$ of critical values of $I_{2_{|S_r^{a,b}}}$ with $d_n \to c_i$ as $n \to \infty$. The fact that d_n are critical values for the restriction of I_2 to $S_r^{a,b}$ implies that for every $n \geq 1$, there exists $\{u_n^1, u_n^2\} \in S_r^{a,b}$ such that

$$I_2(u_n^1, u_n^2) = d_n \text{ and } \lambda^*(u_n^1, u_n^2) = 0,$$

where λ^* is the lower semicontinuous functional defined by

$$\lambda^*(u_1, u_2) := \min\{\|(\xi_1, \xi_2)\|_{V^* \times V^*}; (\xi_1, \xi_2) \in \partial I_{2|S_r^{a,b}}(u_1, u_2)\}. \tag{8.133}$$

Thus one can extract a subsequence, $(u_n^1, u_n^2) \to (u_1, u_2) \in S_r^{a,b}$, strongly in $V \times V$. The continuity of I_2 and the lower semicontinuity of λ^* implies that

$$I_2(u_1, u_2) = c_i \text{ and } \lambda^*(u_1, u_2) = 0, \tag{8.134}$$

which is a contradiction to the initial conditions on c_i. Let us fix now $c_i - \epsilon < a < b < c_i + \epsilon$. By Lemma 8.3, I_2 satisfies the Palais-Smale (PS) condition on $S_r^{a,b}$. Thus for every point $e \in [a, b]$, $(PS)_e$ holds. Accordingly the assumptions of the "Noncritical Interval Theorem" of J. - N. Corvellec, M. Degiovanni and M. Marzocchi (Theorem 2.15 in [7]) hold on the complete metric space $\left(S_r^{a,b}, d(\cdot,\cdot)\right)$. Here we denote by $d(\cdot,\cdot)$ the geodesic distance on $S_r^{a,b}$. We recall here that for points $x, y \in S_r^{a,b}, d(x,y)$ is equal to the infimum of the lengths of all paths on $S_r^{a,b}$ joining x and y. We obtain that there exists a continuous map $\eta : S_r^{a,b} \times [0,1] \to S_r^{a,b}$ such that, for each $(u = (u_1, u_2), t) \in S_r^{a,b} \times [0,1]$, the following conditions are satisfied :

(a) $\eta(u, 0) = u$,

(b) $I_2(\eta(u,t)) \leq I_2(u)$,

(c) $I_2(u) \leq b \Longrightarrow I_2(\eta(u,1)) \leq a$,

(d) $I_2(u) \leq a \Longrightarrow \eta(u,t) = u$.

From these conditions, it follows that the map

$$[I_{2_{|S_r^{a,b}}} \leq b] \ni u \mapsto \eta(u,1) \in [I_{2_{|S_r^{a,b}}} \leq b]$$

is a retraction. We define now the map $\Psi : [I_{2_{|S_r^{a,b}}} \leq b] \times [0,1] \to [I_{2_{|S_r^{a,b}}} \leq b]$ by means of the relation

$$\Psi(u,t) = \eta(u,t).$$

For every $u \in [I_{2_{|S_r^{a,b}}} \leq b]$, we have

$$\Psi(u,0) = u, \ \Psi(u,1) = \eta(u,1), \tag{8.135}$$

and for each $(u,t) \in [I_{2_{|S_r^{a,b}}} \leq a] \times [0,1]$, the equality $\Psi(u,t) = \Psi(u,0)$ holds. Thus it follows that Ψ is $[I_{2_{|S_r^{a,b}}} \leq a]-$ homotopic to the identity of $[I_{2_{|S_r^{a,b}}} \leq a]$. Accordingly Ψ is a strong deformation retraction which implies that the pair

$$\left([I_{2_{|S_r^{a,b}}} \leq b], [I_{2_{|S_r^{a,b}}} \leq a]\right)$$

is trivial. With this argument, we get that c_i is not an essential value of the restriction of I_2 at $S_r^{a,b}$. With this contradiction the proof is complete. □

8.6.2 Applications: Fuzzy Perturbations

Let us put ourselves in the framework of the application (i) in 8.5.1. Let us assume that the superpotential j gives rise to a very irregular graph $[\xi, \partial j(\xi)]$ (i.e. the graph of ∂j has many zig - zag etc.). Then we consider the eigenvalue problem $(P^2_{r,a,b})$, where g^0_y is appropriately chosen in order to "smoothen" the graph $[\xi, \partial j(\xi)]$, i.e. the graph $[\xi, \partial j(\xi) + \partial g(\xi)]$ has a smaller number of irregularities than the graph $[\xi, \partial j(\xi)]$. In the present case we may consider that

$$\partial j(t) + \partial g(t) \subset [\underline{\varphi}(t) + \underline{\varphi}^1(t), \overline{\varphi}(t) + \overline{\varphi^1}(t)], \ \forall t \in \mathbf{R}.$$

In fact, we consider

$$g(t) = \int_0^t \varphi^1(\tau) d\tau, \ t \in \mathbf{R},$$

where $\varphi^1 : \mathbf{R} \to \mathbf{R}$ is locally bounded, measurable and satisfies the subcritical growth condition of Section 8.5.1(i), i.e.

$$|\varphi^1(t)| \leq c_3 + c_4|t|^{p-1}, \ \forall t \in \mathbf{R}$$

Note that we do not need to impose to φ^1 that it is even, as we have assumed on φ. Obviously, Theorem 8.9 applies on every sphere $\|v_1\|^2_V + \|v_2\|^2_V = r^2$ of $V \times V$, with a given small $r > 0$. Analogously we may consider perturbations of all aplications studied in Sections 8.5.1 and 8.5.2.

Let us now consider the perturbed hemivariational inequality eigenvalue problem corresponding to the first buckling problem (i.e. the Timoshenko plate problem) of Section 8.5.4. Let us consider only two plates glued together: Find $\zeta_i \in Z_i$ and $\lambda_i \in \mathbf{R}$, $i = 1, 2$, such that (cf. also Section 6.4)

$$\begin{aligned} & a_1(\zeta_1, z_1) + a_2(\zeta_2, z_2) + \\ & + \int_\Omega [j^0_y(x, (\zeta_1 - \zeta_2)(x); (z_1 - z_2)(x)) + g^0_y(x, (\zeta_1 - \zeta_2)(x); (z_1 - z_2)(x))] dx \\ & \qquad \geq \lambda_1 (L_1\zeta_1, z_1) + \lambda_2 (L_2\zeta_2, z_2), \end{aligned} \tag{8.136}$$

for all $z_i \in Z_i \ \ i = 1, 2$. One can assume that the graph $[\xi, \partial j(\xi) + \partial g(\xi)]$ is much more regular than the graph $[\xi, \partial j(\xi)]$. The case that $f + g$ is monotone is very interesting. Indeed this fact is beneficial in the framework of a numerical calculation.

Further we consider the case of fuzzy effects superimposed on an adhesive contact law. We put ourselves in the framework of the previous example of adhesively connected plates (cf. (8.136)) and we assume that ∂g describes fuzzy effects. We refer to Section 5.7 for the definition of g (cf. also [16] p.77).

Note that existence results related to fuzzy effects have been studied by Naniewicz and Panagiotopoulos in [14] p. 132. Here we can apply our results to the perturbed problem $(P^2_{r,a,b})$. Note that the fuzzy effects do not have any symmetry. Lemma 8.2 shows that if the fuzzy effect is sufficiently small then the energy of the perturbed problem tends to the energy of the initial problem. On the other hand, Theorem 8.9 gives information about the solutions of the perturbed problem.

References

[1] A. Ambrosetti and P. H. Rabinowitz, Dual Variational Methods in Critical Point Theory and Applications, *J. Funct. Anal.* **14** (1973), 349-381.

[2] R. S. Bishop and R. J. Crittenden, *Geometry of Manifolds,* Academic Press, New York, 1964.

[3] M. Bocea, D. Motreanu and P.D. Panagiotopoulos, Multiple Solutions for a Double Eigenvalue Hemivariational Inequality on a Spherelike Type Manifold, *J. Nonl. Anal. T.M.A.* (to appear).

[4] M. Bocea, P.D. Panagiotopoulos and V. Radulescu, A Perturbation Result for a Double Eigenvalue Hemivariational Inequality with Constraints and Applications, *J. of Global Optimiz.* (to appear).

[5] K. C. Chang, Variational Methods for Nondifferentiable Functionals and Applications to Partial Differential Equations, *J. Math. Anal. Appl.* **80** (1981), 102-129.

[6] F. H. Clarke, *Optimization and Nonsmooth Analysis*, John Wiley & Sons, New-York, 1983.

[7] J.N. Corvellec, M. Degiovanni and M. Marzocchi, Deformation Properties for Continuous Functionals and Critical Point Theory, *Top. Meth. Nonl. Anal.* **1** (1993), 151-171.

[8] M. Degiovanni and S. Lancelotti, Perturbations of Even Nonsmooth Functionals, *Differential Integral Equations* **8** (1995), 981-992.

[9] I. Ekeland, On the Variational Principle, *J. Math. Anal. Appl.* **47** (1974), 324-353.

[10] J. Milnor, *Morse Theory* , Princeton University Press, 1963.

[11] D. Motreanu, Existence of Critical Points in a General Setting, *Set-Valued Anal.* **3** (1995), 295-305.

[12] D. Motreanu and P.D. Panagiotopoulos, Double Eigenvalue Problems for Hemivariational Inequalites, *Arch. for Rat. Mech. Anal.* **140** (1997), 225-251.

[13] D. Motreanu and P. D. Panagiotopoulos, Hysteresis: The Eigenvalue Problem for Hemivariational Inequalities, in: Models of Hysteresis (A. Visintin, Editor), *Longman Scientific & Technical*, Harlow (1993), pp. 102-117.

[14] Z. Naniewicz and P. D. Panagiotopoulos, *Mathematical Theory of Hemivariational Inequalities*, Marcel Dekker, New York, 1995.

[15] J. Naumann and H. V. Wenk, On Eigenvalue Problems for Variational Inequalities, *Rendiconti di Matematica* **9** (Serie VI) (1976), 439-463.

[16] P. D. Panagiotopoulos, *Hemivariational Inequalities. Applications in Mechanics and Engineering*, Springer-Verlag, Berlin, 1993.

[17] P. D. Panagiotopoulos, Semicoercive Hemivariational Inequalities. On the delimination of composite Plates. *Quart. of Appl. Math.* **XLVII** (1989), 611-629.

[18]P. D. Panagiotopoulos, *Inequality Problems in Mechanics and Applications. Convex and Nonconvex Energy Functionals*, Birkhäuser Verlag, Basel, 1985.(Russian translation, MIR Publisher, Moscow, 1989).

[19]P. D. Panagiotopoulos and G. Stavroulakis, A Variational-hemivariational Inequality Approach to the Laminated Plate Theory under Subdifferential Boundary Conditions. *Quart. of Appl. Math.* **XLVII** (1989), 611-629.

[20]P. H. Rabinowitz, Minimax Methods in Critical Point Theory with Applications to Differential Equations, *CBMS Reg. Conf. Ser. Math. 65, Amer. Math. Soc.*, Providence, R. J., 1986.

[21]V.D. Rădulescu and P.D. Panagiotopoulos, Perturbations of Hemivariational Inequalities with Constraints and Applications, *J. Global Optimiz.*, (to appear).

[22]E.H. Spanier, *Algebraic Topology*, Mc-Graw-Hill Book Co., New York, 1966.

[23]A. Szulkin, Minimax Principles for Lower Semicontinuous Functions and Applications to Nonlinear Boundary Value Problems, *Ann. Inst. Henri Poincaré, Anal. Non Linéaire* **3** (1986), 77-109.

9. Periodic and Dynamic Problems

The present chapter is devoted to the study of some evolutive or periodic hemivariational inequalities. Specifically, it contains the following three types of problems: hyperbolic hemivariational inequalities modelling nonlinear wave equations with nonlinearities, homoclinic solutions for second order Hamiltonian systems with discontinuous nonlinearities, and multiplicity results for hemivariational inequalities involving periodic energy functionals. The existence of solutions of nonsmooth hyperbolic problems expressed in the form of a hemivariational inequality are studied separately in the nonresonant and the resonant cases. The second section of the chapter contains some existence and multiplicity results for periodic solutions of second order nonautonomous and nonsmooth Hamiltonian systems involving nonconvex superpotentials. These results are proved by showing the existence of homoclinic solutions which are constructed as critical points of the corresponding nonconvex and nonsmooth energy functional. The approach relies on nonsmooth minimax principles, including the nonsmooth Ljusternik-Schnirelman theory, as developed in Chapter 2.

9.1 A Hyperbolic Hemivariational Inequality

Throughout this Section $j : [0,\pi] \times \mathbb{R}^2 \to \mathbb{R}$ denotes a real-valued function $j(x,t,z)$ satisfying the hypothesis

(H_1) $j(x,t,z)$ is 2π-periodic in t, measurable in (x,t), $j(\cdot,\cdot,0) \in L^1((0,\pi) \times (0,2\pi))$, Lipschitz in z, with a bounded generalized gradient

$$|w| \leq M, \quad \forall (x,t,z) \in [0,\pi] \times [0,2\pi] \times \mathbb{R}, \ \forall w \in \partial j(x,t,z).$$

Here $\partial j(x,t,z)$ denotes the generalized gradient of $j(x,t,\cdot)$ with respect to $z \in \mathbb{R}$ (see Definition 1.2 in Chapter 1). The generalized directional derivative of $j(x,t,z)$ with respect to $z \in \mathbb{R}$ (see Definition 1.1 in Chapter 1) will be denoted by $j^0(x,t,z;\cdot)$. For the rest of the Section we fix a number $c \in \mathbb{R}\setminus\{0\}$. We formulate the following hyperbolic hemivariational inequality: Find $u \in H^1$ such that

$$(P) \quad \int_0^{\pi}\int_0^{2\pi}(u_x v_x - u_t v_t - cuv + j^0(x,t,u(x,t);v(x,t))dxdt \geq 0, \forall v \in H^1,$$

where H^1 denotes the Hilbert space equal to the closure with respect to the norm

$$\|u\| := (\|u\|_{L^2}^2 + \|u_x\|_{L^2}^2 + \|u_t\|_{L^2}^2)^{1/2}$$

of the subset of $C^\infty([0,\pi] \times \mathbf{R}, \mathbf{R})$ verifying the boundary conditions

$$u(0,t) = u(\pi,t) = 0, \quad t \in \mathbf{R}, \tag{9.1}$$

and the periodicity conditions

$$u(x,t+2\pi) = u(x,t), \quad (x,t) \in [0,\pi] \times \mathbf{R}. \tag{9.2}$$

A stronger form of Problem (P) is the following nonlinear wave equation with discontinuities

$$(P') \qquad u_{xx} - u_{tt} + cu \in \partial j(x,t,u(x,t)), \quad \text{a.e. } (x,t) \in [0,\pi] \times \mathbf{R},$$

under the boundary and periodicity conditions (9.1), (9.2).

Let us recall that the spectrum of the d'Alembertian $\Box u = u_{tt} - u_{xx}$ is the following

$$\sigma(\Box) = \{k^2 - j^2 : \ k, j \text{ integers}, \ j \neq 0\} \tag{9.3}$$

where the corresponding eigenvalues are

$$\sin jx(\alpha \cos kt + \beta \sin kt), \quad \alpha, \beta \in \mathbf{R}. \tag{9.4}$$

We introduce the unbounded linear operator on H^1 by

$$Lu := \Box u - cu. \tag{9.5}$$

Then the following decomposition holds

$$H^1 = N(L) \oplus N^-(L) \oplus N^+(L), \tag{9.6}$$

where $N(L)$, $N^-(L)$ and $N^+(L)$ denote the null-space, the space on which L is negative definite and positive definite, respectively.

In treating Problem (P') (so (P)) we follow the argument in Goeleven and Motreanu [20] which is a nonsmooth version of the idea in Rabinowitz [39] based on a Galerkin approximation. The finite dimensional approximation procedure is described below.

Lemma 9.1 For any integer $m \geq 1$ let X_m denote the finite dimensional subspace of H^1 defined by

$$X_m := \text{span}\,\{\sin jx \cos kt, \ \sin jx \sin kt\}_{0 \leq j,k \leq m}. \tag{9.7}$$

Assuming condition (H_1) together with

$$c \notin \sigma(\Box), \tag{9.8}$$

there exist $u_m \in X_m$ and $\xi_m \in X_m$ such that

$$\xi_m(x,t) \in \partial j(x,t,u_m(x,t)) \text{ for a.e. } (x,t) \in (0,\pi) \times \mathbb{R} \tag{9.9}$$

and

$$\int_0^\pi \int_0^{2\pi} (u_{mx}\phi_x - u_{mt}\phi_t - cu_m\phi + \xi_m\phi)dxdt = 0, \ \forall \phi \in X_m. \tag{9.10}$$

Proof. From assumption (9.8) it is seen that

$$N(L) = \{0\}. \tag{9.11}$$

Due to (9.6) and (9.11) the space X_m in (9.7) is splitted as follows

$$X_m = X_m^- \oplus X_m^+, \tag{9.12}$$

where $X_m^- := N^-(L) \cap X_m$ and $X_m^+ := N^+(L) \cap X_m$.

We consider the restriction $I|_{X_m} : X_m \to \mathbb{R}$ of the locally Lipschitz functional $I : H^1 \to \mathbb{R}$ given by

$$I(v) = \int_0^\pi \int_0^{2\pi} [\frac{1}{2}(v_x^2 - v_t^2 - cv^2) + j(x,t,v)]dxdt = 0, \ \forall v \in H^1. \tag{9.13}$$

Assumption (H_1) and Lebourg's mean value theorem (see Theorem 1.1 in Chapter 1) imply the estimate

$$|j(x,t,z)| \le M|z| + |j(x,t,0)|, \ \ \forall (x,t,z) \in [0,\pi] \times \mathbb{R}^2. \tag{9.14}$$

It is clear from (9.14) that the functional I in (9.13) is well-defined and locally Lipschitz.

Denote by λ_+ the first positive eigenvalue of the operator L given by (9.5). Then relations (9.7), (9.12), (9.13) and (9.14) show that

$$I(v) \ge \lambda_+ \|v\|_{L^2}^2 - M\|v\|_{L^1} - \int_0^\pi \int_0^{2\pi} |j(x,t,0)|dxdt \tag{9.15}$$

$$\ge c_0\|v\|^2 - c_1\|v\| - c_2, \ \ \forall v \in X_m^+,$$

with positive constants c_0, c_1, c_2, where we have also used that X_m^+ is finite dimensional. Then (9.15) yields the boundedness from below

$$\inf_{X_m^+} I > -\infty. \tag{9.16}$$

Denote now by λ_- the first negative eigenvalue of the linear operator L in (9.5). Then (9.7), (9.12), (9.13) and (9.14) imply the estimate

$$I(v) \le \lambda_- \|v\|_{L^2}^2 + M\|v\|_{L^1} + \int_0^\pi \int_0^{2\pi} |j(x,t,0)|dxdt \tag{9.17}$$

$$\le -d_0\|v\|^2 + d_1\|v\| + d_2, \ \ \forall v \in X_m^-,$$

with positive constants d_0, d_1, d_2, where the fact that X_m^- is finite dimensional was used, too. From (9.17) we obtain

$$I(v) \to -\infty \text{ as } \|v\| \to \infty, \ v \in X_m^-. \tag{9.18}$$

From (9.16) and (9.18) we deduce the existence of a constant $\rho > 0$ such that

$$\sup\{I(v) : \ v \in X_m^-, \ \|v\| = \rho\} < \inf_{X_m^+} I. \tag{9.19}$$

Let us check the Palais-Smale condition in the sense of Chang [14] (see Chapter 2) for the locally Lipschitz functional $I|_{X_m}$. For this let $\{v_n\} \subset X_m$ be a sequence such that $I(v_m)$ is bounded and there exists a sequence $w_n \in \partial(I|_{X_m})(v_n)$ with $w_n \to 0$ in X_m^*. In view of (9.5) and (9.13) we can write

$$w_n = Lv_n + g_n \quad \text{on } X_m,$$

with $g_n \in L^2((0,\pi) \times (0,2\pi))$ such that

$$\langle w_n, v\rangle = \int_0^\pi \int_0^{2\pi} ((Lv_n)v + g_n v) dx dt, \quad \forall v \in X_m,$$

$$g_n(x,t) \in \partial j(x,t,v_n(x,t)) \ \text{ for a.e. } (x,t) \in (0,\pi) \times (0,2\pi).$$

Since $w_n \to 0$ in X_m^* we can estimate the projections $P_m^- v_n$ and $P_m^+ v_n$ of v_n on X_m^- and X_m^+, respectively, as follows

$$\begin{aligned} \|P_m^+ v_n\| &\geq \int_0^\pi \int_0^{2\pi} ((Lv_n)P_m^+ v_n + g_n P_m^+ v_n) dx dt \\ &\geq \lambda_+ \|P_m^+ v_n\|_{L^2}^2 - M\|P_m^+ v_n\|_{L^1} \geq c_0 \|P_m^+ v_n\|^2 - c_3 \|P_m^+ v_n\| \end{aligned}$$

and

$$\begin{aligned} \|P_m^- v_n\| &\geq -\int_0^\pi \int_0^{2\pi} ((Lv_n)P_m^- v_n - g_n P_m^- v_n) dx dt \\ &\geq -\lambda_- \|P_m^- v_n\|_{L^2}^2 - M\|P_m^- v_n\|_{L^1} \geq d_0 \|P_m^- v_n\|^2 - d_3 \|P_m^- v_n\|, \end{aligned}$$

with constants c_3, d_3, and n sufficiently large. The estimates above show that $v_n = P_m^- v_n + P_m^+ v_n \in X_m$ is bounded in X_m. Taking into account that X_m is finite dimensional, we may conclude that $I|_{X_m}$ satisfies the Palais-Smale condition in the sense of Chang [14] (see Definition 2.1 in Chapter 2).

The properties (9.12), (9.16), (9.19) and the Palais-Smale condition allow to apply Saddle-Point Theorem in the nonsmooth version of Chang [14] (see also Corollary 2.6 in Chapter 2) to the locally Lipschitz functional $I|_{X_m} : X_m \to \mathbb{R}$. It ensures the existence of a critical point $u_m \in X_m$ of $I|_{X_m}$, i.e.,

$$0 \in \partial(I|_{X_m})(u_m). \tag{9.20}$$

Then (9.20), together with Theorem 1.3 and Proposition 1.4 in Chapter 1, implies (9.9), (9.10). This completes the proof of Lemma 9.1. □

We focus now on the resonant case, i.e.,

$$c \in \sigma(\Box). \tag{9.21}$$

Taking into account that $c \neq 0$, from (9.21) we obtain that

$$\dim N(L) = p > 0,$$

where

$$N(L) = \text{span } \{\phi_1(x,t), ..., \phi_p(x,t)\}. \tag{9.22}$$

A result, which is analogous to Lemma 9.1, and holds in the resonant case (9.21) for the finite dimensional approximation, is stated below.

Lemma 9.2 Assume that (H_1) and (9.21) hold together with either
(H_2^-) $j(x,t,z) \to -\infty$ as $|z| \to \infty$
or
(H_2^+) $j(x,t,z) \to +\infty$ as $|z| \to \infty$,
uniformly with respect to $(x,t) \in (0,\pi) \times (0,2\pi)$.

Then there exist $u_m \in X_m$ and $\xi_m \in X_m$ such that (9.9) and (9.10) are verified.

Proof. To make a choice we suppose that (H_2^-) is valid. Then (H_2^-) yields a number $a_0 > 0$ such that

$$j(x,t,z) \leq 0, \quad \forall (x,t) \in (0,\pi) \times (0,2\pi), \ |z| \geq a_0, \tag{9.23}$$

and corresponding to each $K > 0$, a number $a_1 = a_1(K) > a_0$ such that

$$j(x,t,z) \leq -K, \quad \forall (x,t) \in (0,\pi) \times (0,2\pi), \ |z| \geq a_1. \tag{9.24}$$

Let $v \in N(L) \setminus \{0\}$. It can be written as

$$v = \alpha\phi, \quad \|\phi\| = 1, \ \alpha > 0. \tag{9.25}$$

Then $I(v)$ can be expressed as follows

$$I(v) = \int_0^{\pi} \int_0^{2\pi} j(x,t,v(x,t))dxdt = \sum_{i=1}^{3} \int_{D_i} j(x,t,v(x,t))dxdt,$$

where, for a fixed $K > 0$ with the corresponding a_1 in (24),

$$\begin{aligned} D_1 &= \{(x,t) \in (0,\pi) \times (0,2\pi) : \ |v(x,t)| \leq a_0\}, \\ D_2 &= \{(x,t) \in (0,\pi) \times (0,2\pi) : \ a_0 < |v(x,t)| < a_1\}, \\ D_3 &= \{(x,t) \in (0,\pi) \times (0,2\pi) : \ |v(x,t)| \geq a_1\}. \end{aligned}$$

In view of (9.25), and since by (9.22) the function ϕ is a linear combination of generators in (9.4), we may find some $\alpha = \alpha(K, \phi)$ large enough to have

$$\text{mes}(D_3) \geq \pi^2. \tag{9.26}$$

The compactness of the unit sphere in $N(L)$ (cf.(9.22)) ensures that in fact (9.26) holds for all v in (9.25) with $\alpha \geq \alpha_0 > 0$, where $\alpha_0 = \alpha_0(K)$ denotes a constant which is independent of $\phi \in N(L)$ with $\|\phi\| = 1$. Then from (9.23)-(9.26) we derive

$$I(v) \le b_0 - K\pi^2, \quad \forall v = \alpha\phi \in N(L),\ \|\phi\| = 1,\ \alpha \ge \alpha_0, \tag{9.27}$$

for a constant $b_0 > 0$. Letting $K \to +\infty$ in (9.27) we conclude that

$$I(v) \to -\infty \ \text{ as } \|v\| \to \infty,\ v \in N(L). \tag{9.28}$$

Let us denote

$$X_m^0 = X_m \cap N(L),$$

and let us consider the splitting

$$X_m = X_m^- \oplus X_m^0 \oplus X_m^+, \tag{9.29}$$

where X_m^- and X_m^+ have the meaning in (9.12). Let $P_m^0 : X_m \to X_m^0$ denote the orthogonal projection of X_m onto X_m^0.

Let $\{v_n\} \subset X_m$ be a sequence such that $I(v_m)$ is bounded and for which there exists a corresponding sequence

$$w_n \in \partial(I|_{X_m})(v_n)$$

with $w_n \to 0$ in X_m^*. By (9.29) we can write

$$v_n = P_m^- v_n + P_m^0 v_n + P_m^+ v_n.$$

The same argument as in the proof of Lemma 9.1 shows that the components $\{P_m^- v_n\}$ and $\{P_m^+ v_n\}$ are bounded in X_m. The boundedness of $I(v_n)$ implies that $I(P_m^0 v_n)$ is bounded (here (H_1) and Lebourg's mean value theorem are used). Then (9.28) ensures that $\{P_m^0 v_n\}$ is bounded. Since X_m is finite dimensional, we obtain that the Palais-Smale condition in the sense of Chang [14] (see Definition 2.1) is satisfied in the case of the locally Lipschitz functional $I|_{X_m}$.

With the notations above and using (9.17) we see that

$$I(v_- + v_0) \le -d_0\|v_-\|^2 + \int_0^\pi \int_0^{2\pi} (j(x,t,v_- + v_0) - j(x,t,v_0))dxdt +$$

$$\int_0^\pi \int_0^{2\pi} j(x,t,v_0)dxdt \le -d_0\|v_-\|^2 + d\|v_-\| + \int_0^\pi \int_0^{2\pi} j(x,t,v_0)dxdt$$

for all $v_- \in X_m^-$ and $v_0 \in X_m^0$, where d_0, d are positive constants. By virtue of (9.28) it follows that

$$I(v) \to -\infty \ \text{ as } \|v\| \to \infty,\ v \in X_m^- \oplus X_m^0. \tag{9.30}$$

Taking into account the Palais-Smale condition and relations (9.29), (9.15), (9.30), we can apply to $I|_{X_m}$ the nonsmooth version of Saddle-Point Theorem of Chang [14], Theorem 3.3 (see Corollary 2.6 for a more general minimax result). Then there exists a critical point $u_m \in X_m$ of $I|_{X_m}$ in the sense of (9.20).

We complete the proof as in Lemma 9.1. □

We present now our main results separately in the nonresonant and resonant case.

Theorem 9.1 Let $c \in \mathbb{R}\backslash\{0\}$ satisfy the nonresonant condition (9.8). Assume that the function $j : [0,\pi] \times I\!R^2 \to I\!R$ verifies (H_1) and

(H_3) for every $r > 0$ there exist constants $c_1 \in I\!R$, $c_2 \in I\!R$, with $c_1 < c$, such that for all $v \in N(\Box) \subset H^1$ with $\|v\|_{L^2} \le r$ and $\xi \in L^2((0,\pi) \times (0,2\pi))$ satisfying

$$\xi(x,t) \in \partial j(x,t,v(x,t)) \text{ for a.e. } (x,t) \in (0,\pi) \times (0,2\pi)$$

one has

$$|\int_0^\pi \int_0^{2\pi} \xi(x,t) v_{tt} dx dt| \le c_1 \|v_t\|^2_{L^2} + c_2.$$

Then Problem (P') admits at least a weak solution. In particular, Problem (P) has a solution.

Proof. Applying Lemma 9.1 one finds a sequence $\{u_m\}$, with $u_m \in X_m$, for which there exists a corresponding sequence $\{\xi_m\} \in X_m$ satisfying (9.10), (9.11). From (9.10) it is clear that, in $L^2((0,\pi) \times (0,2\pi))$, the equality below holds

$$Lu_m = -\xi_m, \quad m \ge 1. \tag{9.31}$$

In view of (9.7) we can write the expansions

$$u_m = \sum_{0 \le j,|k| \le m} f^m_{jk} \sin jx e^{ikt}, \quad f_{jk} = \overline{f}_{j,-k},$$

and

$$\xi_m = \sum_{0 \le j,|k| \le m} h^m_{jk} \sin jx e^{ikt}, \quad h_{jk} = \overline{h}_{j,-k},$$

where the coefficients f^m_{jk} and h^m_{jk} are complex numbers. By (9.8) and (9.31) we arrive at

$$f^m_{jk} = -(j^2 - k^2 - c)^{-1} h^m_{jk}.$$

Since, by virtue of (9.8),

$$|j^2 - k^2 - c| \ge a > 0, \quad \forall j,k,$$

with a constant $a > 0$ which is independent of j and k, it follows that u_m and ξ_m in (9.31) satisfy

$$\|u_m\|_{L^2} \le C_0 \|\xi_m\|_{L^2}, \quad \forall m \ge 1, \tag{9.32}$$

for a constant $C_0 > 0$. On the other hand from (9.9) and (H_1) we derive the boundedness of $\{\xi_m\}$ in $L^2((0,\pi) \times (0,2\pi))$. Consequently, (9.32) yields the L^2-boundedness of $\{u_m\}$, say,

$$\|u_m\|_{L^2} \le C, \quad \forall m > 1. \tag{9.33}$$

Consider the direct sum decomposition

$$H^1 = N(\Box) \oplus N^- \oplus N^+, \tag{9.34}$$

where N^- and N^+ represent the subspaces of H^1 on which the d'Alembertian $\Box$ is negative and positive definite, respectively. Thus each $v \in H^1$ can be expressed uniquely as

$$v = Pv + P^-v + P^+v, \tag{9.35}$$

$$Pv \in N(\Box),\ P^-v \in N^-,\ P^+v \in N^+.$$

Relation (9.31) can be expressed in the form

$$\Box(P^+u_m + P^-u_m) = c(P^+u_m + P^-u_m) - \xi_m, \quad \forall m \geq 1.$$

Then (9.33), (H_1) and a standard estimate for the linear wave equation (see, e.g., Brézis [11]) imply

$$||P^+u_m + P^-u_m|| \leq C, \ \ \forall m \geq 1, \tag{9.36}$$

for a new constant $C > 0$.

It remains to estimate the component $Pu_m \in N(\Box)$ (see (9.35)). Towards this we deduce from (9.10) that

$$\int_0^{\pi}\int_0^{2\pi}(cu_m\phi - \xi_m\phi)dxdt = 0, \ \forall\phi \in X_m \cap N(\Box).$$

If we put in the foregoing equality $\phi = -(Pu_m)_{tt}$ we see that

$$c\int_0^{\pi}\int_0^{2\pi}(Pu_m)_t^2dxdt = -\int_0^{\pi}\int_0^{2\pi}\xi_m(Pu_m)_{tt}dxdt. \tag{9.37}$$

Assumption (H_3) ensures that constants $c_1 > 0$ and $c_2 > 0$ can be found such that

$$|\int_0^{\pi}\int_0^{2\pi}\xi_m(Pu_m)_{tt}dxdt| \leq c_1||(Pu_m)_t||_{L^2}^2 + c_2,$$

with $c_1 < c$ (here we have used also (9.33), as well as the fact that $Pu_m \in N(\Box)$). Then (9.37) yields

$$c||(Pu_m)_t||_{L^2}^2 \leq c_1||(Pu_m)||_{L^2}^2 + c_2.$$

Since $c_1 < c$, this implies the boundedness

$$||(Pu_m)_x||_{L^2} = ||(Pu_m)_t||_{L^2} \leq C, \tag{9.38}$$

for a constant $C > 0$, where the equality in (9.38) is due to the fact that $Pu_m \in N(\Box)$. Combining (9.33), (9.36) and (9.38) we get the boundedness of $\{u_m\}$ in H^1. Thus there is a constant $C > 0$ such that

$$||u_m|| \leq C, \ \ \forall m \geq 1. \tag{9.39}$$

The reflexivity of H^1 and (9.39) allow to assume, after relabeling, that there exists $u \in H^1$ such that

$$u_m \to u \ \text{ weakly in } H^1 \text{ as } m \to \infty. \tag{9.40}$$

Assumption (H_1) implies that, along a subsequence, we have

$$\xi_m \to \xi \ \text{ weakly in } L^2((0,\pi) \times (0,2\pi)) \text{ as } m \to \infty, \tag{9.41}$$

for some $\xi \in L^2((0,\pi) \times (0,2\pi))$. We know that $|\xi_m(x,t)| \leq M$, $\forall (x,t) \in [0,\pi] \times [0,2\pi]$ and thus along a subsequence, we may assume that

$$u_m(x,t) \to u(x,t) \text{ a.e. in } (0,\pi) \times (0,2\pi)$$

and

$$\xi_m(x,t) \to \xi(x,t) \text{ a.e. in } (0,\pi) \times (0,2\pi).$$

The fact that the generalized gradient is weakly*-closed (see (iii) in Proposition 1.4) ensures that

$$\xi(x,t) \in \partial j(x,t,u(x,t)) \ \text{ for a.e. } (x,t) \in (0,\pi) \times (0,2\pi). \tag{9.42}$$

Let us fix $k \geq 1$ and take $\phi \in X_k$. Then $\phi \in X_m$, $\forall m \geq k$, and we can make use of (9.10). Since $k \geq 1$ and $\phi \in X_k$ are arbitrary, we deduce from (9.10), (9.40), (9.41) the formula

$$\int_0^\pi \int_0^{2\pi} (u_x\phi_x - u_t\phi_t - cu\phi + \xi\phi)dxdt = 0, \ \forall \phi \in \cup_{m\geq 1} X_m.$$

By density we derive that $u \in H^1$ is a solution of the equation

$$\int_0^\pi \int_0^{2\pi} (u_x\phi_x - u_t\phi_t - cu\phi + \xi\phi)dxdt = 0, \ \forall \phi \in H^1. \tag{9.43}$$

Therefore u is a solution of Problem (P'). From (9.43), (9.42) and the maximality property of $j^0(x,t,\cdot)$ (see (ii) in Proposition 1.4) one obtains that $u \in H^1$ is a solution of Problem (P) □

We deal further with the resonant case of Problem (P).

Theorem 9.2 Assume that the conditions (H_1), (H_3), (9.21) with $c \neq 0$, and either (H_2^-) or (H_2^+) hold. Then Problem (P') admits at least one solution $u \in H^1$; so Problem (P) has at least a solution.

Proof. To make a choice let us suppose that $c > 0$. Lemma 9.2 leads to sequences $\{u_m\}$, $\{\xi_m\}$ with $u_m \in X_m$ and $\xi_m \in X_m$ satisfying (9.9), (9.10). Then (9.31) is true. Arguing as in the proof of Theorem 9.1 we get the estimate (9.36). We claim that

$$\|Pu_m\|_{L^2} \leq C, \ \ \forall m \geq 1, \tag{9.44}$$

for a constant $C > 0$, where P denotes the projection of H^1 onto $N(\Box)$.

The argument relies on the minimax description of the critical value $I(u_m)$ of $I|_{X_m}$. The minimax characterization of the critical point $u_m \in X_m$ of $I|_{X_m}$ in the nonsmooth version of Saddle-Point Theorem of Chang [14] (see Rabinowitz [39], [41] for the smooth case, and Corollary 2.6 for a more general situation in the nonsmooth case) implies

$$b_m := I(u_m) \geq \inf_{X_m^+} I \geq \inf_{N+} I > -\infty, \quad \forall m \geq 1, \tag{9.45}$$

which is valid because $c > 0$. On the other hand from (9.10) we derive that

$$\begin{aligned} b_m &= \frac{1}{2}\int_0^\pi \int_0^{2\pi} (Lu_m)(P^+u_m + P^-u_m)dxdt \\ &\quad - \int_0^\pi \int_0^{2\pi} (\frac{c}{2}(Pu_m)^2 - j(x,t,u_m))dxdt \\ &= -\frac{1}{2}\int_0^\pi \int_0^{2\pi} \xi_m(x,t)(P^+u_m + P^-u_m)dxdt \\ &\quad - \int_0^\pi \int_0^{2\pi} (\frac{c}{2}(Pu_m)^2 - j(x,t,u_m))dxdt. \end{aligned}$$

The equality above, (9.9) and (H_1) show that

$$b_m \leq M\|P^+u_m + P^-u_m\|_{L^2} - \int_0^\pi \int_0^{2\pi} (\frac{c}{2}(Pu_m)^2 - j(x,t,u_m))dxdt.$$

In view of (9.14), (9.36) and (9.45), we conclude that the claim in (9.44) holds true. From now on the proof proceeds as for Theorem 9.1. □

Remark 9.1 [20] is the first paper studying the nonlinear wave equation with discontinuities without any monotonicity or convexity assumptions. In comparison with the smooth case, i.e., $\xi(x,t)$ in (P) is $g(x,t,u(x,t))$ for $g \in C([0,\pi] \times \mathbb{R}^2)$, no additional hypotheses are imposed (see Rabinowitz [39]). Assumption (g_2) in Rabinowitz [39] is slightly stronger that (H_3), even in the smooth case. For continuous, monotone (in u) nonlinearities in (P), as well as for the case of free vibrations, we refer to Brézis [10]. Theorem 1 in Brézis [10] deals with the nonlinear wave equation $\Box u - g(u) = f(x,t)$, where $g : \mathbb{R} \to \mathbb{R}$ is continuous, monotone, $g(0) = 0$, $|g(u)| \leq \gamma|u| + C$, $\gamma \in [0,3]$, $C \in \mathbb{R}$ and $f \in L^\infty$. In our Theorems 9.1 and 9.2 we drop the restriction on γ, assuming $c \in \mathbb{R}\setminus 0$, but other conditions are added. Necessary conditions of optimality in an optimal control problem subject to the semilinear wave equation above on a bounded domain Ω in $\mathbb{R}^N$ for $g \in C^1(\mathbb{R})$, $g' \in L^\infty(\mathbb{R})$, $\|g'\|_{L^\infty} < \lambda_1(-\Delta)$, with Δ on $H_0^1(\Omega)$, have been obtained in Barbu and Pavel [9]. The use of generalized gradients to derive necessary conditions of optimality can be found in Clarke [15]. (cf. the references in [15] for older papers by Clarke)

Remark 9.2 There is a strong distinction between the nonresonant and resonant cases in Problem (P), i.e., $c \notin \sigma(\Box)$ and $c \in \sigma(\Box)$, respectively. Assumptions (H_2^-), (H_2^+) must hold only in the resonant case. In the smooth nonlinear elliptic boundary value problems they go back to Ahmad, Lazer and Paul [5] (see also Rabinowitz [39]). The first nonsmooth version of Landesman-Lazer result [25] for elliptic problems has been obtained by Chang [14]. An extension of Chang's result allowing unbounded nonlinearities has been obtained by Goeleven, Motreanu and Panagiotopoulos [21].

9.2 Homoclinic solutions

In this Section we study the following boundary value problem, denoted $(\mathcal{P})$, for a second order Hamiltonian system with discontinuous nonlinearities in $I\!R^N$, which reads

$$\begin{aligned}&-\ddot{u}(t) + B(t)u(t) + \xi(t) = 0\\&\xi(t) \in \partial_x j(t, u(t)) \quad \text{a.e. on } [0,T],\\&u(T) - u(0) = \dot{u}(T) - \dot{u}(0) = 0.\end{aligned}$$

Here $T > 0$ is a prescribed period for the matrix $B(\cdot) \in I\!R^{N^2}$ and for the function $j(\cdot, x)$ for all $x \in I\!R^N$. We assume that $B(t)$ is continuous, T−periodic, symmetric and let us suppose that there exists $b > 0$ such that

$$x^T B(t)x \geq b|x|^2, \;\; \forall x \in I\!R^N, \; t \in [0,T].$$

We also suppose that $j(t,x)$ is a Carathéodory real valued function on $I\!R \times I\!R^N$ which is T−periodic with respect to $t \in I\!R$ and locally Lipschitz with respect to $x \in I\!R^N$. The notation $\partial_x j(t,x)$ in the statement of $(\mathcal{P})$ stands for the generalized gradient of $j(t,\cdot)$ with respect to the second variable (see Definition 1.2). For a later use we denote by $j^0(t,x;y)$ the directional derivative in the sense of Clarke of $j(t,\cdot)$ at x in the direction y (see Definition 1.1).

There is a large amount of references dealing with the periodic problem $(\mathcal{P})$ in the smooth case, i.e., when it is supposed at least that $j \in C^1(I\!R \times I\!R^N, I\!R)$ (see for instance Ambrosetti and Bertotti [7], Coti Zelati and Rabinowitz [17], Mawhin and Willem [27], Rabinowitz [40], Yanheng and Girardi [42]).

Following [4], our aim here is to treat the same problem without any differentiability or convexity assumptions required on the superpotential j. More precisely, we will suppose that the function $j(t,x)$ is locally Lipschitz with respect to the second variable and satisfies a polynomial growth condition. This requirement is necessary to assume the applicability of subdifferentiation rule for the integral functionals with generalized gradient (cf. Theorem 1.3). A related result devoted to nonsmooth Hamiltonian systems is the one of Adly and Goeleven [3][2] where it is supposed a global Lipschitz condition for j with respect to the second variable. Since such hypothesis excludes any superlinear growth for j, we work here, as in [4], in a different framework. It is also worth to point out that the homoclinic trajectories for nonsmooth Hamiltonian systems involving locally Lipschitz superpotentials are studied in [2][3] for the first time.

We start by investigating the periodic solutions of problem $(\mathcal{P})$.

We first assume that the function $j : I\!R \times I\!R^N \to I\!R$ satisfies the hypotheses below:

$(\mathcal{H}_1)$ for all $x \in I\!R^N$ the function $j(\cdot, x)$ is measurable and T - periodic. For every $t \in I\!R$ the function $j(t, \cdot)$ is locally Lipschitz and $j(\cdot, 0) \in L^1(0,T)$.

$(\mathcal{H}_2)$ $|w| \leq a + b|x|^{p-1}$ a.e. $t \in I\!R$, $\forall x \in I\!R^N$, $\forall w \in \partial_x j(t,x)$, with $p > 2$, $a \geq 0$ and $b \geq 0$.

Here $|\cdot|$ denotes the euclidean norm in $I\!R^N$.

In view of $(\mathcal{H}_1)$ and $(\mathcal{H}_2)$ the functional $J : L^p(0,T; I\!R^N) \to I\!R$ given by

$$J(u) = \int_0^T j(t, u(t))dt, \quad \forall u \in L^p(0,T; I\!R^N), \tag{9.46}$$

is well defined. Moreover, it is Lipschitz continuous on bounded subsets of $L^p(0,T; I\!R^N)$. Indeed, Lebourg's mean value theorem (see Theorem 1.1) and $(\mathcal{H}_2)$ yield

$$|j(t,u(t)) - j(t,v(t))| \leq (a + b|u(t) - v(t)|^{p-1})(|u(t) - v(t)|$$

a.e. in $[0,T]$ and thus using (9.46) together with Hölder inequality, we obtain

$$|J(u) - J(v)| \leq c\|u - v\|_{L^p}, \tag{9.47}$$

whenever u, $v \in L^p(0,T; I\!R^N)$ belong to a bounded subset of $L^p(0,T; I\!R^N)$. In (9.47), $c > 0$ denotes a constant depending on this bounded set.

Let H_T^1 denote the Hilbert space of $T-$periodic, absolutely continuous, $I\!R^N$ valued functions on $I\!R$ whose derivative is square integrable. The Hilbert space H_T^1 is endowed with the norm

$$\|u\|^2 = \int_0^T \left(|\dot{u}(t)|^2 + B(t)u(t) \cdot u(t)\right)dt, \quad \forall u \in H_T^1. \tag{9.48}$$

Further we make use of the properties of the matrix $B(t)$.

In order to apply a critical point argument for problem $(\mathcal{P})$ we introduce the functional $I : H_T^1 \to I\!R$,

$$I(u) = \frac{1}{2}\|u\|^2 + J(u), \quad \forall u \in H_T^1, \tag{9.49}$$

where (9.46) and (9.48) are used. According to the imbedding $H_T^1 \subset L^p(0,T; I\!R^N)$ and (9.47) we see that I in (9.49) is locally Lipschitz on H_T^1. Thus it makes sense to consider the critical points of the locally Lipschitz functional I in the sense of Chang [14], i.e. the points $u \in H_T^1$ satisfying

$$0 \in \partial I(u), \tag{9.50}$$

(see Definition 1.4), where ∂I represents the generalized gradient of Clarke [15] of I (see Definition 1.2).

In the result below we discuss the relationship between the critical points of I, the solutions of $(\mathcal{P})$ and the solutions to an associated hemivariational inequality in the sense of Panagiotopoulos [34], [35] (see Section 1.4 of Chapter 1).

Proposition 9.1 Assume that assumptions $(\mathcal{H}_1)$ and $(\mathcal{H}_2)$ are satisfied. Then

(i) the critical points of I are weak solutions of problem $(\mathcal{P})$;

(ii) any weak solution $u \in H_T^1$ of $(\mathcal{P})$ satisfies the following hemivariational inequality (HI)

$$\int_0^T \Big(\dot{u}(t) \cdot \dot{v}(t) + B(t)u(t) \cdot v(t) + j^0(t, u(t); v(t))\Big)dt \geq 0, \ \ \forall v \in H_T^1.$$

Proof. The growth condition $(\mathcal{H}_2)$ ensures the following subdifferentiation rule for the generalized gradients of the integral functional J in (9.46) (cf. Theorem 1.3)

$$\partial J(u) \subset \int_0^T \partial_x j(t, u(t))dt, \ \ \forall u \in L^p(0,T; \mathbb{R}^N). \tag{9.51}$$

Then clearly (9.49)-(9.51) imply assertion (i) of Proposition 9.1.

(ii) The fact that $u \in H_T^1$ is a weak solution of $(\mathcal{P})$ shows that

$$\int_0^T \dot{u}(t) \cdot \dot{v}(t)dt + \int_0^T B(t)u(t) \cdot v(t)dt + \int_0^T \xi(t) \cdot v(t)dt = 0, \ \ \forall v \in H_T^1.$$

Therefore

$$\int_0^T \dot{u}(t) \cdot \dot{v}(t)dt + \int_0^T B(t)u(t) \cdot v(t)dt + \int_0^T \max_{z \in \partial_x j(t,u(t))} z \cdot v(t)dt \geq 0,$$

for all $v \in H_T^1$. Hence

$$\int_0^T \dot{u}(t) \cdot \dot{v}(t)dt + \int_0^T B(t)u(t) \cdot v(t)dt + \int_0^T j^0(t, u(t); v(t))dt \geq 0,$$

for all $v \in H_T^1$, which implies that $u \in H_T^1$ solves the hemivariational inequality (HI) and the proof is complete. □

Now we deal with the verification of the Palais-Smale condition in the sense of Chang [14] (see Definition 2.1) for the locally Lipschitz functional I in (9.49). Accordingly, we have to check that any sequence $\{u_n\}_{n\geq 1} \subset H_T^1$ contains a convergent subsequence whenever $I(u_n)$ is bounded and

$$\lambda(u_n) := \inf_{w \in \partial I(u_n)} \|w\| \to 0 \text{ as } n \to +\infty. \tag{9.52}$$

Towards this we need a further assumption:

$(\mathcal{H}_3)$ there are constants $\mu > 2$ and $r > 0$ such that

$$\mu j(t,x) \geq j^0(t,x;x), \ \ \forall t \in \mathbb{R}, \ x \in \mathbb{R}^N, \ |x| \geq r.$$

In the smooth setting this formulation goes back to Ambrosetti and Rabinowitz (see Rabinowitz [41]). For nonsmooth problems, versions of $(\mathcal{H}_3)$ have been used by Adly and Goeleven [3], Chang [14], Motreanu [31], Motreanu and Naniewicz [32] (see Chapter 2 for a general exposition).

Lemma 9.3 Assume that conditions $(\mathcal{H}_1)-(\mathcal{H}_3)$ are fulfilled together with $j(\cdot,0) \in L^1(0,T;\mathbb{R}^N)$. Then the locally Lipschitz functional $I : H^1_T \to \mathbb{R}$ in (9.49) satisfies the Palais-Smale condition in the sense of Chang [14] (see Definition 2.1) .

Proof. Let $\{u_n\}_{n\geq 1}$ be a sequence of H^1_T with the properties (9.52) and

$$I(u_n) \leq M, \quad \forall n \geq 1, \tag{9.53}$$

for a constant $M \geq 0$. From (9.49) and (9.52) it follows the existence of some $w_n \in \partial I(u_n)$ such that

$$w_n \to 0 \text{ as } n \to +\infty. \tag{9.54}$$

Hence, there exists $z_n \in \partial J(u_n) \subset L^{\frac{p}{p-1}}(0,T;\mathbb{R}^N)$ such that

$$w_n = \langle u_n, \cdot\rangle + z_n \quad \to \quad 0 \text{ as } n \to +\infty.$$

Then (9.51), (9.53) and (9.54) ensure that for a sufficiently large n we can write

$$\begin{aligned} M + \frac{1}{\mu}\|u_n\|\,\|w_n\|_* &\geq I(u_n) - \frac{1}{\mu}\langle w_n, u_n\rangle \\ &\geq I(u_n) - \frac{1}{\mu}(\|u_n\|^2 + \langle z_n, u_n\rangle) \\ &\geq (\frac{1}{2} - \frac{1}{\mu})\|u_n\|^2 + \int_0^T (j(t,u_n) - \frac{1}{\mu}j^0(t,u_n;u_n))dt. \end{aligned}$$

Using assumption $(\mathcal{H}_3)$ we obtain

$$\begin{aligned} M + \frac{1}{\mu}\|u_n\|\,\|w_n\|_* &\geq (\frac{1}{2} - \frac{1}{\mu})\|u_n\|^2 + \\ \int_{|u_n(t)|\geq r} \Big(j(t,u_n(t)) &- \frac{1}{\mu}j^0(t,u_n(t);u_n(t))\Big)dt + c_0 \geq \\ &(\frac{1}{2} - \frac{1}{\mu})\|u_n\|^2 + c_0 \end{aligned} \tag{9.55}$$

with $c_0 \in \mathbb{R}$ a constant and each n large enough. To derive the last inequality of (9.55) we use the estimates

$$\begin{aligned} |j(t,x)| &\leq |j(t,x) - j(t,0)| + |j(t,0)| \\ &\leq a + b|x|^{p-1} + |j(t,0)| \end{aligned}$$

and

$$j^0(t,x;x) = \max_{z\in\partial j(t,x)} z\cdot x \leq (a + b|x|^{p-1})|x|$$

for all $t \in [0,T]$ and $x \in I\!R^N$. Taking into account that $\mu > 2$, we deduce from (9.54) and (9.55) that the sequence $\{u_n\}$ is bounded in H^1_T.

Thus we may assume that along a subsequence one has $u_n \rightharpoonup u_0$ weakly in H^1_T and therefore $u_n \to u_0$ in $L^p(0,T;I\!R^N)$ for some $u_0 \in H^1_T$. On the other hand the convergence in (9.54) yields

$$\langle u_n, v-u_n\rangle + J^0(u_n; v-u_n) \geq -\varepsilon_n\|v-u_n\|, \quad \forall v \in H^1_T, \tag{9.56}$$

provided n is sufficiently large, with $\varepsilon_n \to 0$ in $I\!R$. If we set $v = u_0$ in (9.56) we obtain that

$$\|u_n - u_0\|^2 \leq \langle u_0, u_0 - u_n\rangle + \varepsilon_n\|u_n - u_0\| + K\|u_n - u_0\|_{L^p},$$

where $K > 0$ stands for the Lipschitz constant of $J : L^p(0,T;I\!R^N) \to I\!R$ on a suitable bounded set. We conclude that $u_n \to u_0$ strongly in H^1_T, which completes the proof. □

In order to apply an argument based on Mountain Pass Theorem, we impose two new hypotheses

$$(\mathcal{H}_4)\begin{cases} (i) \ \ \liminf\limits_{x\to 0} \dfrac{j(t,x)}{|x|^2} \geq 0 \text{ uniformly with respect to } t \in [0,T]; \\ \\ (ii) \ \ j(t,0) = 0, \ \ \forall t \in [0,T] \end{cases}$$

and

$(\mathcal{H}_5)$ $\int_0^T j(t,x_0)dt < 0$ for some $x_0 \in I\!R^N$ with $|x_0| \geq r$.

The number $r > 0$ entering $(\mathcal{H}_5)$ is the one given in $(\mathcal{H}_3)$.

The main result of this section is stated below.

Theorem 9.3 Under assumptions $(\mathcal{H}_1)$-$(\mathcal{H}_5)$, problem $(\mathcal{P})$ has at least one nontrivial $T-$periodic solution. In particular, the hemivariational inequality (HI) has at least one solution.

Proof. The second assertion of Theorem 9.3 follows directly from (ii) of Proposition 9.1

To prove the first part of the statement we make use of the nonsmooth version of Mountain Pass Theorem for locally Lipschitz functionals (see Chang [14], and Corollary 2.4 for a more general result). By (9.49) and $(\mathcal{H}_4)$ (ii), it is easily verified that

$$I(0) = 0. \tag{9.57}$$

Fix an $\varepsilon > 0$. Assumption $(\mathcal{H}_4)$ (i) insures the existence of a $\delta > 0$ such that

$$j(t,x) \geq -\varepsilon|x|^2, \quad \forall t \in I\!R, \ \ |x| \leq \delta. \tag{9.58}$$

Lebourg's mean value theorem and assumption $(\mathcal{H}_2)$ enable us to write

$$\begin{aligned} j(t,x) &= j(t,x) - j(t,0) \geq -(a + b|x|^{p-1})|x| \\ &\geq -(\frac{a}{\delta^{p-1}} + b)|x|^p, \qquad \forall t \in \mathbb{R},\ |x| \geq \delta. \end{aligned} \tag{9.59}$$

Combining (9.58) and (9.59), we find

$$j(t,x) \geq -\varepsilon|x|^2 - (\frac{a}{\delta^{p-1}} + b)|x|^p,\ \forall t \in \mathbb{R},\ x \in \mathbb{R}^N. \tag{9.60}$$

Then the continuity of the embeddings $H_T^1 \subset L^2(0,T;\mathbb{R}^N)$, $H_T^1 \subset L^p(0,T;\mathbb{R}^N)$ and relations (9.46), (9.60) show that

$$\begin{aligned} J(u) &\geq -\varepsilon\|u\|_{L^2}^2 - (\frac{a}{\delta^{p-1}} + b)\|u\|_{L^p}^p \\ &\geq -C(\varepsilon + (\frac{a}{\delta^{p-1}} + b)\|u\|^{p-2})\|u\|^2,\ \ \forall u \in H_T^1 \end{aligned} \tag{9.61}$$

with $C > 0$ a constant. The assumption $p > 2$ and relation (9.61) allow us to find some $\rho > 0$ such that

$$J(u) \geq -\varepsilon A\|u\|^2,\quad \forall u \in H_T^1 \text{ with } \|u\| \leq \rho, \tag{9.62}$$

where $A > 0$ is a constant which is independent of ε. From (9.62) in conjunction with (9.59) we obtain

$$I(u) \geq (\frac{1}{2} - \varepsilon A)\|u\|^2,\quad \forall u \in H_T^1 \text{ with } \|u\| \leq \rho. \tag{9.63}$$

Take now $0 < \varepsilon < \frac{1}{2A}$. Then (9.63) gives rise to the estimate

$$I(u) \geq \alpha := (\frac{1}{2} - \varepsilon A)\rho^2,\quad \forall \|u\| = \rho. \tag{9.64}$$

The calculus with generalized gradients (see Chapter 1) implies readily the formula

$$\partial_s(s^{-\mu}j(t,sx)) = -\mu s^{-\mu-1}j(t,sx) + s^{-\mu}\partial_x j(t,sx)x, \tag{9.65}$$

for all $t \in \mathbb{R}$, $x \in \mathbb{R}^N$, $s > 0$, where the notation ∂_s has the meaning of generalized gradient of the involved locally Lipschitz function with respect to the variable s. Lebourg's mean value theorem and equality (9.65) imply that for every $s > 1$ there is a $\tau \in]1, s[$ such that for every $t \in \mathbb{R}$, $x \in \mathbb{R}^N$, and $s > 1$

$$s^{-\mu}j(t,sx) - j(t,x) \in (-\mu\tau^{-\mu-1}j(t,\tau x) + \tau^{-\mu}\partial_x j(t,\tau x)x)(s-1).$$

Consequently, we get

$$s^{-\mu}j(t,sx) - j(t,x) \leq \tau^{-\mu-1}(s-1)(j_x^0(t,\tau x;\tau x) - \mu j(t,\tau x)), \tag{9.66}$$

for all $t \in \mathbb{R}$, $x \in \mathbb{R}^N$, $s > 1$. From (9.66) combined with $(\mathcal{H}_3)$ it results

$$j(t,sx) \leq s^{\mu}j(t,x),\ \ \forall t \in \mathbb{R},\ |x| \geq r,\ s > 1. \tag{9.67}$$

By putting $x = x_0$ in (9.67), for the point x_0 with $|x_0| \geq r$ entering $(\mathcal{H}_5)$, we obtain

$$\begin{aligned} I(sx_0) &= \frac{1}{2}s^2\|x_0\|^2 + \int_0^T j(t, sx_0)dt \\ &\leq \frac{1}{2}s^2\|x_0\|^2 + s^\mu \int_0^T j(t, x_0)dt, \quad \forall s > 1. \end{aligned}$$

Since $\mu > 2$, we derive from $(\mathcal{H}_5)$ and the relation above that

$$\lim_{s \to +\infty} I(sx_0) = -\infty.$$

Thus we can determine a number $s > 0$ to satisfy

$$I(sx_0) \leq 0 \text{ with } \|sx_0\| > \rho. \tag{9.68}$$

Lemma 9.3 and properties (9.57), (9.64), (9.68) imply that the requirements of the Mountain Pass Theorem are satisfied for the locally Lipschitz functional $I : H_T^1 \to I\!R$ given by (9.49). We conclude that a nontrivial critical point $u \in H_T^1$ of I can be found. Proposition 9.1 completes the proof. □

In order to obtain multiplicity information regarding $T-$periodic solutions to problem $(\mathcal{P})$ we strengthen assumption $(\mathcal{H}_5)$ and add a symmetry condition. Namely, we assume that

$$j(t, x) \leq -d, \ \ \forall t \in [0, T], \ \ |x| = r \text{ for a constant } d > 0; \tag{$\mathcal{H}_5'$}$$

and

$$j(t, -x) = j(t, x), \ \ \forall t \in [0, T], \ \ x \in I\!R^N. \tag{$\mathcal{H}_6$}$$

Our multiplicity result in solving problem $(\mathcal{P})$ is formulated as follows:

Theorem 9.4 Assume that assumptions $(\mathcal{H}_1)$-$(\mathcal{H}_4)$, $(\mathcal{H}_5')$ and $(\mathcal{H}_6)$ hold. Then problem $(\mathcal{P})$ possesses infinitely many distinct pairs $(u, -u)$ of weak solutions.

Proof. Let E be any linear subspace of H_T^1 with a finite dimension m. We claim that

$$\lim_{\substack{u \in E \\ \|u\| \to +\infty}} I(u) = -\infty. \tag{9.69}$$

Since E is a finite dimensional subspace, there is a constant $K > 0$ such that

$$\|u\|_{L^\mu} \geq K\|u\|, \ \ \forall u \in E. \tag{9.70}$$

Inequality (9.70) holds due to the equivalence of norms on a finite dimensional subspace. We have also

$$\|u\|_{L^\infty} \geq K'\|u\|, \ \ \forall u \in E, \ \ (K' > 0),$$

and thus for $\|u\|$ large enough, $\text{mes}\{t \in [0, T] \ : \ u(t) > r\} > 0$ and from (9.67), we obtain

$$\int_0^T j(t,u(t))dt \le c + \int_{|u(t)|>r} r^{-\mu}|u(t)|^\mu j(t, \frac{ru(t)}{|u(t)|})dt \tag{9.71}$$

with a constant $c > 0$. By $(\mathcal{H}_5')$ and (9.71), we arrive at

$$\begin{aligned} I(u) &= \frac{1}{2}\|u\|^2 + \int_0^T j(t,u(t))dt \\ &\le \frac{1}{2}\|u\|^2 - r^{-\mu} d\|u\|_{L^\mu}^\mu + c, \ \ \forall u \in H_T^1. \end{aligned} \tag{9.72}$$

Then by (9.70) and (9.72), we obtain

$$I(u) \le \frac{1}{2}\|u\|^2 - d_0\|u\|^\mu + c, \ \ \forall u \in H_T^1 \tag{9.73}$$

where $d_0 > 0$, and $c \in \mathbb{R}$ is a constant. Taking into account estimate (9.73), conditions $d_0 > 0$ and $\mu > 2$, property (9.69) is established.

Now, let us recall that the locally Lipschitz functional $I : H_T^1 \to \mathbb{R}$ fulfills the Palais-Smale condition as shown in Lemma 9.3 and the properties stated by (9.57), (9.64), (9.69) and $(\mathcal{H}_6)$. So we are in position to apply to I the symmetric version of Mountain Pass Theorem for locally Lipschitz functionals (see Corollary 2.7, and Rabinowitz [41] for the smooth case). Hence, there exists $m :=$ dim E – codim H_T^1 pairs $(u, -u)$ of symmetric solutions to problem $(\mathcal{P})$. As $m \ge 1$ is an arbitrary positive integer, the desired conclusion follows. □

With the foregoing data we turn to the homoclinic problem in $\mathbb{R}^N$ which corresponds to problem $(\mathcal{P})$, namely

$$(HP) \begin{cases} -\ddot{u}(t) + B(t)u(t) + \xi(t) = 0, \\ \xi(t) \in \partial_x j(t,u(t)) \quad \text{a.e.} \ \ t \in \mathbb{R} \\ u(t) \to 0 \quad \text{as} \quad |t| \to +\infty \\ \dot{u}(t) \to 0 \quad \text{as} \quad |t| \to +\infty \end{cases}$$

The first equation in (HP) is understood in the weak sense, i.e.

$$\int_{\mathbb{R}} \Big(\dot{u}(t)\cdot\dot{\varphi}(t) + B(t)u(t)\cdot\varphi(t) + \xi(t)\cdot\varphi(t)\Big)dt = 0 \tag{9.74}$$

for all $\varphi \in C_0^\infty(\mathbb{R}, \mathbb{R}^N)$, where the solution u belongs to $H^1(\mathbb{R}, \mathbb{R}^N)$ and some $\xi \in L^1(\mathbb{R}, \mathbb{R}^N)$. It is well known that $H^1(\mathbb{R}, \mathbb{R}^r) = H_0^1(\mathbb{R}, \mathbb{R}^N)$, so (9.74) formulates a right concept of weak solution. The solutions to problem (HP) are called homoclinics. The existence of nontrivial homoclinics u, i.e. $u(t) \not\equiv 0$, is given by the next result.

Theorem 9.5 Assume that conditions $(\mathcal{H}_1)$- $(\mathcal{H}_5)$ are satisfied together with the following additional property: for all sequences $\{s_k\} \subset \mathbb{R}$, $\{x_k\} \subset \mathbb{R}^N$, $\{y_k\} \subset \mathbb{R}^N$ such that $|s_k| \le kT$, $y_k \cdot x_k < 0$, $x_k \to 0$ and $y_k \in \partial_x j(s_k, x_k)$ one has

$$\frac{y_k \cdot x_k}{|x_k|^2} \to 0 \text{ as } k \to +\infty. \tag{9.75}$$

Then problem (HP) has at least one nontrivial homoclinic solution.

Proof. For any positive integer $k \geq 1$, let H^1_{2kT} denote the Hilbert space of $2kT$–periodic functions from $\mathbb{R}$ to $\mathbb{R}^N$ which are absolutely continuous with square integrable derivative. The norm of H^1_{2kT} is taken to be

$$\|u\|_k^2 := \int_{-kT}^{kT} \left(|\dot{u}(t)|^2 + B(t)u(t) \cdot u(t)\right) dt. \tag{9.76}$$

Let us relate problem (HP), for each $k \geq 1$, to the functional $I_k : H^1_{2kT} \to \mathbb{R}$ defined by

$$I_k(u) := \frac{1}{2}\|u\|_k^2 + \int_{-kT}^{kT} j(t, u(t))dt. \tag{9.77}$$

As shown for the functional I before, we can prove that the functional I_k in (9.77) is locally Lipschitz and its critical points solve in the weak sense the $2kT$–periodic problem

$$(P_k) \begin{cases} -\ddot{u}(t) + B(t)u(t) + \xi(t) = 0, \\ \xi(t) \in \partial_x j(t, u(t)) \quad \text{a.e. on } [-kT, kT], \\ u(kT) - u(-kT) = \dot{u}(kT) - \dot{u}(-kT) = 0. \end{cases}$$

In view of the assumptions $(\mathcal{H}_1) - (\mathcal{H}_5)$ it is clear that an analogue of Theorem 9.3, with the interval $[-kT, kT]$ in place of $[-T, T]$, ensures the existence of a nontrivial critical point $u_k \in H^1_{2kT}$ of I_k in (9.77), and thus a nontrivial weak solution of problem (P_k). Since each u_k is obtained by means of the Mountain Pass Theorem, the corresponding critical value $c_k := I_k(u_k)$ admits the following minimax description

$$c_k := I_k(u_k) = \inf_{g \in \Gamma_k} \max_{s \in [0,1]} I_k(g(s)), \tag{9.78}$$

where

$$\Gamma_k = \{g \in C([0,1], H^1_{2kT}) \ : \ g(0) = 0, \ g(1) = e_k\} \tag{9.79}$$

Here $e_k \in H^1_{2kT}$ is constructed as it follows: By $(\mathcal{H}_5)$ we can easily construct a function $\varphi \in H^1_{2T} \setminus \{0\}$ such that $I_1(\varphi) < 0$ and $\varphi(T) = \varphi(-T) = 0$. Then $e_k \in H^1_{2kT} \setminus \{0\}$ is taken to be $e_k(t) = \varphi(t)$ on $[-T, T]$ and $e_k(t) = 0$ otherwise. Now from the minimax characterization (9.78)-(9.79), we derive the estimate

$$c_k \leq \max_{0 \leq s \leq 1} I_k(se_k) = \max_{0 \leq s \leq 1} I_1(s\varphi) = M, \quad \forall k \geq 1 \tag{9.80}$$

with $M > 0$ independent of k. Since $0 \in \partial I_k(u_k)$, assumption $(\mathcal{H}_3)$, estimate (9.80) and formula (9.61) lead to

$$\begin{aligned} M &\geq c_k = I_k(u_k) \geq I_k(u_k) - \frac{1}{2} I_k^0(u_k; u_k) \\ &\geq \frac{1}{2} \int_{-kT}^{kT} \left(2j(t, u_k(t)) - j^0(t, u_k(t); u_k(t))\right) dt \\ &= \frac{1}{2} \int_{-kT}^{kT} \left(\mu j(t, u_k(t)) - j^0(t, u_k(t); u_k(t))\right) dt \end{aligned}$$

$$+(1-\frac{\mu}{2})\int_{-kT}^{kT} j(t,u_k(t))dt$$
$$\geq (1-\frac{\mu}{2})\int_{-kT}^{kT} j(t,u_k(t))dt + c \tag{9.81}$$

for each integer $k \geq 1$ and for a constant $c > 0$. Then (9.76), (9.77), (9.78) and (9.81) imply the a priori estimate

$$\|u_k\|_k^2 = 2c_k - 2\int_{-kT}^{kT} j(t,u_k(t))dt \leq \tag{9.82}$$

$$2M + \frac{4}{\mu-2}M - \frac{4}{\mu-2}c \leq \bar{c}, \ \ \forall k \geq 1$$

where $\bar{c}$ is a positive constant which is independent of k. Let us now recall the following relation from Rabinowitz [40] (see also Yanheng and Girardi [42])

$$\|v\|_{L^\infty([-kT,kT],\mathbb{R}^N)} \leq a_0\|v\|_k, \ \ \forall v \in H^1_{2kT} \tag{9.83}$$

with $a_0 > 0$ a constant. Combining (9.82) and (9.83), we get

$$\|u_k\|_{L^\infty([-kT,kT],\mathbb{R}^N)} \leq a_0\sqrt{\bar{c}} =: \bar{a}, \ \ \forall k \geq 1. \tag{9.84}$$

The second relation of (P_k), $(\mathcal{H}_2)$ and (9.84) show that $\xi(t) = \xi_k(t)$ satisfies

$$|\xi_k(t)| \leq a + b|u_k(t)|^{p-1} \leq a + b\bar{a}^{p-1} =: \beta, \ \ \forall t \in [-kT,kT]. \tag{9.85}$$

By (9.84) and (9.85) we find subsequences of $\{u_k\}$ and $\{\xi_k\}$, denoted again by $\{u_k\}$ and $\{\xi_k\}$ respectively, such that

$$u_k(t) \to u(t) \text{ and } \xi_k(t) \to \xi(t) \text{ for a.e. } t \in \mathbb{R} \tag{9.86}$$

Properties (9.84)-(9.86) and Lebesgue Dominated Convergence Theorem tell us that $u \in L^\infty_{\text{loc}}(\mathbb{R},\mathbb{R}^N)$ and $\xi \in L^{\frac{p}{p-1}}_{\text{loc}}(\mathbb{R},\mathbb{R}^N)$. From (P_k) we know that

$$\xi_k(t) \in \partial j(t,u_k(t)) \text{ for a.e. } t \in [-kT,kT].$$

Then (9.86) and the upper semi-continuity of the generalized gradient (see Proposition 1.4) ensure

$$\xi(t) \in \partial_x j(t,u(t)) \text{ a.e. } t \in \mathbb{R}. \tag{9.87}$$

Estimate (9.82) and Fatou's Lemma entail

$$b^{-1}\bar{c} \geq \liminf_{k\to+\infty}\int_{-s}^{s}|u_k(t)|^2dt \geq \int_{-s}^{s}|u(t)|^2dt, \ \ \forall s > 0,$$

which shows that $u \in L^2(\mathbb{R},\mathbb{R}^N)$. For each $\varphi \in C_0^\infty(\mathbb{R},\mathbb{R}^N)$, there exists $k_0 > 0$ such that supp $\varphi \subset [-kT,kT]$, for all $k \geq k_0$. We have

$$|\int_{\mathbb{R}} u_k(t)\varphi'(t)dt| = |\int_{kT}^{kT} u_k'(t)\varphi(t)dt| \leq \bar{c}\|\varphi\|_{L^2}. \tag{9.88}$$

Moreover

$$u_k(t)\varphi'(t) \to u(t)\varphi'(t), \quad \text{a.e. } t \in \mathbb{R}$$

and

$$\begin{aligned} |u_k(t)\varphi'(t)| &\leq \bar{a}|\varphi'(t)| \text{ a.e. } t \in [-kT, kT], \\ |u_k(t)\varphi'(t)| &= 0, \text{ a.e. } t \in \mathbb{R} \setminus [-kT, kT]. \end{aligned}$$

The function

$$h(t) = \begin{cases} \bar{a}|\varphi'(t)| & \text{if} \qquad t \in [-kT, kT] \\ 0 & \text{otherwise} \end{cases}$$

belongs to $L^2(\mathbb{R})$. We may use Lebesgue Theorem and take the limit as $k \to +\infty$ in (9.88) to obtain

$$|\int_{\mathbb{R}} u(t)\varphi'(t)dt| \leq \bar{c}\|\varphi\|_{L^2}.$$

Then using Proposition IX.3, p. 153 in Brézis [11], we obtain $u \in H^1(\mathbb{R}, \mathbb{R}^N)$. Using (9.85), the second convergence in (9.86) and property (13.9.19) in Dieudonné [18], we may conclude that ξ is measurable. By $(\mathcal{H}_2)$ and (9.87) one obtains that $\xi \in L^{\frac{p}{p-1}}(\mathbb{R}, \mathbb{R}^N)$. Moreover, for every $\varphi \in C_0^\infty(\mathbb{R}, \mathbb{R}^N)$ as in (9.88), we derive from (9.85) and (9.86) the convergence

$$\int_{\mathbb{R}} \xi_k(t)\varphi(t)dt \to \int_{\mathbb{R}} \xi(t)\varphi(t)dt \text{ as } k \to +\infty. \tag{9.89}$$

Similarly,

$$\int_{\mathbb{R}} B(t)u_k(t)\varphi(t)dt \to \int_{\mathbb{R}} B(t)u(t)\varphi(t)dt \text{ as } k \to +\infty \tag{9.90}$$

and

$$\begin{aligned} \int_{\mathbb{R}} u_k'(t)\varphi'(t)dt = -\int_{\mathbb{R}} u_k(t)\varphi''(t)dt &\longrightarrow \\ \int_{\mathbb{R}} u(t)\varphi''(t)dt = \int_{\mathbb{R}} u'(t)\varphi'(t)dt.& \end{aligned} \tag{9.91}$$

Fix now again $\varphi \in C_0^\infty(\mathbb{R}, \mathbb{R}^N)$. Since the first equation in (P_k) holds in the weak sense, it is permitted to write

$$\int_{\mathbb{R}} \Big(\dot{u}_k(t) \cdot \varphi'(t) + B(t)u_k(t) \cdot \varphi(t) + \xi_k(t) \cdot \varphi(t)\Big)dt = 0, \tag{9.92}$$

for all $k \geq 1$, supp $\varphi \subset [-kT, kT]$, where (9.88) has been also used. Then (9.89)-(9.92) show that u is a weak solution of the first equation in (HP).

The Hilbert space $H^1(\mathbb{R}, \mathbb{R}^N)$ is continuously embedded in the space of continuous functions from $\mathbb{R}$ to $\mathbb{R}^N$ converging to 0 at infinity. Since $u \in H^1(\mathbb{R}, \mathbb{R}^N)$, it follows that

$$u(t) \to 0 \text{ as } |t| \to \infty. \tag{9.93}$$

Knowing that $u \in H^1(\mathbb{R}, \mathbb{R}^N)$ is a weak solution of the first equation in (HP) which corresponds to some $\xi \in L^{\frac{p}{p-1}}(\mathbb{R}, \mathbb{R}^N)$, we have

$$\int_{\mathbb{R}} \Big(\dot{u}(t) \cdot \varphi'(t) + B(t)u(t) \cdot \varphi(t) + \xi(t) \cdot \varphi(t)\Big)dt = 0, \quad \forall \varphi \in C_0^\infty(\mathbb{R}, \mathbb{R}^N).$$

Therefore $\dot{u}$ has a weak derivative

$$\ddot{u} = B(\cdot)u + \xi \in L^{\frac{p}{p-1}}(\mathbb{R}, \mathbb{R}^N) \tag{9.94}$$

and

$$-\ddot{u}(t) + B(t)u(t) + \xi(t) = 0 \text{ a.e. } t \in \mathbb{R}. \tag{9.95}$$

Then by (9.94), (9.95) and the the $T-$periodicity of B, we obtain

$$\begin{aligned}
|\dot{u}(t)| &\leq |\dot{u}(\tau)| + |\int_\tau^t \ddot{u}(s)ds| \\
&\leq |\dot{u}(\tau)| + K|\int_\tau^t |u(s)|ds| + |\int_\tau^t |\xi(s)|ds|, \quad \forall t, \ \tau \in \mathbb{R},
\end{aligned}$$

for a constant $K > 0$. Integrating the inequality above over $[t - \frac{1}{2}, t + \frac{1}{2}]$ and then using Fubini Theorem and Hölder inequality, we see that

$$\begin{aligned}
|\dot{u}(t)| \leq \ & K_1\Big[\Big(\int_{t-\frac{1}{2}}^{t+\frac{1}{2}} |u(\tau)|^2 d\tau\Big)^{1/2} + \Big(\int_{t-\frac{1}{2}}^{t+\frac{1}{2}} |\dot{u}(\tau)|^2 d\tau\Big)^{1/2} + \\
& + \Big(\int_{t-\frac{1}{2}}^{t+\frac{1}{2}} |\xi(\tau)|^{\frac{p}{p-1}} d\tau\Big)^{\frac{p-1}{p}}\Big], \qquad \forall t \in \mathbb{R}
\end{aligned} \tag{9.96}$$

with a new constant $K_1 > 0$. Using the regularity properties that $u \in H^1(\mathbb{R}, \mathbb{R}^N)$ and $\xi \in L^{\frac{p}{p-1}}(\mathbb{R}, \mathbb{R}^N)$, and passing to the limit in (9.96) as $|t| \to +\infty$, we conclude that

$$\dot{u}(t) \to 0 \text{ as } |t| \to \infty. \tag{9.97}$$

Then (9.87), (9.93), (9.94) and (9.97) express that $u \in H^1(\mathbb{R}, \mathbb{R}^N)$ solves the homoclinic problem (HP).

It remains to establish that the homoclinic orbit u is nontrivial, i.e. $u(t) \not\equiv 0$. Our argument relies on the approximate problem (P_k). Since u_k solves weakly (P_k), we know that

$$\begin{aligned}
b\|u_k\|_{L^2}^2 &\leq \|u_k\|_k^2 = -\int_{-kT}^{kT} \xi_k(t) \cdot u_k(t)dt \\
&\leq -\min_{t \in [-kT, kT]} \frac{\xi_k(t) \cdot u_k(t)}{|u_k(t)|^2} \|u_k\|_{L^2}^2.
\end{aligned}$$

By Theorem 9.3, we have $\|u_k\|_k \neq 0$, so the computation above implies

$$\min_{t \in [-kT, kT]} \frac{\xi_k(t) \cdot u_k(t)}{|u_k(t)|^2} \leq -b < 0, \ \ \forall k \geq 1. \tag{9.98}$$

Thus (9.98) produces a sequence $s_k \in [-kT, kT]$ such that

$$\frac{\xi_k(s_k) \cdot u_k(s_k)}{|u_k(s_k)|^2}$$

is bounded away from zero. On the other hand if $u \equiv 0 \in H^1(I\!R, I\!R^N)$, we would have $u_k \to 0$ in $L^\infty(I\!R, I\!R^N)$ as $k \to \infty$. Hence one obtains that $u_k(s_k) \to 0$ as $k \to \infty$ which allows us to apply assumption (9.75) for $x_k := u_k(s_k)$ and $y_k := \xi_k(s_k)$. Thus we arrive at a contradiction. This completes the proof of the theorem. □

Remark 9.3 Assumption (9.75) has been used only to show that the obtained homoclinic solution is nontrivial.

9.3 Periodic solutions

In this section we discuss the existence of critical orbits for locally Lipschitz functionals invariant with respect to a discrete group, and some applications to multiple eigensolutions for dynamic hemivariational inequalities involving symmetric functionals. Notice that the problems with symmetries treated here differ essentially from the ones of Chapter 2, where the action of a compact group has been considered. This fact requires to reconsider the framework of Chapter 2 in the present setting of a discrete group of symmetries. In this part of the chapter we follow Goeleven, Motreanu and Panagiotopoulos [23] relying on Chang [14] and the approach exposed by J. Mawhin and M. Willem in [27] (see also Rabinowitz [41]). As already mentioned the given applications concern the multiple eigensolutions for symmetric functionals and nonsmooth dynamic problems. One obtains extensions of the results of Mawhin and Willem [27] for simple pendulum problems and of Fournier and Willem [19] for double pendulum problems. The class of dynamic hemivariational inequalities described here can also be used to study boundary value problems for some forced pendulum type equations involving discontinuous nonlinearities (see Adly and Goeleven [3] for more details). Concerning the multiple eigensolutions for an eigenvalue problem for a hemivariational inequality, the buckling problems in Mechanics are typical examples which illustrate this issue.

We start by studying the orbits of critical points. Let G be a discrete subgroup of a Banach space X and let $\pi : X \to X/G$ be the canonical surjection. A subset C of X is said to be G-invariant if $\pi^{-1}(\pi(C)) = C$. A function $f : X \to I\!R$ is called G- invariant if

$$f(u + g) = f(u), \ \forall u \in X, \ \forall g \in G.$$

Lemma 9.4 Assume that the function $f : X \to I\!R$ is G-invariant and locally Lipschitz. Then one has

$$\partial f(x) = \partial f(x+g),\ \forall x \in X,\ \forall g \in G.$$

In particular, the set K of critical points of f, i.e.,

$$K = \{x \in X : \ 0 \in \partial f(x)\}$$

is G-invariant.

Proof. The result is derived from the equivalences below

$$w \in \partial f(x) \Leftrightarrow$$

$$f^0(x;v) = \limsup_{\substack{u \to x \\ \lambda \downarrow 0}} \frac{1}{\lambda}(f(u+\lambda v) - f(u)) \ \geq \ \langle w, v\rangle, \forall v \in X \Leftrightarrow$$

$$f^0(x+g;v) = \limsup_{\substack{u \to x \\ \lambda \downarrow 0}} \frac{1}{\lambda}(f(u+g+\lambda v) - f(u+g)) \geq \langle w, v\rangle,\ \forall v \in X$$

$$\Leftrightarrow w \in \partial f(x+g).$$

□

We recall from Chapter 2 some terminology for the critical point theory in the locally Lipschitz setting. For a locally Lipschitz function $f : X \to \mathbb{R}$ and any $c \in \mathbb{R}$. we use the notations

$$f_c := \{x \in X : \ f(x) \leq c\},$$

$$K_c := \{x \in X : \ 0 \in \partial f(x) \text{ and } \ f(x) = c\} = K \cap f_c.$$

Any element $x \in K$ is called a critical point of f. If the function f is G-invariant, Lemma 9.4 reveals that, for each $x \in K$, $x + G = \pi^{-1}(\pi(x))$ is a set of critical points of f which is called a critical orbit of f. The function on X

$$x \to \lambda(x) := \min_{w \in \partial f(x)} \|w\|_{X^*}$$

is known to be lower semicontinuous (see Chang [14] and Chapter 1). Lemma 9.4 gives rise to the following

Lemma 9.5 Under the assumptions of Lemma 9.4, the function λ is G-invariant.

We shall use an appropriate compactness condition of Palais-Smale type (see Definition 2.1). A G-invariant, locally Lipschitz function $f : X \to \mathbb{R}$ is said to satisfy the $(PS)_{c,G}$-condition if, for every sequence $\{u_k\}_{k\geq 1}$ in X such that $f(u_k) \to c$ and $\lambda(u_k) \to 0$, there exists $v_k \in \pi^{-1}(\pi(u_k))$ such that $\{v_k\}_{k\geq 1}$ contains a convergent sequence in X. If the $(PS)_{c,G}$-condition is satisfied for all $c \in \mathbb{R}$, we simply say that f verifies the $(PS)_G$-condition. The usual Palais-Smale condition is obtained when $G = \{0\}$ (see Definition 2.1).

For a later use we review the notion of Ljusternik- Schnirelman category. More details can be found, e.g., in Ambrosetti [6], Browder [12], Mawhin and

Willem [27]. A subset A of a topological space Y is contractible in Y if there exists a homotopy $h \in C([0,1] \times A, Y)$ and a point $y \in Y$ such that $h(0,u) = u$ and $h(1,u) = y$ for all $u \in A$. We say that a subset A of a topological space Y has the category k in Y, if k is the smallest positive integer such that A can be covered by k closed and contractible sets in Y. If no such k exists, we say that A is of category $+\infty$. The category of A in Y is denoted by $\mathrm{cat}_Y(A)$. We set $\mathrm{cat}_Y(\emptyset) = 0$. We summarize now some basic properties of this notion.

Proposition 9.2 Let $A, B \subset Y$.

(i) If $A \subset B$, then $\mathrm{cat}_Y(A) \le \mathrm{cat}_Y(B)$.

(ii) $\mathrm{cat}_Y(A \cup B) \le \mathrm{cat}_Y(A) + \mathrm{cat}_Y(B)$.

(iii) If A is closed and $B = \eta(1, A)$, where $\eta \in C([0,1] \times A, Y)$ is such that $\eta(0,u) = u$, $\forall u \in A$, then $\mathrm{cat}_Y(A) \le \mathrm{cat}_Y(B)$. □

We recall some additional notations. By $B_\delta(x)$ we denote the closed ball in the Banach space X with center $x \in X$ and radius $\delta > 0$. For a subset C of X, $N_\delta(C)$ denotes the open δ -neighborhood of C in X.

Lemma 9.6 Suppose that $f : X \to \mathbb{R}$ is a locally Lipschitz and G-invariant functional on the Banach space X satisfying the $(PS)_{c,G}$-condition with a given $c \in \mathbb{R}$. Then, for every $\delta > 0$, there exist constants $\bar{\varepsilon} > 0$, $b > 0$ and a locally Lipschitz X-valued mapping v on

$$M(f, \bar{\varepsilon}, \delta, c) := f^{-1}([c - \bar{\varepsilon}, c + \bar{\varepsilon}]) \backslash N_\delta(K_c)$$

provided

(i) $\|v(x)\| \le 1$,

(ii) $\langle x^*, v(x) \rangle > \frac{b}{2}$,

(iii) $v(x+g) = v(x)$,

for all $x \in M(f, \bar{\varepsilon}, \delta, c)$, $x^* \subset \partial f(x)$ and $g \in G$.

Proof. Fix $\delta > 0$. The $(PS)_{c,G}$-condition implies that there exist $b > 0$, $\bar{\varepsilon} > 0$ such that

$$\lambda(x) \ge b, \ \forall\, x \in M(f, \bar{\varepsilon}, \delta, c). \tag{9.99}$$

In order to check (9.99), suppose the contrary. Then there exists a sequence $\{x_n\}_{n \ge 1}$ in $X \backslash N_\delta(K_c)$ such that $f(x_n) \to c$ and $\lambda(x_n) \to 0$ as $n \to \infty$. By $(PS)_{c,G}$-condition we can assume, passing if necessary to a subsequence, that there is a sequence $\{y_n\}_{n \ge 1}$ in X and an element $x \in X$ with the properties

$$\pi(y_n) = \pi(x_n), \ \forall n \ge 1, \tag{9.100}$$

$$y_n \to x \ in \ X \ as \ n \to \infty. \tag{9.101}$$

Equality (9.100) implies that

$$y_n = x_n + g_n \text{ for some } g_n \in G, \ \forall n \geq 1.$$

Since $x_n \notin N_\delta(K_c)$, Lemma 9.4 yields

$$\text{dist }(y_n, K_c) = \text{ dist }(x_n + g_n, K_c) = \text{ dist }(x_n, K_c) \geq \delta.$$

Taking into account (9.101) we deduce that $x \notin N_\delta(K_c)$. The lower semicontinuity of λ and Lemma 9.5 show that

$$\lambda(x) \leq \liminf_{n\to\infty} \lambda(y_n) = \liminf_{n\to\infty} \lambda(x_n) = 0,$$

so $0 \in \partial f(x)$. In addition, we obtain from (9.101) that $f(x) = c$. It follows that $x \in K_c$. The contradiction establishes our claim in (9.99). Given $x_0 \in M(f, \bar{\varepsilon}, \delta, c)$ there exists $w_0 \in \partial f(x_0)$ such that $\|w_0\|_{X^*} = \lambda(x_0)$. The set $\partial f(x_0)$ is nonempty, w^*-compact and convex and the open ball $\overset{o}{B}_*(0, \|w_0\|_{X^*})$ in X^* is nonempty, convex and

$$\partial f(x_0) \cap \overset{o}{B}_* (0, \|w_0\|_{X^*}) = \emptyset.$$

According to the Eberlein Separation Theorem there exists $h_0 \in X$ such that $\|h_0\| = 1$ and

$$\langle x^*, h_0 \rangle \ \geq \ \langle w, h_0 \rangle, \ \forall \ w \in B_*(0, \|w_0\|_{X^*}), \ \forall x^* \in \partial f(x_0). \tag{9.102}$$

The Hahn-Banach Theorem implies

$$\max_{w \in B_*(0, \|w_0\|_{X^*})} \langle w, h_0 \rangle = \|w_0\|_{X^*} \|h_0\| = \|w_0\|_{X^*}.$$

Then, by (9.99) and (9.102), we obtain

$$\langle x^*, h_0 \rangle \geq \|w_0\|_{X^*} = \lambda(x_0) \geq b, \ \forall x^* \in \partial f(x_0).$$

The upper semicontinuity of ∂f at x_0 as stated in part (iv) of Proposition 1.4 ensures that for any $\varepsilon > 0$ there is a $\eta_0 = \eta(x_0) > 0$ such that to each $x^* \in \partial f(x)$ with $\|x - x_0\| < \eta_0$ a $z_0 \in \partial f(x_0)$ corresponds satisfying

$$|\langle x^* - z_0, h_0 \rangle| < \varepsilon.$$

It turns out that

$$\langle x^*, h_0 \rangle = \langle x^* - z_0, h_0 \rangle + \langle z_0, h_0 \rangle \geq b - \varepsilon,$$

and thus, taking a small ε,

$$\langle x^*, h_0 \rangle > \frac{b}{2}, \ \forall x^* \in \partial f(x), \ \forall x \in \overset{o}{B} (x_0, \eta_0). \tag{9.103}$$

Letting x_0 run in $M(f, \bar{\varepsilon}, \delta, c)$, we consider the open covering

$$\mathcal{N} = \{\overset{o}{B}(x_0, \eta(x_0)) : x_0 \in M(f, \bar{\varepsilon}, \delta, c)\}$$

of $M(f, \bar{\varepsilon}, \delta, c)$. The assumption upon G to be discrete assures that for each $u \in X$ we can find an open neighborhood A_u of u in X so that each orbit $\pi(x) = x + G$ intersects A_u at finitely many points whose number is locally constant with respect to x. We form the open covering

$$\mathcal{A} = \{A_u : u \in M(f, \bar{\varepsilon}, \delta, c)\}$$

of $M(f, \bar{\varepsilon}, \delta, c)$. Then there exists an open covering $\{U_j\}_{j \in J}$ of $M(f, \bar{\varepsilon}, \delta, c)$ which is locally finite and finer than $\mathcal{N}$ and $\mathcal{A}$. We assign to each $j \in J$ an $x \in M(f, \bar{\varepsilon}, \delta, c)$ with

$$U_j \subset \overset{o}{B}(x_j, \eta(x_j)),$$

and let us define

$$\varphi_j(x) = \frac{\sum_{g \in G} \operatorname{dist}(x + g, X \backslash U_j)}{\sum_{i \in J} \sum_{g \in G} \operatorname{dist}(x + g, X \backslash U_i)} \quad \text{if } x \in \bigcup_{i \in J} U_i.$$

We set

$$v(x) := \sum_{j \in J} \varphi_j(x) h_j, \ \forall x \in \bigcup_{j \in J} U_j,$$

where h_j plays the same role for x_j as h_0 for x_0 previously. It follows that v is defined and locally Lipschitz. We see that

$$\|v(x)\| \leq \sum_{j \in J} \varphi_j(x) \|h_j\| = \sum_{j \in J} \varphi_j(x) = 1$$

and, by (9.103),

$$\langle x^*, v(x) \rangle = \sum_{j \in J} \varphi_j(x) \langle x^*, h_j \rangle > \frac{b}{2},$$

for all $x \in \cup_{j \in J} U_j$ and $x^* \in \partial f(x)$. Additionally, one has

$$v(x + g) - \sum_{j \in J} \varphi_j(x + g) h_j - \sum_{j \in J} \varphi_j(x) h_j,$$

for all $x \in M(f, \bar{\varepsilon}, \delta, c)$ and $g \in G$ observing that $x + g$ belongs to $M(f, \bar{\varepsilon}, \delta, c)$. Assertions (i)-(iii) are thus established. □

We proceed to the construction of a suitable equivariant deformation following the argument in Chang [14] (see Chapter 2 for a general equivariant construction in the case of the action of a compact group). Given $\varepsilon \in (0, \bar{\varepsilon})$ choose locally Lipschitz functions $\varphi : X \to [0, 1]$ and $\psi : X \to [0, 1]$ satisfying

$$\varphi(x) = 1 \text{ if } x \in f^{-1}([c - \varepsilon, c + \varepsilon]),$$

$$\varphi(x) = 0 \text{ if } x \in f^{-1}([c - \bar{\varepsilon}, c + \bar{\varepsilon}])$$

and

$$\psi(x) = 1 \text{ if } x \in N_{4\delta}(K_c),$$
$$\psi(x) = 0 \text{ if } x \in N_{2\delta}(K_c).$$

Then we introduce $V : X \to X$ by setting

$$V(x) = \varphi(x)\psi(x)v(x),\ x \in X,$$

with the $v(x)$ of Lemma 9.6. Due to the choice of functions φ, ψ, function V is well defined, locally Lipschitz and bounded. Consequently, the Cauchy problem

$$\frac{d}{dt}\eta(t,x) = -V(\eta(t,x)), \tag{9.104}$$

$$\eta(0,x) = x$$

determines a unique solution $\eta : [0,1] \times X \to X$.

Lemma 9.7 The global flow $\eta : [0,1] \times X \to X$ of (9.104) fulfills the requirements

(i) $\|\eta(t,x) - x\| \leq t,\ \forall (t,x) \in [0,1] \times X$;

(ii) $f(x) - f(\eta(t,x)) > \frac{b}{2}t$ if $\eta(s,x) \in M(f,\bar{\varepsilon},\delta,c)$ for $s \in [0,1]$;

(iii)$\eta(t,x+g) = \eta(t,x) + g,\ \forall (t,x) \in [0,1] \times X,\ \forall g \in G$.

Proof. Property (i) follows directly from (9.104) and the boundedness of V. To justify (ii) we note that Lebourg's mean value theorem (see Theorem 1.1) and (ii) of Lemma 9.6 imply

$$f(x) - f(\eta(t,x)) = -\int_0^t \frac{d}{ds} f(\eta(s,x))ds > \frac{b}{2}t$$

for t and x as required in (ii). The equivariant property (iii) holds because we obtain from (iii) of Lemma 9.6, that

$$\frac{d}{dt}(\eta(t,x)+g) = \frac{d}{dt}\eta(t,x) = -V(\eta(t,x)) = -V(\eta(t,x)+g)$$

and

$$\eta(0,x) + g = x + g,$$

whenever $g \in G$. □

We are now in a position to state the following deformation result.

Theorem 9.6 (Equivariant Deformation Lemma). Let f be a locally Lipschitz function on a real Banach space X which is invariant with respect to a discrete subgroup G of X and satisfies the $(PS)_{c,G}$-condition for a fixed $c \in \mathbb{R}$. Then, for any neighborhood $\mathcal{N}$ of K_c and $\varepsilon_0 > 0$ there exist $\varepsilon \in (0,\varepsilon_0)$ and a homeomorphism $\eta_0 : X \to X$ such that

(i) $\eta_0(x) = x, \quad \forall x \notin f^{-1}([c-\varepsilon_0, c+\varepsilon_0])$;

(ii) $\eta_0(f_{c+\varepsilon}\backslash N) \subset f_{c-\varepsilon}$;

(iii) if $K_c = \emptyset$, then $\eta_0(f_{c+\varepsilon}) \subset f_{c-\varepsilon}$;

(iv) $\eta_0(x+g) = \eta_0(x) + g, \ \forall x \in X, \ \forall g \in G.$

Proof. By $(PS)_{c,G}$-condition, K_c is a compact set. Hence we can choose $\delta > 0$ such that $N_{6\delta}(K_c) \subset \mathcal{N}$. Lemma 9.6 provides numbers $\bar{\varepsilon} > 0$, $b > 0$ and a locally Lipschitz vector field v on $M(f, \bar{\varepsilon}, \delta, c)$ such that the assertions (i)-(iii) of Lemma 9.6 are verified. Moreover, it is clear that if $0 < \varepsilon_1 \leq \varepsilon_2$ then $M(f, \varepsilon_1, \delta, c) \subset M(f, \varepsilon_2, \delta, c)$, and this allows to impose $\bar{\varepsilon} \leq \min\{\varepsilon_0, b\delta/4\}$. Fixing $\varepsilon \in (0, \bar{\varepsilon})$ let $\eta \in C([0,1] \times X, X)$ be defined by (9.104). Putting

$$t_0 = \frac{4\bar{\varepsilon}}{b},$$

we introduce $\eta_0 : X \to X$ by

$$\eta_0(x) = \eta(t_0, x), \forall x \in X.$$

If $x \notin f^{-1}([c-\varepsilon_0, c+\varepsilon_0])$, then we get $x \notin f^{-1}([c-\bar{\varepsilon}, c+\bar{\varepsilon}])$ and $V(x) = 0$ which implies (i).

Concerning (ii), since f is non-increasing along $\eta(.,x)$, we may restrict to $x \in M(f, \varepsilon, 6\delta, c)$. Arguing by contradiction suppose that there exists $x \in M(f, \varepsilon, 6\delta, c)$ such that $\eta_0(x) \notin f_{c-\varepsilon}$. It is seen that $\eta(s,x) \in f_{c+\varepsilon}\backslash f_{c-\varepsilon}$ for all $s \in [0, t_0]$. We claim that

$$\eta(s,x) \notin N_{4\delta}(K_c), \ \forall s \in [0, t_0].$$

If not, there is $s_0 \in [0, t_0]$ such that $\eta(s,x) \in N_{4\delta}(K_c)$ for all $s \in [0, s_0)$, and $\eta(s_0, x) \notin N_{4\delta}(K_c)$. Since then

$$\eta(s,x) \in M(f, \varepsilon, 4\delta, c), \ \forall s \in [0, s_0),$$

we derive from (i), (ii) of Lemma 9.7 where t, $\bar{\varepsilon}$, δ are replaced by s, ε, 4δ, respectively,

$$\|\eta(s_0,x) - x\| \leq s_0 \leq \frac{2}{b}(f(x) - f(\eta(s_0,x))) \leq \frac{2}{b}(c+\varepsilon-(c-\varepsilon)) = \frac{4\varepsilon}{b} < \delta.$$

We arrive at a contradiction with $x \notin N_{6\delta}(K_c)$. We thus establish that

$$\eta_0(x) \in M(f, c, \varepsilon, 4\delta).$$

Making use once more of (ii) of Lemma 9.7 we get

$$2\varepsilon = c + \varepsilon - (c-\varepsilon) \geq f(x) - f(\eta_0(x)) > \frac{b}{2}t_0 = 2\bar{\varepsilon}.$$

This contradicts the choice of ε. The statement (ii) is valid. Then it results that (iii) is also true. Property (iv) follows directly from (iii) of Lemma 9.7. This completes the proof. □

The main result of the Section is formulated below.

Theorem 9.7 Let $f : X \to \mathbb{R}$ be a G - invariant and locally Lipschitz functional which satisfies the $(PS)_G$ -condition. If f is bounded from below and if the dimension N of the linear subspace of X spanned by G is finite, then f has at least $N+1$ critical orbits.

Proof. For each $1 \leq j \leq N+1$ we set

$$\mathcal{A}_j = \{A \subset X : \; A \text{ is compact and } \operatorname{cat}_{\pi(X)}\pi(A) \geq j\}.$$

Due to the fact that $\dim(\operatorname{span} G) = N$ one has that $\operatorname{cat}_{\pi(X)}(\pi([0,1]^N \times\{0\})) = N+1$ (see Mawhin and Willem [27]). Therefore every set $\mathcal{A}_j$ is nonempty. We introduce for every $1 \leq j \leq N+1$ the minimax value

$$c_j = \inf_{A\in\mathcal{A}_j} \max_{x\in A} f(x).$$

Since $\mathcal{A}_{j+1} \subset \mathcal{A}_j$ it follows that

$$-\infty < \inf_X f = c_1 \leq c_2 \leq ... \leq c_{N+1} < +\infty.$$

We will prove that, if $c := c_j = c_k$ for some $1 \leq j \leq k \leq N+1$, then the set of critical points K_c contains at least $k-j+1$ critical orbits.

Suppose the contrary. Thus K_c contains $0 \leq n \leq k-j$ distinct critical orbits $\pi(u_1), ..., \pi(u_n)$. Let $\delta > 0$ be such that π restricted to $B(u_i, \delta)$ is a homeomorphism, $1 \leq i \leq n$. Then

$$\mathcal{N} = \bigcup_{i=1}^{n} \bigcup_{g\in G} \overset{o}{B}(u_i + g, \delta)$$

is an open neighborhood of K_c (in the case where $K_c = \emptyset$ one takes $\mathcal{N} = \emptyset$). Corresponding to $\mathcal{N}$, Theorem 9.6 supplies an $\varepsilon > 0$ and the homeomorphism η_0 of X satisfying (i)-(iv) of Theorem 9.6. The definition of $c = c_k$ shows the existence of $A \in \mathcal{A}_k$ such that $A \subset f_{c+\varepsilon}$. Let $B = A \backslash \mathcal{N}$. By Proposition 9.2 we infer that

$$\begin{aligned} k &\leq \operatorname{cat}_{\pi(X)}\pi(A) \leq \operatorname{cat}_{\pi(X)}(\pi(B) \cup \pi(\mathcal{N})) \\ &\leq \operatorname{cat}_{\pi(X)}\pi(B) + n \leq \operatorname{cat}_{\pi(X)}(\pi(B) + k - j. \end{aligned}$$

It follows that $B \in \mathcal{A}_j$ and $B \subset f_{c+\varepsilon} \backslash \mathcal{N}$. From (ii) of Theorem 9.6 we deduce that $\eta_0(B) \subset f_{c-\varepsilon}$. The equivariance of the homeomorphism η_0 ensures

$$\operatorname{cat}_{\pi(X)}(\pi(\eta_0(B)) = \operatorname{cat}_{\pi(X)}\pi(B) \geq j,$$

so $\eta_0(B) \in \mathcal{A}_j$.Consequently, we find that

$$c \le \max_{\eta_0(B)} f \le c - \varepsilon.$$

This contradiction completes the proof. □

We deal now with multiple eigensolutions for symmetric functionals. We argue on an infinite dimensional Hilbert space X with the scalar product $(.,.)$ whose associated norm is $\|.\|$. Now G stands for a discrete subgroup of the group of linear isometries of X. For every fixed $r > 0$ we denote by S_r the sphere in X centered at 0 and of radius r, i.e.,

$$S_r = \{x \in X : \ \|x\| = r\}.$$

We are concerned with a functional $f : X \to \mathbb{R}$ which is G-invariant, that is,

$$f(gx) = f(x), \ \forall x \in X, \ \forall g \in G.$$

We denote the restriction of f to S_r by $\tilde{f}$, thus $\tilde{f} : S_r \to \mathbb{R}$. Assuming that $f : X \to \mathbb{R}$ is locally Lipschitz, the generalized gradient $\partial\tilde{f}$ of $\tilde{f}$ is determined by ∂f as follows

$$\partial\tilde{f}(x) = \{z - r^{-2}\langle z, x\rangle \Lambda x : \ z \in \partial f(x)\}, \tag{9.105}$$

at an arbitrary $x \in S_r$, where $\Lambda : X \to X^*$ denotes the duality map of X, namely, $\langle \Lambda x, y\rangle = (x, y)$, $\forall x, y \in X$ (see Definition 1.3 and Example 1.1).

Lemma 9.8 If $f : X \to \mathbb{R}$ is G-invariant and locally Lipschitz, then the generalized gradient $\partial\tilde{f}$ of $\tilde{f} = f_{|S_r}$ satisfies

$$\partial\tilde{f}(x) = g^*\partial\tilde{f}(gx), \ \forall x \in S_r, \ \forall\, g \in G.$$

In particular, the set $\tilde{K}$ of critical points of $\tilde{f}$, i.e.,

$$\tilde{K} = \{x \in S_r : \ 0 \in \partial\tilde{f}(x)\}$$

is G-invariant, i.e.

$$g\tilde{K} = \tilde{K}, \ \forall g \in G.$$

Proof. Notice that

$$gS_r = S_r, \ \forall g \in G,$$

because G consists of isometries of X. Let us check the formula

$$\partial f(x) = g^*\partial f(gx), \ \forall x \in X, \ \forall g \in G. \tag{9.106}$$

The fact that $w \in \partial f(x)$ is equivalent to

$$\begin{aligned}
\langle w, v\rangle &\le \limsup_{\substack{u \to x \\ \lambda \downarrow 0}} \tfrac{1}{\lambda}(f(u + \lambda v) - f(u)) \\
&= \limsup_{\substack{u \to x \\ \lambda \downarrow 0}} \tfrac{1}{\lambda}(f(gu + \lambda gv) - f(gu)) \\
&= f^0(gx; gv), \ \forall v \in X, \ \forall g \in G.
\end{aligned}$$

Equality (9.106) follows. Combining (9.105) and (9.106) we see that the conclusion of Lemma 9.8 is true. □

Lemma 9.9 Under the assumptions of Lemma 9.8, the function $\tilde{\lambda} : S_r \to [0, +\infty)$ defined by

$$\tilde{\lambda}(x) = \min_{w \in \partial \tilde{f}(x)} \|w\|_{X*}, \ x \in S_r,$$

is G-invariant.

Proof. The result is deduced from Lemma 9.8 as follows,

$$\tilde{\lambda}(gx) = \min_{w \in \partial \tilde{f}(gx)} \|w\|_{X^*} = \min_{z \in \partial \tilde{f}(x)} \|g^{*-1}z\|_{X^*} = \min_{z \in \partial \tilde{f}(x)} \|z\|_{X^*} = \tilde{\lambda}(x)$$

for all $x \in S_r$ and $g \in G$. □

In the following the mapping $\pi : X \to X/G$ denotes the quotient map $\pi(x) = Gx$, $x \in X$.

We introduce a compactness condition of Palais-Smale type. Given $r > 0$ and $c \in \mathbb{R}$, a G-invariant, locally Lipschitz function $f : X \to \mathbb{R}$ is said to satisfy the $(PS)_{r,c,G}$-condition if for every sequence $\{x_k\} \subset S_r$ with $f(u_k) \to c$ and $\tilde{\lambda}(u_k) \to 0$, where $\tilde{\lambda}$ is as in Lemma 9.9, there exist a subsequence $\{u_{kn}\}$ and a sequence $\{x_n\} \subset S_r$ with $\pi(u_{kn}) = \pi(x_n)$ such that $\{x_n\}$ converges. We say that the $(PS)_{r,G}$-condition is satisfied if the preceding holds for every $c \in \mathbb{R}$.

Let us denote

$$\tilde{K}_c := \tilde{K} \cap f^{-1}(c) = \{x \in S_r : \ 0 \in \partial \tilde{f}(x) \ \text{ and } \ f(x) = c\}.$$

A result analogous to Lemma 9.7 is stated below.

Lemma 9.10 Assume that $f : X \to \mathbb{R}$ is locally Lipschitz, G-invariant and satisfies the $(PS)_{r,c,G}$-condition. Then, for each $\delta > 0$, there exist constants $\bar{\varepsilon} > 0$, $b > 0$ and a locally Lipschitz map $v : \tilde{M}(f, \bar{\varepsilon}, \delta, c) \to X$, where

$$\tilde{M}(f, \bar{\varepsilon}, \delta, c) := \tilde{f}^{-1}([c - \bar{\varepsilon}, c + \bar{\varepsilon}]) \backslash N_\delta(\tilde{K}_c),$$

such that

(i) $(v(x), x) = 0$,

(ii) $\|v(x)\| \leq 1$,

(iii) $\langle x^*, v(x) \rangle > \frac{b}{2}$,

(iv) $v(gx) = gv(x)$,

for all $x \in \tilde{M}(f, \bar{\varepsilon}, \delta, c)$, $x^* \in \partial \tilde{f}(x)$ and $g \in G$.

Proof. For a fixed $\delta > 0$ there exist constants $b > 0$ and $\bar{\varepsilon}$ such that

$$\tilde{\lambda}(x) \geq b, \ \forall x \in \tilde{M}(f, \bar{\varepsilon}, \delta, c). \tag{9.107}$$

The proof of (9.107) relies on the $(PS)_{r,c,\tilde{G}}$ - condition, the G–invariance in Lemma 9.9 and the lower semicontinuity of $\tilde{\lambda}$. Since it is similar to the proof of (9.99) we omit the details.

Take now a fixed $x_0 \in \tilde{M}(f, \bar{\varepsilon}, \delta, c)$. Arguing in the tangent space of the sphere S_r at x_0, i.e.,

$$T_{x_0} S_r = \{x \in X : \ (x, x_0) = 0\},$$

as we have done before in the proof of Lemma 9.6 in the space X, with $\partial \tilde{f}(x_0)$ in place of $\partial f(x_0)$, we find $h_0 = h(x_0) \in T_{x_0} S_r$ satisfying $\|h_0\| = 1$ and

$$\langle x^*, h_0 \rangle \geq b, \ \forall x^* \in \partial \tilde{f}(x_0).$$

In addition, we can suppose that

$$h(gx_0) = gh(x_0), \ \forall g \in G,$$

because by Lemma 9.7 we have

$$\langle \partial \tilde{f}(gx_0), gh(x_0) \rangle = \langle g^* \partial \tilde{f}(gx_0), h(x_0) \rangle = \langle \partial \tilde{f}(x_0), h(x_0) \rangle, \forall g \in G,$$

whose minimum is $\geq b$. The upper semicontinuity of ∂f in the weak sense of Chang [14] (assertion (iv) of Proposition 1.4) and the fact that h_0 belongs to $T_{x_0} S_r$ assure that for every $\varepsilon > 0$ there is $\eta_0 = \eta(x_0) > 0$ provided for each $x^* \in \partial \tilde{f}(x)$ with $\|x - x_0\| < \eta_0$ and $x \in S_r$ there exists $z_0 \in \partial \tilde{f}(x_0)$ with

$$|\langle x^* - z_0, h_0 \rangle| < \varepsilon.$$

Choosing $\varepsilon > 0$ small enough we conclude that

$$\langle x^*, h_0 \rangle \ > \ \frac{b}{2}, \ \forall x^* \in \partial \tilde{f}(x), \tag{9.108}$$

whenever $\|x - x_0\| < \eta_0$ and $x \in S_r$.

Corresponding to the open covering

$$\mathcal{N} = \{\overset{o}{B}\,(x_0, \eta_0) : x_0 \in \tilde{M}(f, \bar{\varepsilon}, \delta, c)\}$$

of $\tilde{M}(f, \bar{\varepsilon}, \delta, c)$ there exists a locally finite open covering $\{U_j\}_{j \in J}$ of $\tilde{M}(f, \bar{\varepsilon}, \delta, c)$ which is a refinement of $\mathcal{N}$ and has the property that any orbit $\pi(u) = Gu$ meets an arbitrary U_j at a finite number of points, that is locally constant with respect to $u \in X$. We assign to each $j \in J$ a point $x_j \in \tilde{M}(f, \bar{\varepsilon}, \delta, c)$ such that $U_j \subset \overset{o}{B}\,(x_j, \eta_j)$. For every $j \in J$ we define

$$\varphi_j(x) = \frac{\sum_{g \in G} \text{ dist } (gx, S_r \backslash U_j)}{\sum_{i \in J} \sum_{g \in G} \text{ dist } (gx, S_r \backslash U_i)},$$

and then

$$v(x) = \sum_{i \in J} \varphi_i(x)(h_i - r^{-2}(h_i, x)x)$$

for all $x \in \cup_{i\in J} U_i \cap S_r$, where $h_i = h(x_i)$, $i \in J$. Then v is well defined and locally Lipschitz. A direct computation yields (i),(ii). From (9.105) and (9.110) we infer that (iii) holds. A preceding remark yields

$$\begin{aligned} v(gx) &= \sum_{j\in J} \varphi_j(gx)(h_j - r^{-2}(h_j, gx)gx) \\ &= \sum_{j\in J} \varphi_j(gx)(gh_j - r^{-2}(gh_j, gx)gx) \\ &= g\sum_{j\in J} \varphi_j(gx)(h_j - r^{-2}(h_j, x)x) = gv(x) \end{aligned}$$

whenever $x \in \tilde{M}(f, \bar{\varepsilon}, \delta, c)$ and $g \in G$. Thus (iv) is verified. □

Choosing $\varepsilon \in (0, \bar{\varepsilon})$ we consider the locally Lipschitz functions $\varphi : S_r \to [0,1]$ and $\psi : S_r \to [0,1]$ having the same properties as required in the construction of V in (9.104), where we replace f and K_c by $\tilde{f}$ and $\tilde{K}_c$, respectively. We define $\tilde{V} : S_r \to X$ by means of the mapping v of Lemma 9.10 as follows

$$\tilde{V}(x) = \varphi(x)\psi(x)v(x), \ \forall x \in S_r.$$

Due to Lemma 9.10 and because of the properties of the functions φ and ψ, $\tilde{V}$ is well defined, locally Lipschitz and tangent to S_r, i.e., $\tilde{V}(x) \in T_x S_r$, $\forall \in S_r$. The tangency to S_r and the boundedness of $\tilde{V}$ ensure the existence of the global flow $\tilde{\eta} : [0,1] \times S_r \to S_r$ described by

$$\begin{aligned} \frac{d}{dt}\tilde{\eta}(t,x) &= -\tilde{V}(\tilde{\eta}(t,x)) \\ \tilde{\eta}(0,x) &= x \in S_r. \end{aligned}$$

Lemma 9.11 The flow $\tilde{\eta} \in C([0,1] \times S_r, S_r)$ has the properties

(i) $\|\tilde{\eta}(t,x) - x\| \le t$, $\forall x \in S_r$,

(ii) $f(x) - f(\tilde{\eta}(t,x)) > \frac{b}{2}t$, $\forall t \in [0,1]$, $\forall x \in \tilde{M}(f, \bar{\varepsilon}, \delta, c)$,

(iii) $\tilde{\eta}(t, gx) = g\tilde{\eta}(t,x)$, $\forall t \in [0,1]$, $\forall\, x \in S_r$, $\forall g \in G$.

Proof. The argument is the same as in the proof if Lemma 9.7 involving the vector field $\tilde{V}$ on the sphere S_r in place of V on X. □

By means of the flow $\tilde{\eta}$ one obtains the equivariant deformation on the sphere S_r.

Theorem 9.8 Assume that $f : X \to \mathbb{R}$ satisfies the hypotheses of Lemma 9.10 corresponding to a discrete group G of linear isometries of X and with fixed numbers $c \in \mathbb{R}$ an $r > 0$. Given a neigborhood $\mathcal{N}$ of $\tilde{K}_c$ and a number $\varepsilon_0 > 0$ there exist $\varepsilon \in (0, \varepsilon_0)$ and a homeomorphism $\tilde{\eta}_0 : S_r \to S_r$ such that

(i) $\tilde{\eta}_0(x) = x, \forall x \in \tilde{f}^{-1}([c-\varepsilon_0, c+\varepsilon_0])$;

(ii) $\tilde{\eta}_0(\tilde{f}_{c+\varepsilon}\backslash\mathcal{N}) \subset \tilde{f}_{c-\varepsilon}$;

(iii) if $\tilde{K}_c = \emptyset$, then $\tilde{\eta}_0(\tilde{f}_{c+\varepsilon}) \subset \tilde{f}_{c-\varepsilon}$;

(iv) $\tilde{\eta}_0(gx) = g\tilde{\eta}_0(x), \forall\, x \in S_r,\ \forall g \in G$.

Proof. The proof is analogous to the proof of Theorem 9.6 with the difference that we apply the Lemma 9.10 and 9.11 instead of the Lemma 9.6 and 9.7. It is necessary to replace in the proof of Theorem 9.6 the data K_c and $\eta \in C([0,1] \times X, X)$ by $\tilde{K}_c$ and $\tilde{\eta} \in C([0,1] \times S_r, S_r)$, respectively. Comparing with Theorem 9.6 we must also take into account the different way of action (on S_r) of the group G. □

Following an idea in Browder [12] we study the category of $\pi(S_r)$ over compact sets, i.e., cat $\pi(S_r) = \sup\{$ cat $_{\pi(S_r)}\pi(A) :$ A is compact subset of $S_r\}$, where $\pi : X \to X/G$ is the quotient projection.

Lemma 9.12 For any sphere S_r in X one has

$$\text{cat } \pi(S_r) = +\infty$$

in the case, where X is infinite dimensional and G is cyclic of prime order.

Proof. This follows from Proposition 8.3 of Browder [12].

The next result establishes the existence of infinitely many orbits of solutions to the abstract eigenvalue problem formulated for a symmetric functional f.

Theorem 9.9 Let G be a discrete subgroup of linear isometries of a Hilbert space X with the usual action on X and let $f : X \to \mathbb{R}$ be a G-invariant and locally Lipschitz functional which is bounded from below on S_r and satisfies the $(PS)_{r,G}$-condition for some $r > 0$. Then there exist cat $\pi(S_r)$ many distinct orbits $Gu_j \subset S_r$ of eigenelements of f on S_r, $j \geq 1$, in the sense that for each $j \geq 1$ there are $\lambda_j \in \mathbb{R}$ and $u_j \in S_r$ such that

$$\lambda_j \Lambda x_j \in \partial f(x_j),\ \forall x_j \in Gu_j.$$

Under the additional assumptions of Lemma 9.12 there exist infinitely many orbits Gu_j.

Proof. With the sequence of sets

$$\tilde{\mathcal{A}}_j = \{A \subset S_r :\ A \text{ is compact and } \text{cat }_{\pi(S_r)}\pi(A) \geq j\},\ j \geq 1,$$

we construct the sequence of real numbers

$$\tilde{c} = \inf_{A \in \tilde{\mathcal{A}}_j} \max_{x \in A} f,\quad 1 \leq j \leq \text{ cat } \pi(S_r).$$

Lemma 9.12 ensures that every collection $\tilde{\mathcal{A}}_j$ is nonempty, so each $\tilde{c}_j$ is finite. Our goal is to show that if $c := \tilde{c}_k = \tilde{c}_j$ for $j \le k$ then $\tilde{K}_c$ contains at least $k - j + 1$ critical orbits of $\tilde{f} = f|_{S_r}$. If not, one would find at most $n = k - j$ distinct critical orbits $\pi(u_1), ..., \pi(u_n)$ with representatives $u_1, ..., u_n \in S_r$. In view of the fact that G is discrete we may choose mutually disjoint, open and contractible neigborhoods $U_1, ..., U_n$ of $u_1, ..., u_n$ in S_r, respectively, such that π is a homeomorphism on each U_i, $i = 1, ..., n$. Then

$$\mathcal{N} = \cup_{j=1}^{n} \cup_{g \in G} gU_j$$

is an open neighborhood of $\tilde{K}_c$ in S_r. Corresponding to $\mathcal{N}$, Theorem 9.6 supplies an $\varepsilon > 0$ and a homeomorphism η_0 of S_r with the properties stated there. Let us take $A \in \tilde{\mathcal{A}}_K$ such that $A \subset \tilde{f}_{c+\varepsilon}$ and set $B = A \backslash \mathcal{N}$. It turns out that

$$\text{cat}_{\pi(S_r)} \pi(B) \ge k - \text{cat}_{\pi(S_r)} \pi(\mathcal{N}) = k - n \ge j,$$

and thus $B \in \tilde{\mathcal{A}}_j$. Theorem 9.8 implies further that $\tilde{\eta}_0(B) \subset \tilde{f}_{c-\varepsilon}$. The equivariance property (iv) in Theorem 9.8 for $\tilde{\eta}_0$ and (ii) in Proposition 9.2 permit us to write

$$\text{cat}_{\pi(S_r)} \pi(\tilde{\eta}_0(B)) \ge \text{cat}_{\pi(S_r)} \pi(B) \ge j.$$

Hence we get $\tilde{\eta}_0(B) \in \tilde{\mathcal{A}}_j$. This leads to the contradiction

$$c \le \max_{\tilde{\eta}_0(B)} f \le c - \varepsilon.$$

Therefore the existence of infinitely many critical orbits for $\tilde{f}$ is justified. Finally, we note that $u \in \tilde{K}$ if and only if there is $x^* \in \partial f(u)$ such that

$$x^* = r^{-2} \langle x^*, u \rangle \Lambda u.$$

This remark completes the proof. □

We apply now our multiplicity results to study the periodic solutions of a hemivariational inequality.

For a given number $T > 0$ we assume that

(H_1) $M(t, x)$ is a symmetric matrix of order n continuously differentiable on $[0, T] \times I\!R^N$ such that

$$\langle M(t, x)y, y \rangle \ge \alpha |y|^2, \forall (t, x, y) \in [0, T] \times I\!R^N \times I\!R^N,$$

for some constant $\alpha > 0$;

(H_2) $j(t, x)$ is a real-valued function which is measurable in t, for every $x \in I\!R^N$, and for which there exists $K \in L^2([0, T])$ such that

$$|j(t, x) - j(t, y)| \le K(t)|x - y|, \ \forall x, y \in I\!R^N, \ \forall t \in [0, T];$$

(H_3) there exists $h \in L^1([0,T])$ such that

$$|j(t,x)| \leq h(t), \ \forall x \in I\!R^N \text{ and a.e. } t \in [0,T].$$

(H_4) $M(t,x)$ and $j(t,x)$ are T_i-periodic in x_i for prescribed $T_i > 0, \ i = 1, ..., N$, where $x = (x_1, ..., x_N)$;

(H_5) $g(t)$ is an $I\!R^N$-valued function on $[0,T]$ with $g \in L^1(0,T,I\!R^N)$ and $\int_0^T g(t)dt = 0$.

Using the data above we formulate the next problem in the form of a hemivariational inequality: find $u \in H_T^1$ such that

$$\int_0^T (\frac{1}{2}\langle M(t,u)\dot{u}, \dot{v}\rangle \ + \ \frac{1}{2}\sum_{i=1}^N \langle D_{x_i} M(t,u)\dot{u}, \dot{u}\rangle v)dt$$
$$+ \int_0^T j^0(t,u;v)dt \geq \int_0^T \langle g(t), v(t)\rangle dt, \ \forall v \in H_T^1.$$

Here H_T^1 stands for the Sobolev space obtained as the completion of the set of C^∞, $I\!R^N$-valued, T-periodic functions on $\mathbb{R}$ with respect to the H^1-norm.

To put the problem in an abstract setting we consider the following discrete subgroup G of H_T^1 defined by

$$G = \{\sum_{i=1}^N k_i T_i e_i : \ k_i \text{ integer }, \ 1 \leq i \leq N\},$$

where $\{e_i\}_{1 \leq i \leq N}$ denotes the canonical basis of $I\!R^N$.

Our existence result is the following.

Theorem 9.10 Under hypotheses $(H_1) - (H_5)$ the foregoing hemivariational inequality possesses at least $N+1$, T - periodic solutions which are distinct.

Proof In view of (H_2) the functional

$$J(u) := \int_0^T j(t,u(t))dt$$

is well defined and is Lipschitz continuous on $L^2(0,T;I\!R^N)$. We introduce the functional $f : H_T^1 \to I\!R$ where

$$f(u) := \int_0^T (\langle \frac{1}{2}M(t,u)\dot{u}, \dot{u}\rangle - \langle g(t), u(t)\rangle)dt + J(u), \ \forall u \in H_T^1.$$

It is clear that f is locally Lipschitz. A straightforward computation based on (H_4), (H_5) yields

$$f(u + \sum_{i=1}^N k_i T_i e_i) = f(u), \ \forall u \in H_T^1, \ \forall \text{ integer } k_i, \ i = 1, ..., N.$$

Therefore f is G-invariant.

For every $u \in L^1(0,T,\mathbb{R}^N)$ we write $u = \bar{u} + \tilde{u}$, where

$$\bar{u} = \frac{1}{T}\int_0^T u(t)dt.$$

By $(H_1), (H_3), (H_5)$ and the usual imbedding theorems the estimate below holds

$$f(u) \geq \alpha\|\dot{u}\|_{L^2}^2 - \int_0^T h(t)dt + \int_0^T \langle g(t), u(t)\rangle dt \geq$$

$$\alpha\|\tilde{u}\|_{L^2}^2 - c_1 - \|g\|_{L^1}\|\tilde{u}\|_{L^\infty} \geq \tilde{\alpha}\|\tilde{u}\|_{H_T^1}^2 - c_1 - c_2\|\tilde{u}\|_{H_T^1}, \ \forall u \in H_T^1, \tag{9.109}$$

with positive constants $\tilde{\alpha}, c_1, c_2$. It follows that f is bounded from below.

Our next aim is to justify that f satisfies the $(PS)_G$-condition. Towards this end we need the following: If $u \in H_T^1$ has the decomposition above $u = \bar{u} + \tilde{u}$, then there exist uniquely determined integers k_i such that

$$(\bar{u}, e_i) - k_iT_i \in [0, T_i),\ 1 \leq i \leq N.$$

We set

$$\bar{u}^0 = ((\bar{u}, e_1) - k_1T_1, ..., (\bar{u}, e_N) - k_NT_N).$$

Then $v = \bar{u}^0 + \tilde{u}$ is a representative of $\pi(u) \in H_T/G$.

In order to check the $(PS)_G$-condition let $\{u_n\}$ be a sequence in H_T^1 such that $f(u_n)$ is bounded and $\lambda(u_n) \to 0$. Firstly, we shall prove that

$$v_n := \bar{u}_n^0 + \tilde{u}_n$$

is bounded. Since

$$\langle \bar{u}_n^0, e_i\rangle \leq T_i,\ 1 \leq i \leq N,\ \forall n \geq 1,$$

the sequence $\bar{u}_n^0$ is bounded. Taking into account the boundedness of $f(u_n)$, it follows from (9.109) that $\{\tilde{u}_n\}$ is bounded in H_T^1. Hence $\{v_n\}$ is bounded in H_T^1.

There exists a subsequence of $\{v_n\}$, denoted again by $\{v_n\}$, such that $v_n \to v$ weakly in $H_T^1, v_n \to v$ strongly in $L^2(0,T;\mathbb{R}^N)$ and $v_n \to v$ strongly in $C([0,T];\mathbb{R}^N)$. By Lemma 9.5 we know that $\lambda(v_n) \to 0$. Thus there exists $w_n \in \partial f(v_n)$ with $w_n \to 0$. Putting $\varepsilon_n := \|w_n\|$ we get

$$\int_0^T (\langle Mt, v_n)\dot{v}_n, \dot{z}\rangle + \frac{1}{2}\sum_{i=1} N\langle D_{x_i}M(t,v_n)\dot{v}_n, \dot{v}_n\rangle z_i)dt +$$

$$\int_0^T j^0(t, v_n; z)dt \geq -\varepsilon_n\|z_n\|_{H_T^1} + \int_0^T \langle g(t), z\rangle dt, \tag{9.110}$$

for all $z = (z_1, ..., z_N) \in H_T^1$.

If we take $z := v - v_n$ in (9.110), by employing (H_1) and standard inequalities we obtain

$$\varepsilon_n\|v - v_n\|_{H_T^1} + |\int_0^T j^0(t, v_n; z)dt| -$$

$$-\int_0^T \langle M(t,v_n)\dot v, \dot v_n - \dot v\rangle dt + \int_0^T \langle g(t), v_n - v\rangle dt \geq$$

$$\geq \alpha\|\dot v_n - \dot v\|_{L2}^2 - \frac{1}{2}\sum_{i=1}^N \|\langle D_{xi}M(t,v_n)\dot v_n, \dot v_n\rangle\|_{L^1}\|v_{n,i} - v_i\|_{L^\infty}, \qquad (9.111)$$

where we have used the element $v = (v_1, ..., v_N) \in H_T^1$. The boundedness of

$$\|v - v_n\|_{H_T^1}$$

implies that

$$\varepsilon_n\|v - v_n\|_{H_T^1} \to 0.$$

Hypothesis (H_2) ensures that

$$\|\int_0^T j^0(t, v_n; v - v_n)dt\| \leq \|K\|_{L^2}\|v - v_n\|_{L^2},$$

and therefore

$$\int_0^T j^0(t, v_n; v - v_n)dt \to 0 \text{ as } n \to \infty.$$

The weak convergence of $\{v_n\}$ to v in H_T^1 implies that

$$\int_0^T \langle f(t), v_n - v\rangle dt \to 0 \text{ as } n \to \infty.$$

The uniform convergence of $\{v_n\}$ to v and the weak convergence of $\{\dot v_n\}$ to $\dot v$ in $L^2(0,T;\mathbb{R}^N)$ imply that

$$\int_0^T \langle M(t,v_n)v, \dot v_n - \dot v\rangle dt \to 0 \text{ as } n \to \infty.$$

Since the partial derivatives $D_{xi}M$ are continuous, the sequence $\{v_n\}$ is uniformly convergent and the sequence $\{\dot v_n\}$ is bounded in $L^2(0,T;\ \mathbb{R}^N)$, it turns out that $\langle D_{xi}M(t,v_n)\dot v_n, \dot v_n\rangle$ is bounded in $L^1(0,T;\ \mathbb{R}^N)$. Hence we derive that

$$\frac{1}{2}\sum_{i=1}^N \|\langle D_{xi}M(t,v_n)\dot v_n, \dot v_n\rangle\|_{L^1}\|v_{n,i} - v_i\|_{L^\infty} \to 0 \text{ as } n \to \infty.$$

Then, from (9.111), we deduce the strong convergence of $\{\dot v_n\}$ to $\dot v$ in $L^2(0,T;\mathbb{R}^N)$. Moreover, due to $v_n \to v$ in $L^2(0,T;\mathbb{R}^N)$ we conclude that $v_n \to v$ in H_T^1. Recalling now the equality $\pi(u_n) = \pi(v_n),\ n \geq 1$, we deduce that the $(PS)_G$-condition is verified.

We note also that $\dim\{\text{ span }(G)\} = N$. Therefore, for the functional f on H_T^1 and the group G, all the assumptions of Theorem 9.6 are satisfied. Then Theorem 9.7 insures that at least $N+1$ critical orbits of f exist. It suffices now to observe that each critical point of f is a solution of the stated hemivariational inequality. The essential point here is that one has

$$\partial J(u) \subset \int_0^T \partial_u j(t,u)dt,\ \forall u \in L^2(0,T,\mathbb{R}^N),$$

because (H_2) allows to apply Theorem 1.3. This completes the proof. □

We close the chapter with the study of multiple eigensolutions for a hemivariational inequality. Precisely, we treat the following eigenvalue problem for a hemivariational inequality:

(P_r) find $u \in X$ with $\|u\| = r$ and $\lambda \in \mathbb{R}$

such that

$$a(u,v) + \int_\Omega j^0(x,u;v)dx \geq \lambda(u,v), \ \forall v \in X.$$

In the formulation of (P_r), X denotes an infinite dimensional real Hilbert space with the scalar product $(.,.)$ and the associated norm $\|.\|$, which is densely and compactly imbedded in $L^p(\Omega;\mathbb{R}^N)$ for a bounded domain Ω in $\mathbb{R}^m$, $m \geq 1$, and some $p \geq 2$ and an integer $N \geq 1$. Problem (P_r) is stated for a prescribed number $r > 0$. The data of (P_r) have the following meaning: $a : X \times X \to \mathbb{R}$ is a continuous, symmetric bilinear form whose corresponding self-adjoint operator is denoted by $A : X \to X$, i.e.,

$$(Au, v) = a(u,v), \ \forall u, v \in X,$$

and $j : \Omega \times \mathbb{R}^N \to \mathbb{R}$ is a function of Carathéodory type for which one imposes the assumption (A_1):

(i) $j(.,y) : \Omega \to \mathbb{R}$ is measurable for all $y \in \mathbb{R}^N$,

(ii) $j(x,.) : \mathbb{R}^N \to \mathbb{R}$ is locally Lipschitz for all $x \in \Omega$,

(iii) $j(.,0) : \Omega \to \mathbb{R}$ is essentially bounded.

We denote by $j^0(x,\xi;\eta)$ Clarke's generalized directional derivative of j with respect to the second variable

$$j^0(x,\xi;\eta) = \limsup_{h\to 0, t\downarrow 0} \frac{1}{t}(j(x,\xi+h+t\eta) - j(x,\xi+h)), \ \forall x \in \Omega, \ \forall \xi, \eta \in \mathbb{R}^N,$$

and accordingly the generalized gradient of $j(x,.)$ is denoted by ∂j. Thus

$$\partial j(x,y) = \{z \in \mathbb{R}^N : \ \langle z, \eta\rangle \leq j^0(x,y;\eta), \ \forall \eta \in \mathbb{R}^N\}$$

at any $(x,y) \in \Omega \times \mathbb{R}^N$.

Moreover it is given a discrete subgroup G of the group of linear isometries of X with properties as in Lemma 9.12. The three following assumptions are assumed to hold

(A_2) $|z| \leq c(1 + |y|^{p-1})$, $\forall (x,y) \in \Omega \times \mathbb{R}^N$, $\forall z \in \partial j(x,y)$, with a constant $c > 0$.

(A_3) Whenever a sequence $\{v_n\} \subset X$ with $\|v_n\| = r$, a number $b \in [-r^2\|A\|, r^2\|A\|]$ and a measurable map $z : \Omega \to \mathbb{R}^N$ are such that $v_n \to v$ in $L^p(\Omega,\mathbb{R}^N)$, $a(v_n,v_n) \to \alpha$ and

$$z(x) \in \partial j(x, v(x)) \text{ for a.e. } x \in \Omega,$$

then

$$\inf_{\|w\|=1} \{a(w,w)\} - \frac{1}{r^2}(\alpha + \int_\Omega \langle z(x), v(x)\rangle dx) > 0.$$

(A_4) a and j are G-invariant in the sense that

$$a(gu, gv) = a(u,v), \ \forall u, v \in X, \ \forall g \in G,$$

and

$$j(x, (gu)(x)) = j(x, u(x)), \ \forall u \in X, \ \forall g \in G \text{ and for a.e. } x \in \Omega.$$

Our existence and multiplicity result for the solutions of problem (P_r) is formulated as follows.

Theorem 9.11 Assume that conditions (A_1) – (A_4) are satisfied and let $I(v) = \int_\Omega j(x, v(x))dx$, $v \in L^p(\Omega; \mathbb{R}^N)$. Then the eigenvalue problem (P_r) admits infinitely many pairs $(\lambda_n, Gu_n), n \geq 1$, such that $\lambda = \lambda_n$ and every $u \in Gu_n$ solves (P_r). In addition, the sequence $I(u_n)$ is bounded.

Proof. Hypotheses (A_1) and (A_2) imply that the functional $I : L^p(\Omega; \mathbb{R}^N) \to \mathbb{R}$

$$I(v) = \int_\Omega j(x, v(x))dx, \ v \in L^p(\Omega; \mathbb{R}^N),$$

is well defined and locally Lipschitz. Then the same properties are valid for the functional $f : X \to \mathbb{R}$ defined by

$$f(v) = a(v,v) + I(v), \ \forall v \in X.$$

The growth condition (A_2), Lebourg's mean value theorem (see Theorem 1.1) and (iii) of (A_1) guarantee the estimate

$$|j(x,y)| \leq c_1 + c_2|y|^p, \ \forall (x,y) \in \Omega \times \mathbb{R}^N,$$

with positive constants c_1, c_2. Then using the continuous imbedding $X \subset L^p(\Omega, \mathbb{R}^N)$ it results that

$$|f(v)| \leq \frac{1}{2}\|A\| \, \|v\|^2 + c_1 + c_2\|v\|^p, \ \forall v \in X,$$

for possibly new positive constants c_1, c_2. We conclude that the functional f is bounded on the sphere S_r

$$|f(v)| \leq M, \ \forall v \in S_r. \tag{9.112}$$

Clearly hypothesis (A_4) shows that f is G- invariant

$$f(gv) = f(v),\ \forall v \in X,\ \forall g \in G. \tag{9.113}$$

We claim that the functional f satisfies the $(PS)_{r,G}$-condition. According to this let $\{v_n\} \subset S_r$ be such that $\tilde{\lambda}(v_n) \to 0$ (due to (9.112) the condition concerning the convergence of $f(v_n)$ is not relevant). Denoting by $\Lambda : X \to X^*$ the duality mapping

$$\langle \Lambda u, v \rangle = (u, v),\ \forall u, v \in X,$$

we find a sequence $\{z_n\} \subset X^*$ that verifies

$$z_n \in \partial I(v_n),\ \forall n \geq 1,$$

and

$$Av_n - \frac{1}{r^2}\langle \Lambda A v_n + z_n, v_n \rangle v_n \to 0 \text{ in } X \text{ as } n \to \infty.$$

The boundedness of $\{v_n\}$ and the compactness of the imbedding $X \subset L^p(\Omega;,\mathbb{R}^N)$ allow to assume that a subsequence again denoted by $\{v_n\}$ can be extracted such that

$$a(v_n, v_n) \to b \text{ in } \mathbb{R},$$

$$v_n \to v \text{ weakly in } X$$

and

$$v_n \to v \text{ strongly in } L^p(\Omega; \mathbb{R}^N),$$

for a number b. The fact that I is locally Lipschitz and the compactness of the imbedding $L^p(\Omega; \mathbb{R}^N)^* \subset X^*$ enable us to write

$$z_n \to z \text{ weakly in } L^p(\Omega; \mathbb{R}^N)$$

and

$$z_n \to z \text{ strongly in } X^*.$$

It turns out that

$$z \in \partial I(v) \tag{9.114}$$

(cf. (iii) of Proposition 1.4) and that

$$Av_n - \frac{1}{r^2}(b + \langle z, v \rangle)v_n \text{ converges in } X. \tag{9.115}$$

Relation (9.114) can be interpreted as the condition required for z in (A_3). This expresses that the assumption (A_3) can be used. On the other hand one can write the inequality

$$\inf_{\|w\|=1}\{a(w,w\} - \frac{1}{r^2}(b + \int_\Omega \langle z(x), v(x) \rangle dx)\|v_n - v_k\|^2 \leq$$

$$\leq \|A(v_n - v_k) - \frac{1}{r^2}(b + \int_\Omega \langle z(x), v(x) \rangle dx))(v_n - v_k)\|\ \|v_n - v_k\|$$

for all $n, k \geq 1$.

From (A_3) and (9.115) it follows that $\{v_n\}$ contains a strongly convergent subsequence. Consequently, the Palais-Smale condition for $\tilde{f} = f|_{S_r}$ holds. In particular, it follows that the $(PS)_{r,G}$-condition is true. The hypothesis of Theorem 9.7 are satisfied. The conclusion of Theorem 9.7 leads to the existence of infinitely many pairs (λ_n, Gu_n), $n \geq 1$, such that

$$\Lambda(\lambda_n x_n - Ax_n) \in \partial I(x_n), \ \forall x_n \in Gu_n, \ \forall n \geq 1. \tag{9.116}$$

The final step in the proof is to make use of the relation

$$\partial I(u) \subset \int_\Omega \partial j(x,u)dx, \ \forall u \in L^p(\Omega; I\!R^N). \tag{9.117}$$

The inclusion above is derived from (A_2) because we can apply Theorem 1.3. Combining (9.116) and (9.117) we see that $\lambda = \lambda_n$ and $u = x_n$ for $x_n \in Gu_n$, $n \geq 1$, forms an eigensolution to problem (P_r). The last assertion of Theorem follows from property (9.112). The proof is thus complete. □

Remark 9.4 Theorem 9.11 extends Theorem 2 in Motreanu and Panagiotopoulos [29] which can be obtained when $G = \{id_X, -id_X\}$. Hence we can extend related results in Chang [14] and Rabinowitz [39] (see Motreanu and Panagiotopoulos [29] for details). A different variational approach to problem (P_r) is given in Goeleven, Motreanu and Panagiotopoulos [22]. Specific problems in Mechanics can be discussed in the framework of (P_r) and of the indicated assumptions $(A_1)-(A_4)$ (see [35],[38],[28],[33]. For analogous problems and related mathematical methodologies we refer also to [1],[8],[13],[16],[24],[26],[30],[36],[37]). An advantage of Theorem 9.11 is that it points out the relationship between a group of symmetries and the multiple solutions. The class of eigenvalue problems (P_r), where groups of symmetries are present, is large.

References

[1]S. Adly, G. Buttazzo and M. Théra, Critical Points for Nonsmooth Energy Functions and Applications, *Nonlinear Anal. T.M.A.*, (to appear).

[2]S. Adly and D. Goeleven, Periodic Solutions for a Class of Hemivariational Inequalities, *Comm. Appl. Nonlinear Anal.* **2** (1995), 45-57.

[3]S. Adly and D. Goeleven, Homoclinic Orbits for a Class of Hemivariational Inequalities, *Applicable Anal.* **58** (1995), 229-240.

[4]S. Adly, D. Goeleven and D. Motreanu, Periodic and Homoclinic Solutions for a Class of Unilateral Problems, *Discrete Cont. Dyn. Systems* **3** (1997), 579-590.

[5]S. Ahmad, A. C. Lazer and J. L. Paul, Elementary Critical Point Theory and Perturbations of Elliptic Boundary Value Problems at Resonance, *Indiana Univ. Math. J.* **25** (1976), 933-944.

[6]A. Ambrosetti, Critical Points and Nonlinear Variational Problems, *Suppl. Bull Math.* France, **120** (1992), fasc. 2, Memoire n° 49.

[7]A. Ambrosetti and M. L. Bertotti, Homoclinics for Second Order Conservative Systems, in: Partial Differential Equations and Related Subjects (Proc. Conf. in honor of L. Nirenberg), *Longman Scientific & Technical*, Harlow, England, 1992, pp. 21-37.

[8]C.C. Baniotopoulos and P.D. Panagiotopoulos, Variational Methods for Structures with Adhesive Joints, in Variational Methods in Engineering, Proceedings of the 2nd international conference University of Southampton, England, Springer-Verlag, 1985.

[9]V. Barbu and N. H. Pavel, Periodic Optimal Control in Hilbert Spaces, *Appl. Math. Optimiz.* **33** (1996), 169-188.

[10]H. Brézis, Periodic Solutions of Nonlinear Vibrating Strings and Duality Principles, *Bull. Amer. Math. Soc.* **8** (1983), 409-426.

[11]H. Brézis, *Analyse Fonctionnelle, Théorie et Applications*, Masson, Paris, 1987.

[12]F.E. Browder, *Nonlinear Eigenvalue Problems and Group Invariance. In: Functional Analysis and Related Fields*, Springer-Verlag, Berlin, 1970.

[13]A. Canino, Multiplicity of Solutions for Quasi Linear Elliptic Equations, Preprint, Universita' Degli Studi Della Calabria, 1994.

[14]K. C. Chang, Variational Methods for Non-Differentiable Functionals and their Applications to Partial Differential Equations, *J. Math. Anal. Appl.* **80** (1981), 102-129.

[15]F. H. Clarke, Nonsmooth Analysis and Optimization, Wiley & Sons, New York, 1984.

[16]J. N. Corvellec, M. Degiovanni and M. Marzocchi, Deformation Properties for Continuous Functionals and Critical Point Theory, *Topological Methods in Nonlinear Analysis* **1** (1993), 151-171.

[17]V. Coti Zelati and P.L. Rabinowitz, Homoclinic Orbits for Second Order Hamiltonian Systems Possessing Superquadratic Potentials, *J. Amer. Math. Soc.* **4** (1992), 693-727.

[18]J. Dieudonné, *Eléments d'Analyse*, Gauthier-Villars, Paris, 1968.

[19]G. Fournier and M. Willem, Multiple Solutions of the Forced Pendulum Equation, Ann. Inst. H. Poincaré, *Analyse Nonlinéaire* **6** (1989), 259-282.

[20]D. Goeleven and D. Motreanu, Hyperbolic Hemivariational Inequality and Nonlinear Wave Equation with Discontinuities, in "From Convexity to Nonconvexity" (Ed. by R. Gilbert, P.D. Panagiotopoulos, P. Pardalos). A volume dedicated to the memory of Gaetano Fichera, Kluwer Academic Publishers, Boston (to appear).

[21]D. Goeleven, D. Motreanu and P. D. Panagiotopoulos, Eigenvalue Problems for Variational-Hemivariational Inequalities at Resonance, *Nonlin. Anal. TMA*, **33** (1998), 161-180.

[22]D. Goeleven, D. Motreanu and P.D. Panagiotopoulos, Multiple Solutions for a Class of Eigenvalue Problems in Hemivariational Inequalities, *Nonlinear Anal. T.M.A.* **29** (1997), 9-26 .

[23]D. Goeleven, D. Motreanu and P.D. Panagiotopoulos, Multiple Solutions for a Class of Hemivariational Inequalities Involving Periodic Energy Functionals, *Math. Methods Appl. Sci.* **20** (1997), 548-568.

[24]D. Goeleven, V.H. Nguyen and M. Willem, Existence and Multiplicity Results for Noncoercive Unilateral Problems, *Bull. Australian Math. Soc.* **49** (1994), 489-498.

[25]E. M. Landesman and A. C. Lazer, Nonlinear Perturbations of Linear Elliptic Boundary Value Problems at Resonance, *J. Math. Mech.* **19** (1970), 609-623.

[26]J. Mawhin and M. Willem, Multiple Solutions for the Periodic Boundary Value Problem for some Forced Pendulum Type Equations, *J. Differential Equations* **52** (1984), 264-287.

[27]J. Mawhin and M. Willem, *Critical Point Theory and Hamiltonian Systems*, Springer-Verlag, New-York, 1989.

[28]J.J. Moreau and P.D. Panagiotopoulos (eds.), *Nonsmooth Mechanics and Applications*, Springer-Verlag, New York, CISM Lecture Notes vol. **302**, 1988.

[29]D. Motreanu and P.D. Panagiotopoulos, On the Eigenvalue Problem for Hemivariational Inequalities: Existence and Multiplicity of Solutions, *J. Math. Anal. Appl.* **197** (1996), 75-89.

[30]D. Motreanu and P.D. Panagiotopoulos, Hysteresis: The Eigenvalue Problem for Hemivariational Inequalities, in: Models of Hysteresis (A. Visitin, Editor), *Pitman Research Notes in Mathematics*, vol. 286, 1993, pp. 102-117.

[31]D. Motreanu, Existence of Critical Points in a General Setting, *Set-Valued Anal.* **3** (1995), 295-305.

[32]D. Motreanu and Z. Naniewicz, Discontinuous Semilinear Problems in Vector-Valued Functions Spaces, *Differ. Int. Equations* **9** (1996), 581-598.

[33]Z. Naniewicz and P. D. Panagiotopoulos, *The Mathematical Theory of Hemivariational Inequalities and Applications*, Marcel Dekker, New York, 1994.

[34]P.D. Panagiotopoulos, *Inequality Problems in Mechanics and Applications, Convex and Nonconvex Energy Functions*, Birkhäuser, Basel, 1985.

[35]P. D. Panagiotopoulos, *Hemivariational Inequalities. Applications in Mechanics and Engineering*, Springer-Verlag, Berlin, 1993,

[36]P. D. Panagiotopoulos, Nonconvex Energy Function, Hemivariational Inequalities and Substationarity Principles, *Acta Mech.* **48** (1983), 160-183.

[37]P.D. Panagiotopoulos, Nonconvex Problems of Semipermeable Media and Related Topics, *ZAMM* **65** (1995), 29-36.

[38]P.D. Panagiotopoulos and G.E. Stavroulakis, The Delamination Effect in Laminated von Kármán Plates under Unilateral Boundary Conditions. A Variational-Hemivariational Inequality Approach, *J. of Elasticity* **23** (1990), 69-96.

[39]P. H. Rabinowitz, Some Minimax Theorems and Applications to Nonlinear Partial Differential Equations, in: Nonlinear Analysis: A collection of papers in honor of E. Röthe, *Academic Press*, New York, 1978, pp. 161-177.

[40]P.H. Rabinowitz, Homoclinic Orbits for a Class of Hamiltonian Systems, *Proc. Royal Soc. Edinburg* **114A** (1990), 33-38.

[41]P. H. Rabinowitz, Minimax Methods in Critical Point Theory with Applications to Differential Equations, CBMS Reg. Conf. Ser. Math. 65, *Amer. Math. Soc.*, Providence, R. I., 1986.

[42]D. Yanheng and M. Girardi, Periodic and Homoclinic Solutions to a Class of Hamiltonian Systems with the Potentials Changing Sign, *Dynamic Syst. Appl.* **2** (1993), 131-145.

MIX
Papier aus verantwortungsvollen Quellen
Paper from responsible sources
FSC® C105338

If you have any concerns about our products,
you can contact us on
ProductSafety@springernature.com

In case Publisher is established outside the EU,
the EU authorized representative is:
Springer Nature Customer Service Center GmbH
Europaplatz 3, 69115 Heidelberg, Germany

Printed by Libri Plureos GmbH
in Hamburg, Germany